NIGHTCLUB

设计速递 **夜店设计**

精品文化工作室 / 编
赵欣 / 译

大连理工大学出版社
Dalian University of Technology Press

图书在版编目(CIP)数据

夜店设计：汉英对照 / 精品文化工作室编；赵欣
译． 一大连：大连理工大学出版社，2012.8
　　（设计速递）
　　ISBN 978-7-5611-7079-3

　　Ⅰ．①夜… Ⅱ．①精…②赵… Ⅲ．①文娱活动－文
化建筑－建筑设计－世界－图集 Ⅳ．①TU242.4-64

　　中国版本图书馆CIP数据核字（2012）第149050号

出版发行：大连理工大学出版社
　　　　　　（地址：大连市软件园路80号　邮编：116023）
印　　　刷：精一印刷（深圳）有限公司
幅面尺寸：225mm×300mm
印　　张：15
插　　页：4
出版时间：2012年8月第1版
印刷时间：2012年8月第1次印刷
责任编辑：刘　蓉
责任校对：王丹丹
封面设计：连　帅

ISBN 978-7-5611-7079-3
定　　价：210.00元

电　话：0411-84708842
传　真：0411-84701466
邮　购：0411-84703636
E-mail: designbooks_dutp@yahoo.cn
URL: http://www.dutp.cn

如有质量问题请联系出版中心：（0411）84709246　84709043

NIGHTCLUB

夜店设计

Roaming in Fashion and Managing the Art

1. Sunny design and sunny entertainment

Speaking of China's fashion of night, we will naturally think of crazy and romantic bars, and some of the entertainment bases are my designing works. The night revelers are lingering and indulged in the artistic charm of sound effect of light and shadow, experiencing the cutting-edge technology and aesthetic of design.

What I think more about is how to feature my designs with personality and culture, and I convince that the entertainment should be sunny. The theme of culture is refined, combining with Eastern and Western culture to create themed private rooms, and to upgrade the commercial entertainment space onto the level of culture and art. We will do the design in accord with different partitions and different themes, endowing them with distinctive personality and feature. In the past few years, I have designed many kinds of projects in different styles and themes.

2. Designers and operators

I am not only a designer, but also a consultant, planner and practitioner. I manage bars and restaurants, and I am often hired by my customers as technical advisor or invited to buy shares as shareholder. My idea is that developing is diversified instead of a point, and I combine the elements close to life together, integrating my own system and standing in different positions and angles, to understand the mentality of the owners and investors, including risk control and business model, to let them complement with each other and keep pace. So we constantly focus on new consumption models, new business models and new management models, in order to more accurately grasp the demand point of market, investors and consumers, which will endow our business works with more market competitiveness and influence.

I think the commercial constraints in design are necessary, and the designers need to be familiar with multiple areas, edge knowledge and supporting systems involved in design. From a commercial point of view, we help our customers make business decision to succeed. I deeply realize the commercial operating skills from my actual operating experience, thus to further expand to the customers' application, forming a positive cycle of development to become customers' long-term technical advisor. This brings me to establish my own hotel management team, audio creation team and other plan agencies, to extend new services in design industry, which has become the feature and advantage incomparable by other similar design companies.

3. Fashion mainstream and comprehensive design

I walk freely between theme and non-theme, mainstream and non-mainstream, extending many meaningful areas and introducing the fashion of various artistic industries into space design. From my point of view, the non-themed theme features many elements, such as music theme, science and technology theme, modern life theme, color theme and theme of the state the owner like, and we can properly combine the new idea of space design with cultural entertainment, which is thoroughly shown in Shanghai CICI CLUB. The design is integrated with night fashion, through the understanding of the stylish interior design, to interpret the international modeling clothing fashion, entertainment hair salon fashion, sound and light tech fashion, modern creative fashion and other trends, showing out the strong artistic fashion atmosphere of new entertainment space. We advocate that the entertainment design space should feature advance, to lead the industrial and public awareness of the future entertainment trends.

The comprehensive design I defined must feature a comprehensive knowledge, life experience, and the understanding of operating, and it needs to integrate different elements together. I communicate with friends in fashion industry, graphic industry and construction industry about the direction of the mainstream, and which elements we can share. Behind the interactive form, shared spirit and never-ending enthusiasm for science and technology, it prompts the needs of people. Through the win-win way of cross-border cooperation, I experience the creativity of mainstream, to improve myself and inspire others, forming a positive circle among industries.

4. Field of vision and value

In early years, I have been to many Western countries, gaining a very keen and unique insight on the international fashion culture, and put the fashion, art, culture, science and technology into perfect integration, interpreting the post-modern decoration and design style to achieve many classic works with mix and match of eastern and western cultures. We specialize in multi-integrated mix and match, like to fully integrate theater scene, mainstream fashion, industrial products, furnishing works of art, and other multimedia technology, diversified comprehensive elements into our design work, and we cooperate with designers of different areas from all over the world to successfully launch

漫游时尚 经营艺术

extremely creative diversified works of culture, beyond the definition of consuming goods, bringing our consumers with more strong sense of art and spiritual enjoyment.

The level of vision determines that whether a designer is unique, and the space we design highlights the participation, respects individuality and shows the application of science and technology. Interaction and sharing are positive and wonderful states, allowing us to treat others and ourselves in a broader and opener area. Integrating with others to share the beauty of fashion and the beauty of design, to provide mutual penetration and impact on design elements, style and concept.

Chen Wu

（一）阳光设计 阳光娱乐

说起中国的夜时尚，自然地便会想到疯狂、浪漫的酒吧，其中一些欢乐根据地便是我的作品。夜色中狂欢的人们流连、沉醉于光影音效的艺术魅力，体验前沿的科技与设计美感。

我思考更多的是如何把自己的作品做出个性与文化来，坚信娱乐就应该是阳光的，将文化的主题进行提炼，结合东西方文化打造主题包房，把商业的娱乐空间提升到文化艺术的层面上。我们会将作品分区分主题地进行设计，令其皆具鲜明的个性特征。这几年中，我设计了多种风格和主题的作品。

（二）设计者 经营者

我不仅是设计师，也是咨询师、策划人和实践者。自己经营酒吧和餐厅，经常被客户聘为技术顾问或是被邀去做股东。我的观念是：发展是多元性而非一个点，把很贴近生活的元素结合在一起，融入自己的体系，站在不同的立场与角度，了解业主与投资者的心态，包括风险的控制、经营的模式，使其相辅相成，并驾齐驱。这样不断地关注新的消费模式、新的经营模式、新的管理模式，从而更准确地把握市场、投资者与消费者的需求点，使我们的商业作品更具市场竞争力和影响力。

我认为设计中的商业约束是有必要的。设计涉及的多元领域、边缘知识、配套系统，设计师都有必要通晓。要站在一个商业的角度，从经营的决策上给予客户帮助，帮助客户成功。我从实际经营经验中更深切地体会到商业的运作技巧，从而再扩大到客户中，形成良性的循环发展，成为客户长期的技术顾问，也由此我建立起了自己的酒店经营团队、音乐创作团队等策划机构，延伸出设计行业中的新型服务，这也成为别的同类设计公司无法比拟的特点与优势。

（三）时尚主流 设计共和

在主题与非主题、主流与非主流之间，我自由游走，延伸出许多有意义的领域，并将各个艺术行业的时尚纳入空间设计中。在我看来，非主题的主题有很多元素，如音乐主题、科技主题、现代生活主题、色彩主题及业主希望的某种状态的主题，我们可以恰当地将空间设计新思维与文化娱乐进行有机结合，这一点在上海 CICI CLUB 中表现得淋漓尽致。把设计与夜时尚相结合，通过对时尚室内设计的理解，诠释了国际造型服饰时尚、娱乐发艺时尚、声光科技时尚与现代创意时尚等潮流趋势，展现出新娱乐空间浓郁的艺术时尚氛围。我们倡导在娱乐设计空间上要具有超前性，更极力引导行业、大众对未来娱乐趋势的认知。

我定义中的设计共和，必须有综合的知识、生活的阅历，还要有对经营的认识，需要把不同的元素组合在一起。跟服装界、平面界、建筑界的朋友一起探讨主流方向是什么，有哪些元素可以共享。这种互动的形式、共享的精神以及对科技永不停止的热情背后是人的需求。通过这种跨界合作共赢的方式我领略到了主流的创造力，从而提高了自己，启发了别人，形成了业界间的良性循环。

（四）眼界 价值

早年我曾游历西方多国，对国际时尚文化有着极为敏锐的观察和独到的领悟，并将时尚、艺术、文化和科技完美结合，演绎后现代主义的装饰设计风格，成就了许多东西方文化混搭的经典作品。我们擅长于多元综合混搭，喜欢将戏剧场景、主流时装、工业产品、装饰艺术品等多媒体科技和多元化的综合元素充分融合到设计作品中，与世界各国不同领域的设计师合作，从而成功推出极具创意、极具多元化的文化作品，超越了消费品本身的定义，更多地是带给消费群强烈的艺术观感与精神享受。眼界高低决定了设计师是否与众不同，我们所设计的空间突出参与性、尊重个性、体现科技的使用角度。互动与共享是两种积极美妙的状态，它们让人在更宽广更开放的领域内看待自己和他人。联合起来与人们分享时尚之美、设计之美，让彼此在设计元素、风格和理念上相互渗透、互相影响。

陈武

Contents 目录

"ArKADIA" Fontainebleau Hotel and Resort Leisure Center

—— 枫丹白露酒店与度假村休闲中心 "ArKADIA"

In October 2010, an adults' entertainment place is opened in Fontainebleau Miami, where the brilliant lighting and high-quality materials create an attractive space together, and the stylish environment and sexy lines feature the most intimate contact, making people relaxed and indulged in the space...

The designer Francois Frossard introduces the concept of mirror image jewelry box to create the space. The gold mirror on ceiling is cut by laser, coupled with the all-round reflection from diamond-shaped mirrors on bar and DJ booth, shining light and mirror image effect to make the space more brilliant and bright. Different from nightclubs of the same type, the space is designed according to its size and equipped with customized furniture. The beige sofa, silver and gold tea table, white side tale...stand quietly among the swaggering clusters of silver "pillars", describing the elegance and fashion. LED lights timely debut to cover the space with a layer of blurred color. The atmosphere is just right, waiting for the protagonist debut...

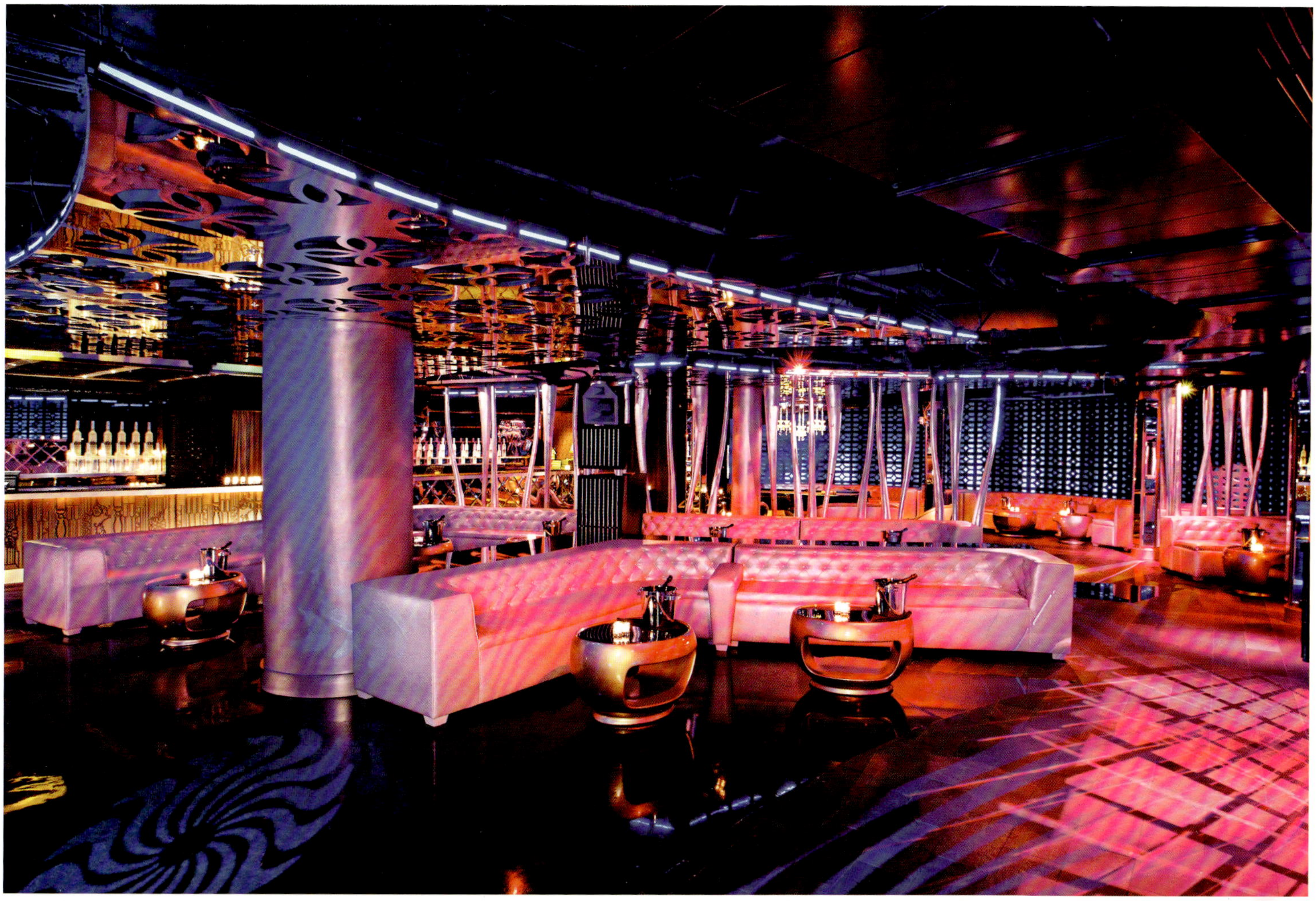

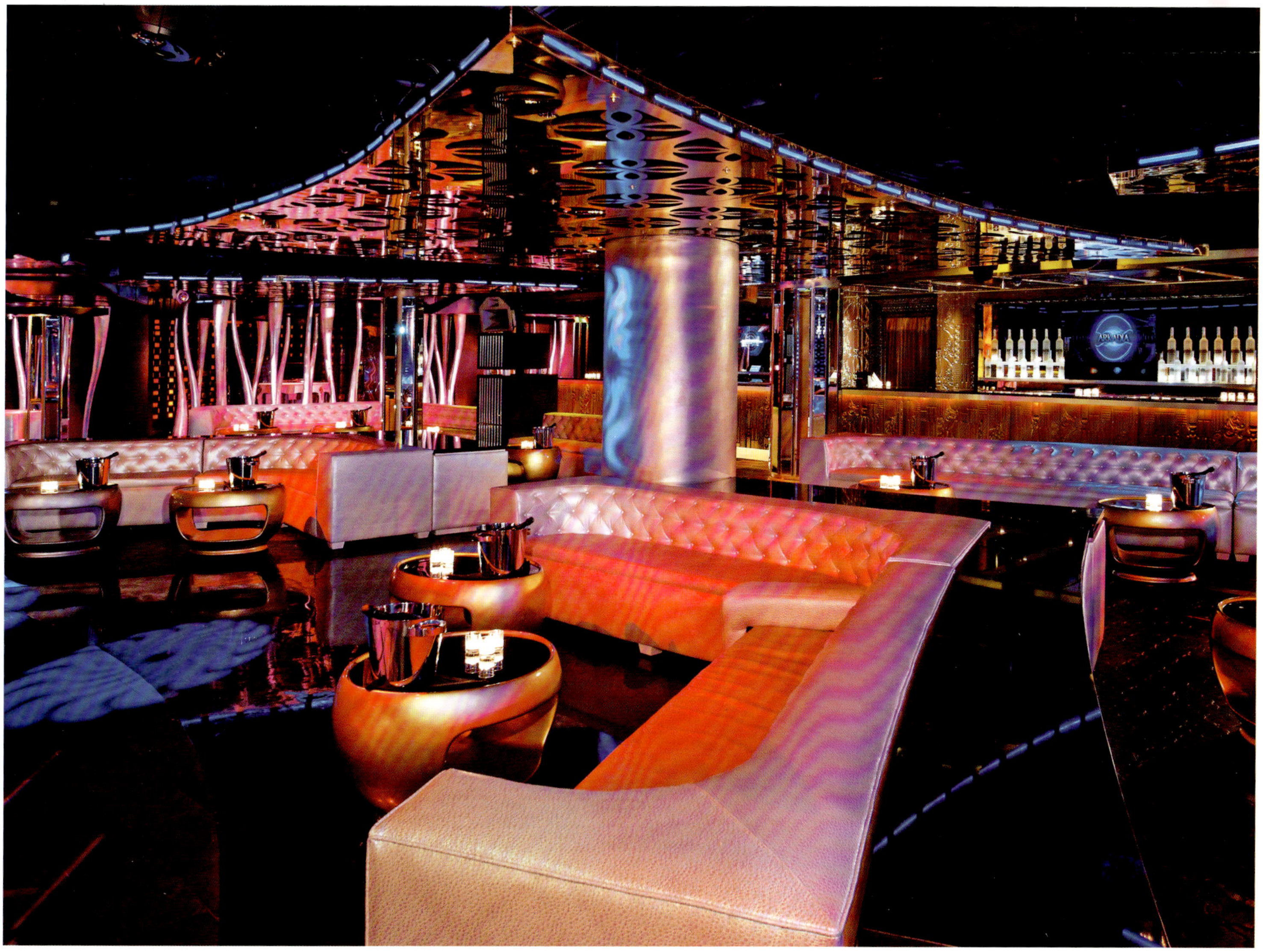

2010 年 10 月，一家成年人的娱乐场所在迈阿密枫丹白露酒店开张，绚丽的灯光与高品质的材料共同营造了一个诱人的空间，时尚的环境与性感的线条带给人最亲密的接触，让人放松、沉溺其中……

设计师 Francois Frossard 以镜像首饰盒的概念来打造这个空间。对天花板上的黄金镜子实施激光切割，加上吧台、DJ 台菱形镜子的多面反射、灯光辉映及镜像效果，让空间更添璀璨光芒。不同于其他相同类型的夜总会，设计师为之量身设计、定制了家具。米色的沙发，银色和金色的茶几，白色的边几……在招摇丛生的银色"柱子"间静静侍立，书写着优雅与时尚。LED 灯适时登场，为空间蒙上了一层迷离的色彩。氛围刚刚好，只等主角登场……

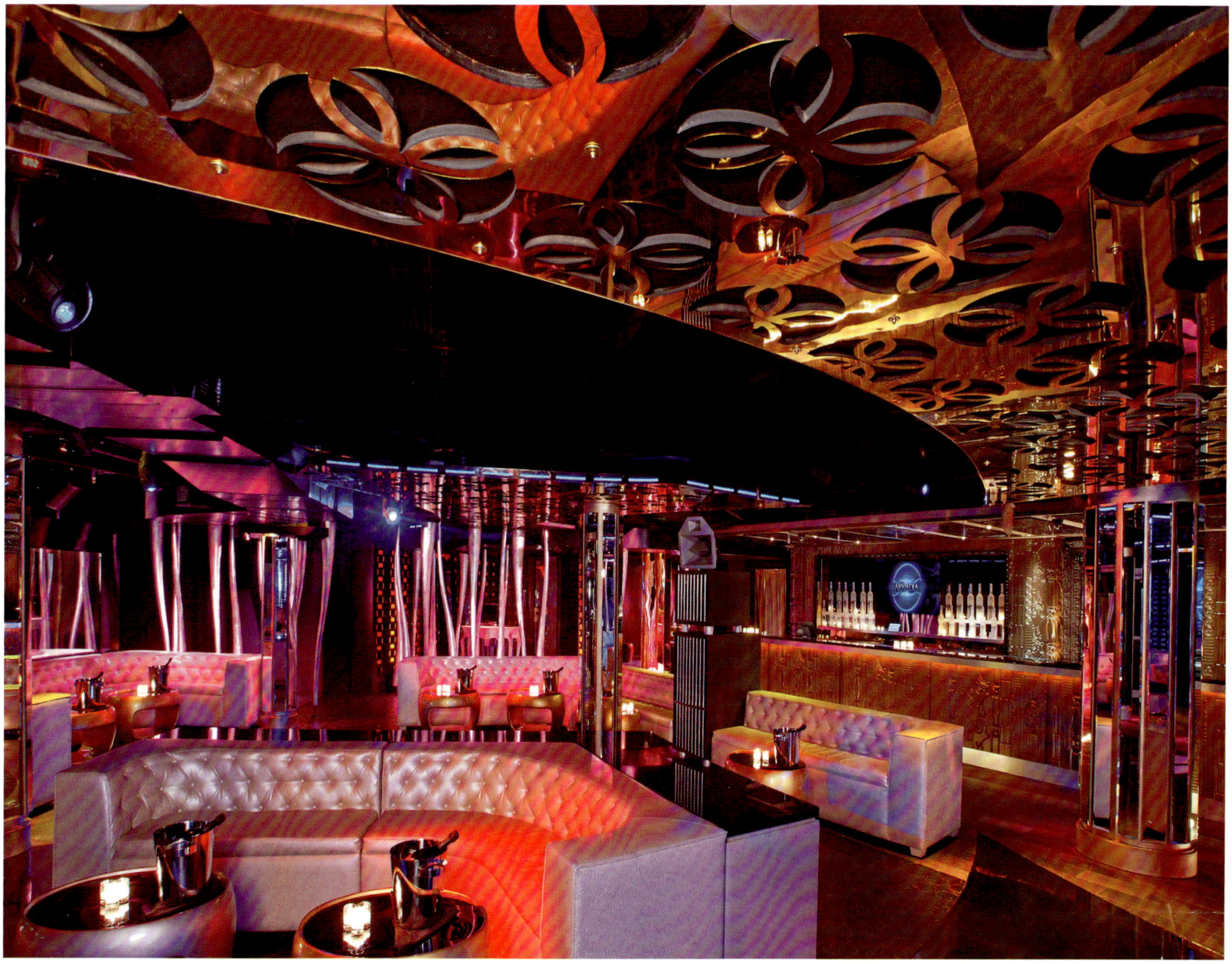

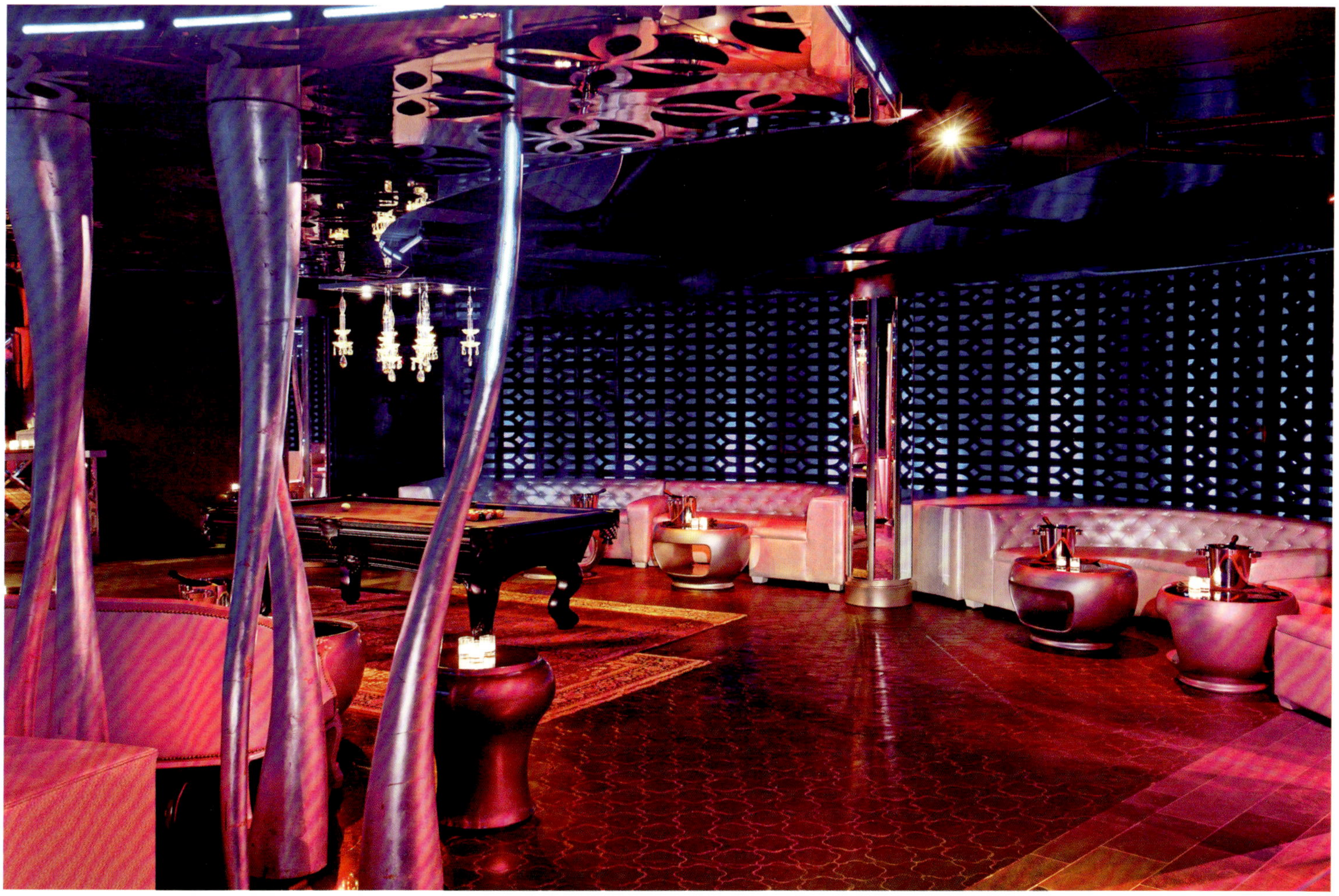

地点	面积	设计师	设计公司
/ 美国拉斯维加斯	/ 1116m²	/ Francois Frossard	/ FFD Inc

"PRIVE" LAS VEGAS

拉斯维加斯 "PRIVE"

"PRIVE" is a subsidiary bar of Las Vegas Hollywood Planet Hotel, located in Las Vegas Street, next to BELLAGIO fountain, and is a stylish and elegant place for relaxing at night in the hotel.

At the entrance of the space, the first thing catches customers' eyes is the magnificent and generous main bar. The concave-shaped ceiling over the bar area is designed with granite, and the bar and wine cabinet are high-gloss wooden products. The inverted pyramid-shaped crystal chandelier and kaleidoscope-style lighting at the entrance describe an elegant and generous atmosphere. Turn right is a VIP private enjoyment area, where customers can enjoy the comfortable chairs and intimate dialogue, with a private bathroom, presenting a blend of private and low-key.

In the luxurious space of more than 1000 square meters, the designer Francois Frossard creates a most typical Las Vegas. The ethereal fantasy, dramatic elements, sleek intentional decorations and cutting-edge lighting design perfect the space expression, creating a unique atmosphere to allow customers to have an expectation of surprising performances, celebrity guests and first-class service.

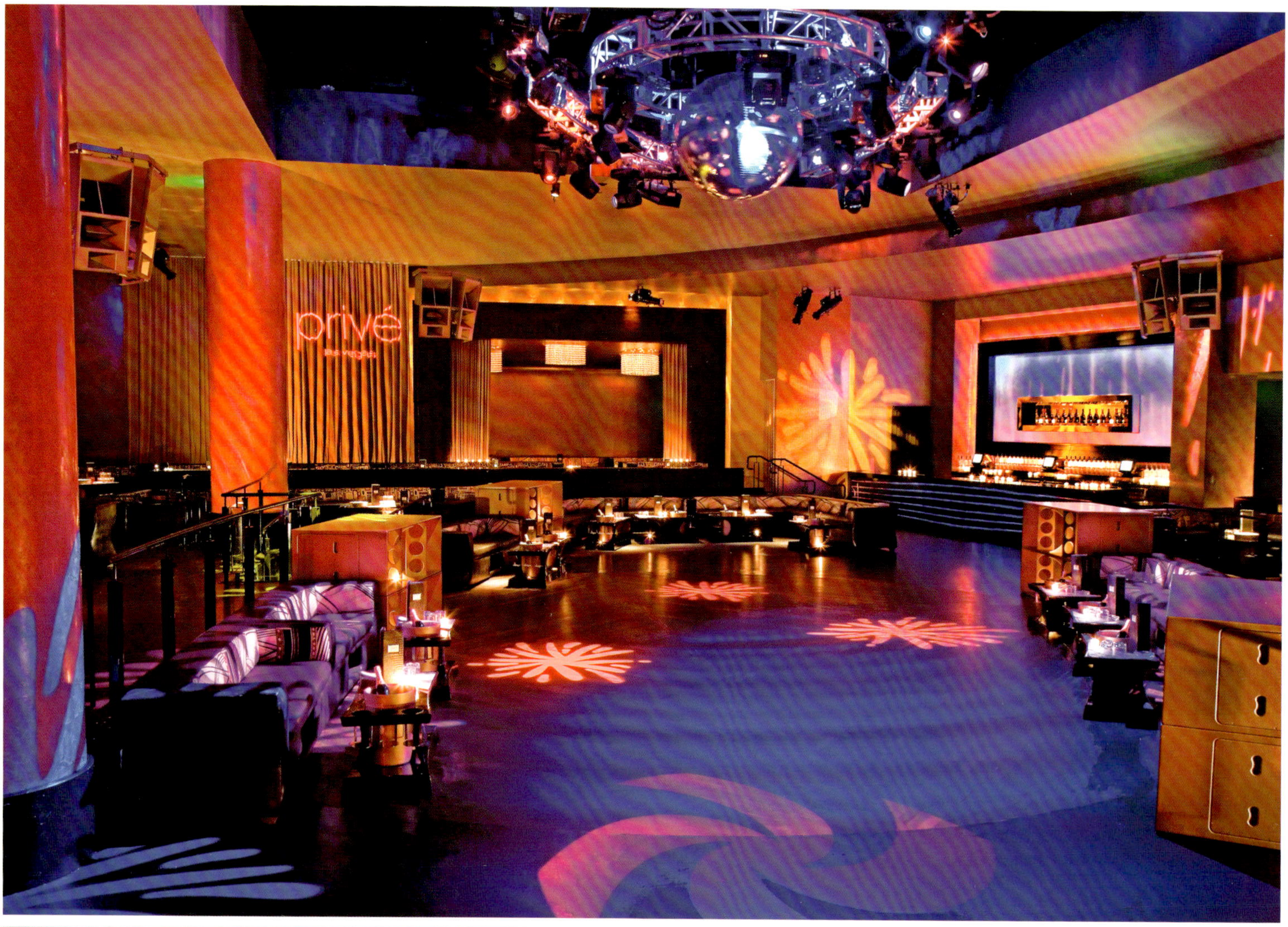
privé

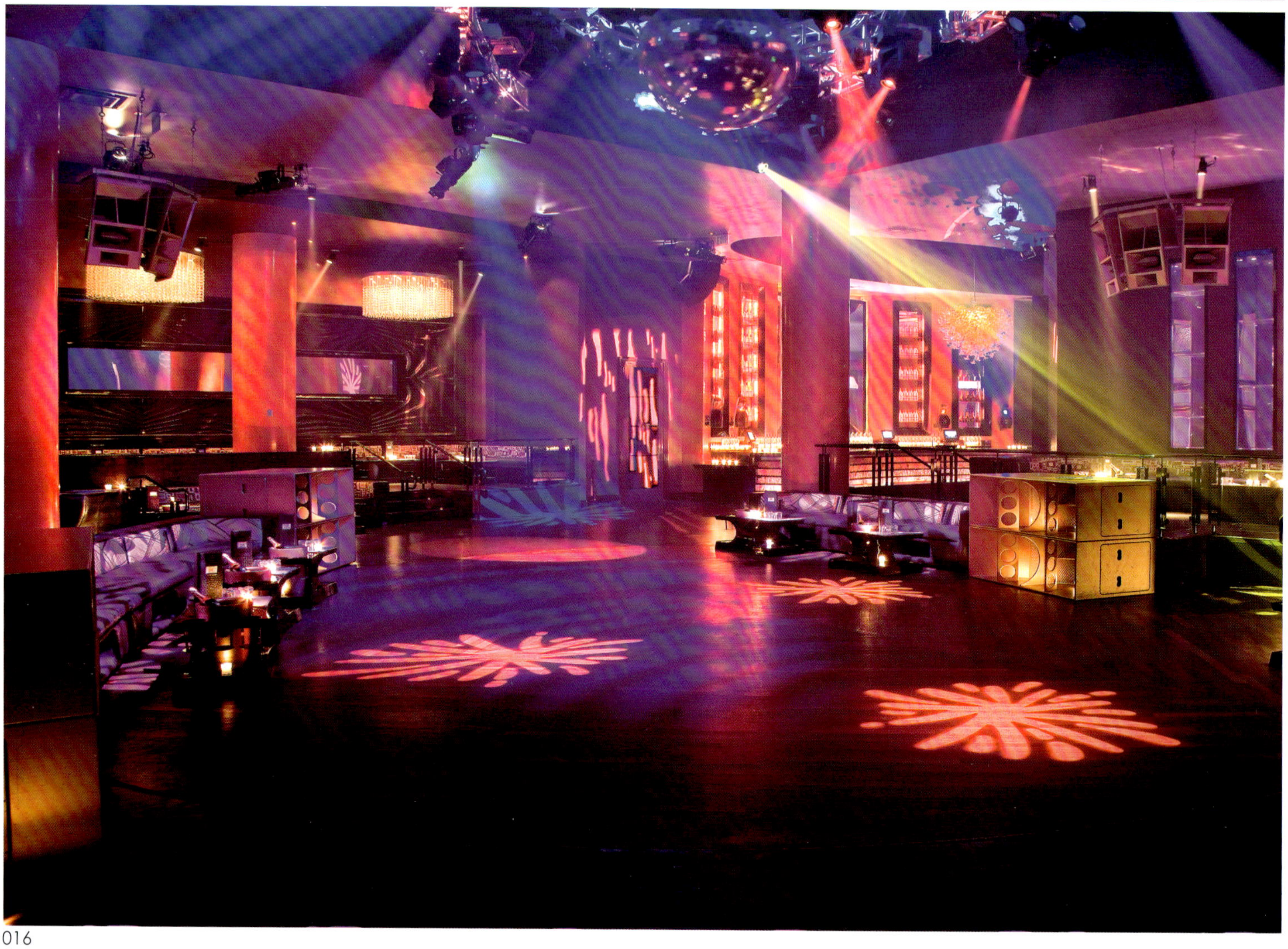

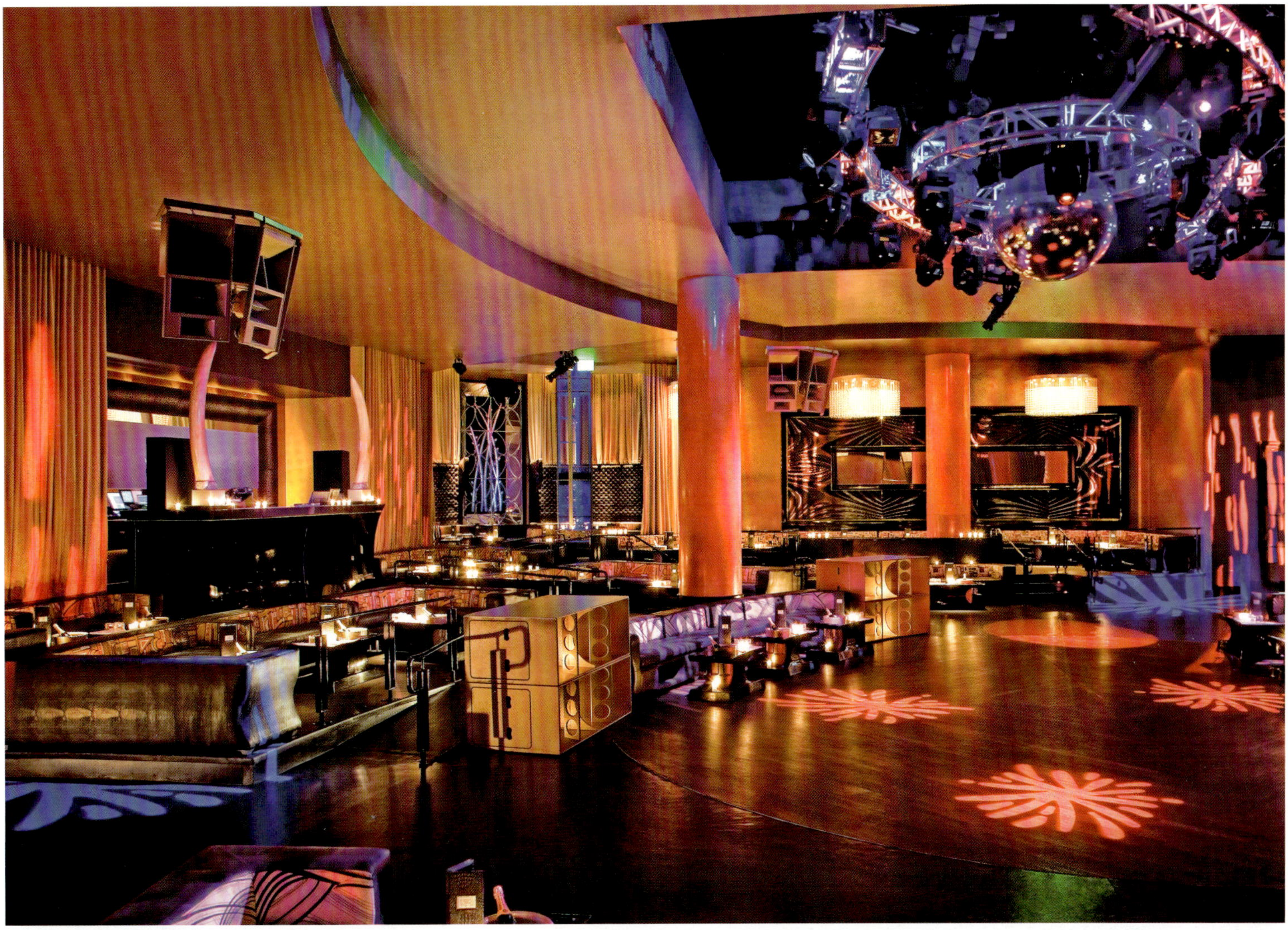

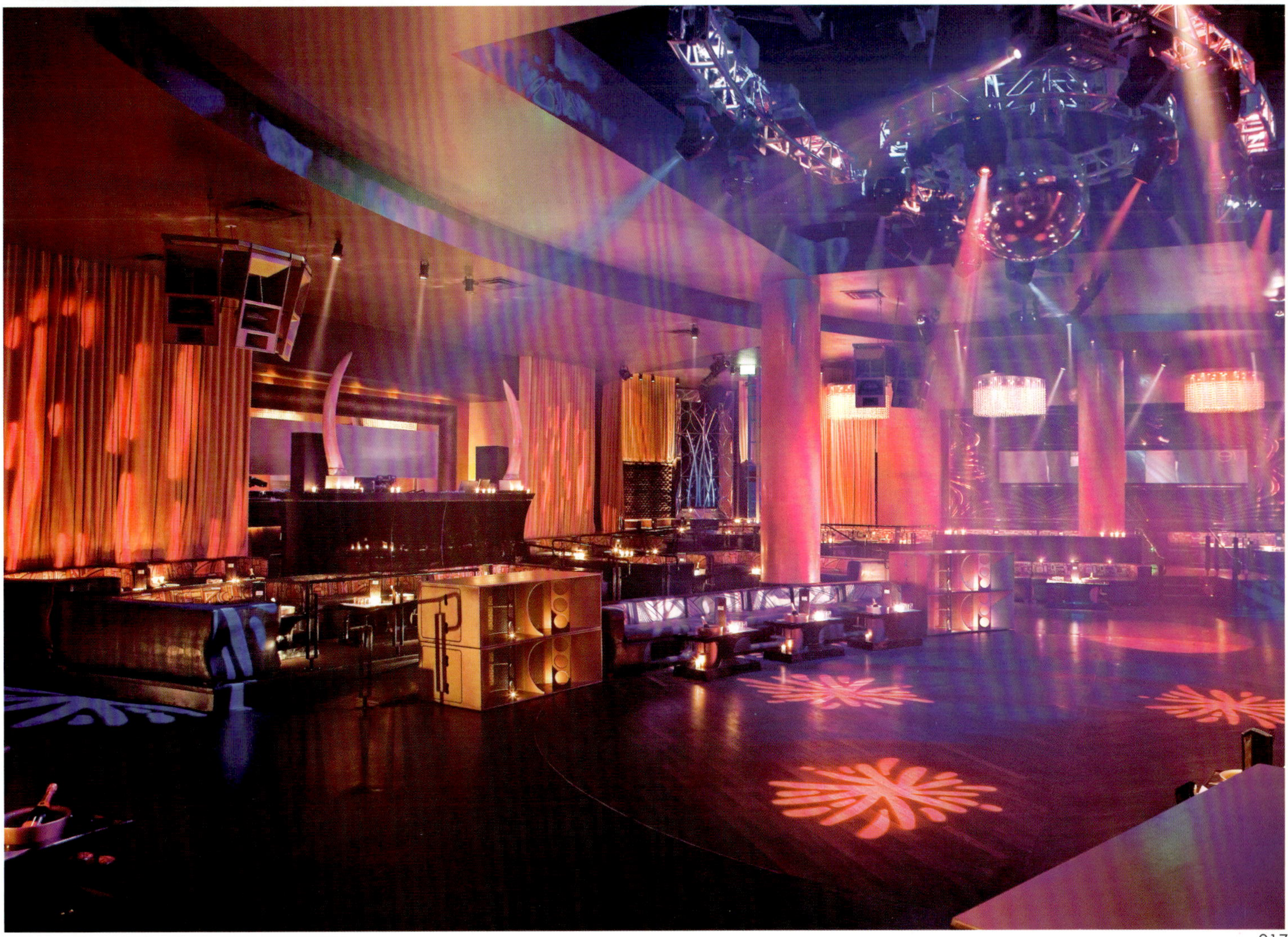

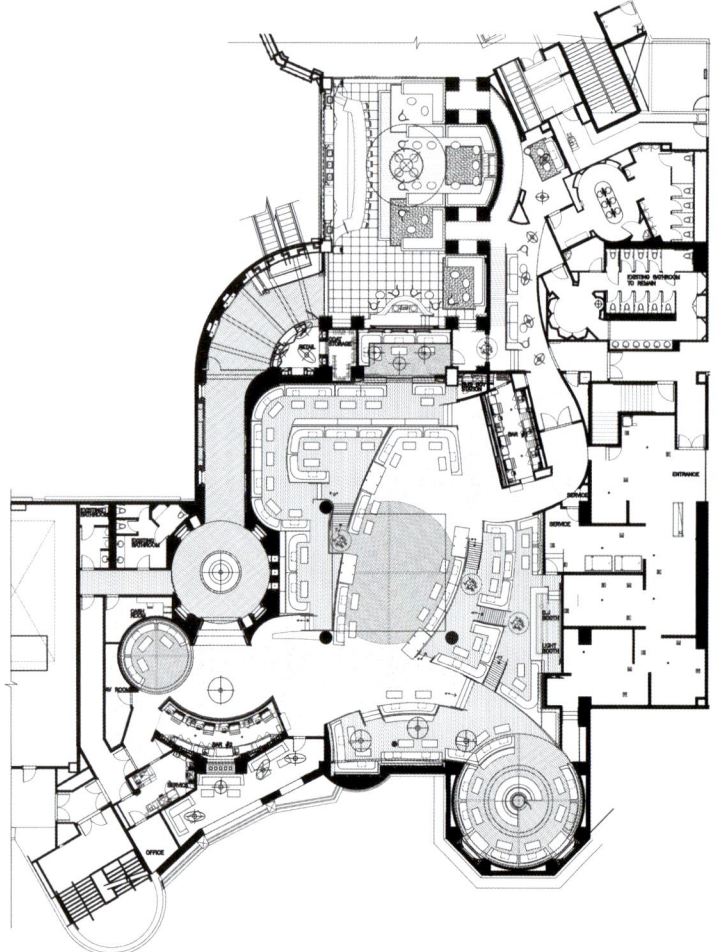

"PRIVE"是拉斯维加斯好莱坞星球酒店的附属酒吧，位于拉斯维加斯大道，毗邻BELLAGIO喷泉，是酒店时尚、高雅的夜间休闲场所。

进入空间，首先看到的是恢弘大气的主吧台。凹面造型的吧台区顶部以花岗岩打造，吧台及酒柜是高光泽的木制品，倒金字塔式的水晶吊灯和万花筒式的入口照明书写出一派优雅、大气。向右转，是贵宾私享区，可以享受舒适的座位和亲密对话，同时具有私人浴室，是私密和低调的交融。

在这个1000多平方米的豪华空间里，设计师Francois Frossard创造出了一个最具典型性的拉斯维加斯。空灵的想象、戏剧性的元素、圆滑的意向装饰和尖端的照明设计都完善了空间表情，营造出独特的氛围，让客人充满期待，期待充满惊喜的表演、名人嘉宾和一流的服务。

OTHER TENANT SPACE
STORAGE
ELECT. ROOM
BAR
CLEANING CLOSET
EXIT TO STREET
WOMEN
STORAGE
FOYER
ENTRY HALL
BAR
MEN
BAR
OTHER TENANT SPACE
D.J. BOOTH
SERVICE
SERVICE ROOM
ELECTRICAL ROOM
EXIT TO STREET

MENS RESTROOM
SERVICE
D.J.
SERVICE
LIGHT BOTH VIDEO BOOTH
STORAGE
LIQUOR ROOM
BAR
BAR
PRIVATE ROOM
BAR
WOMENS RESTROOM
OFFICE
RESTROOM
STORAGE

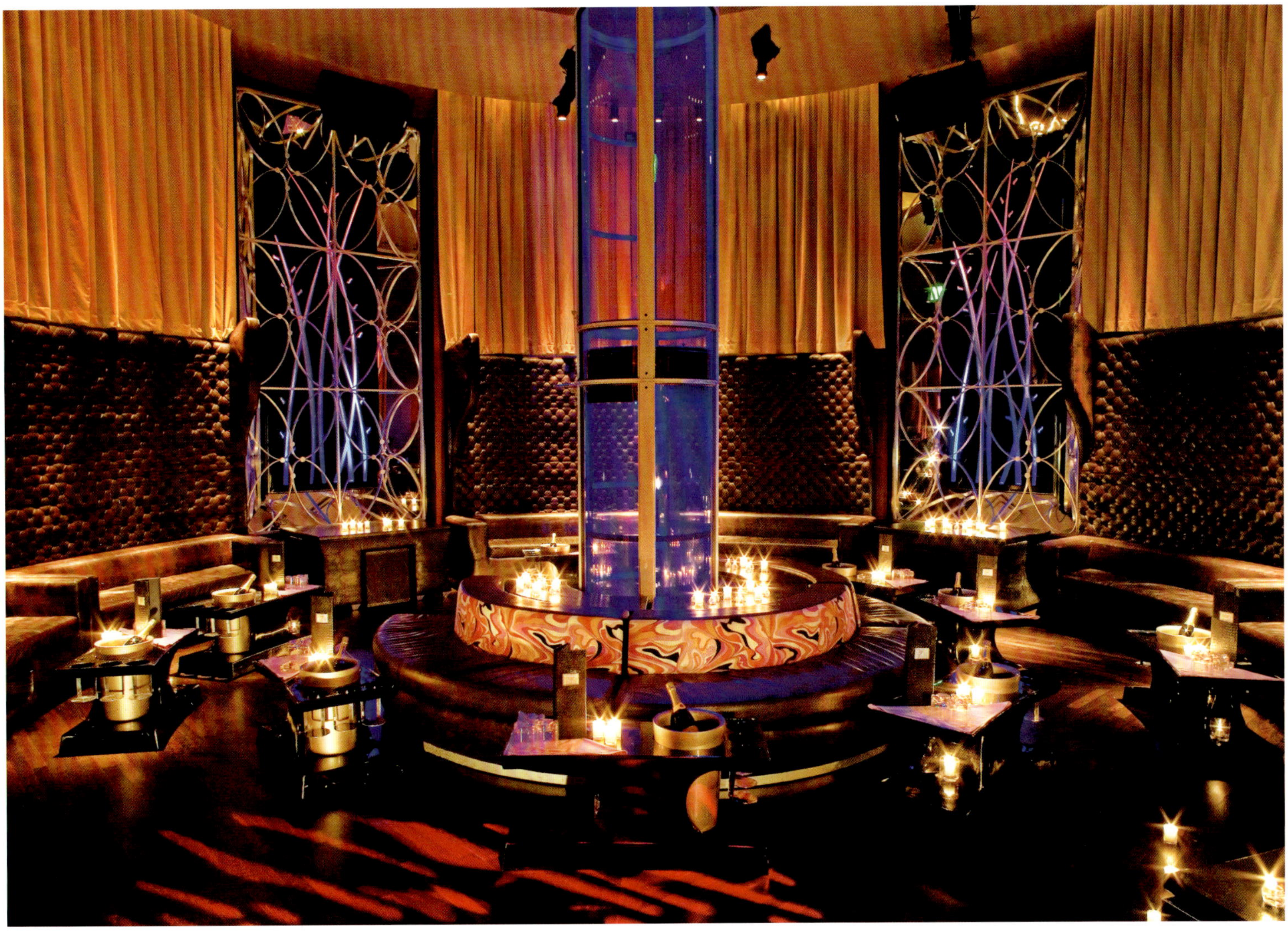

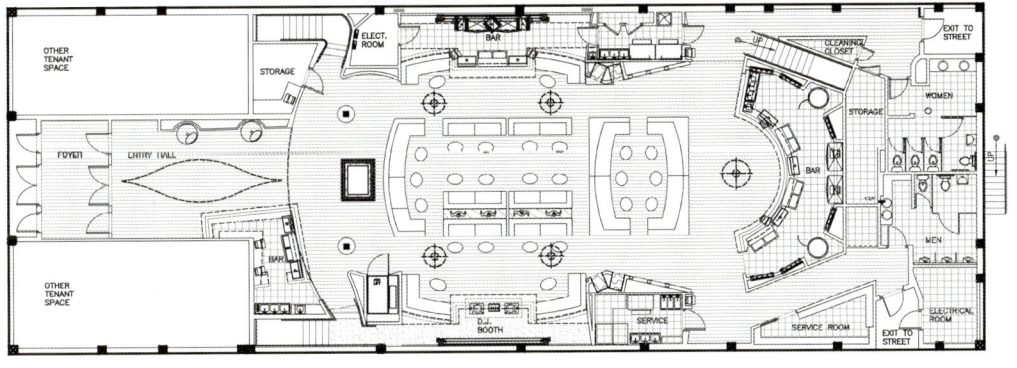

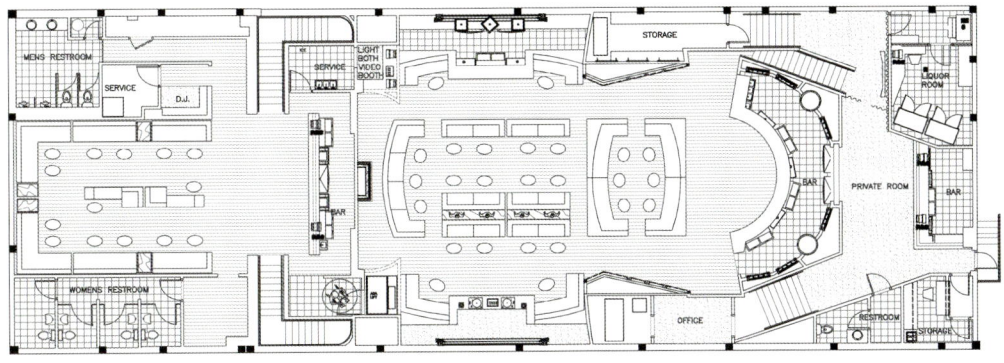

地 点	面 积	设计师	设计公司
/ 美国拉斯维加斯	/ 930m²	/ Francois Frossard	/ FFD Inc

SET Miami Beach

——迈阿密海滩上的 "SET"

"SET" is a nightclub project on Miami Beach. The designer Francois Frossard endows the whole club with "gorgeous and charming, noble and elegant" style, to create what he called "Hollywood luxurious mansion atmosphere in 1940s."

The space of the small dance floor is arranged and decorated with fireplace, decorative ivory, fabrics and other items. The heightened space is equipped with pneumatic lift, and the wall around the dance floor is designed with special movable stage for dancers. The lifting stage is made of metal frames and glass surfaces, and the walls on both sides are designed with neat wine shelves. In the echo of lighting and the reflective effect of glass, the red wall increases the charm of the space. The disco balls of widespread use are replaced by four glittering Swarovski crystal chandeliers, providing a sense of level to enrich the color and form of the space, and with the combination of European style and modern elements, the space seems luxurious and elegant, full of appeal.

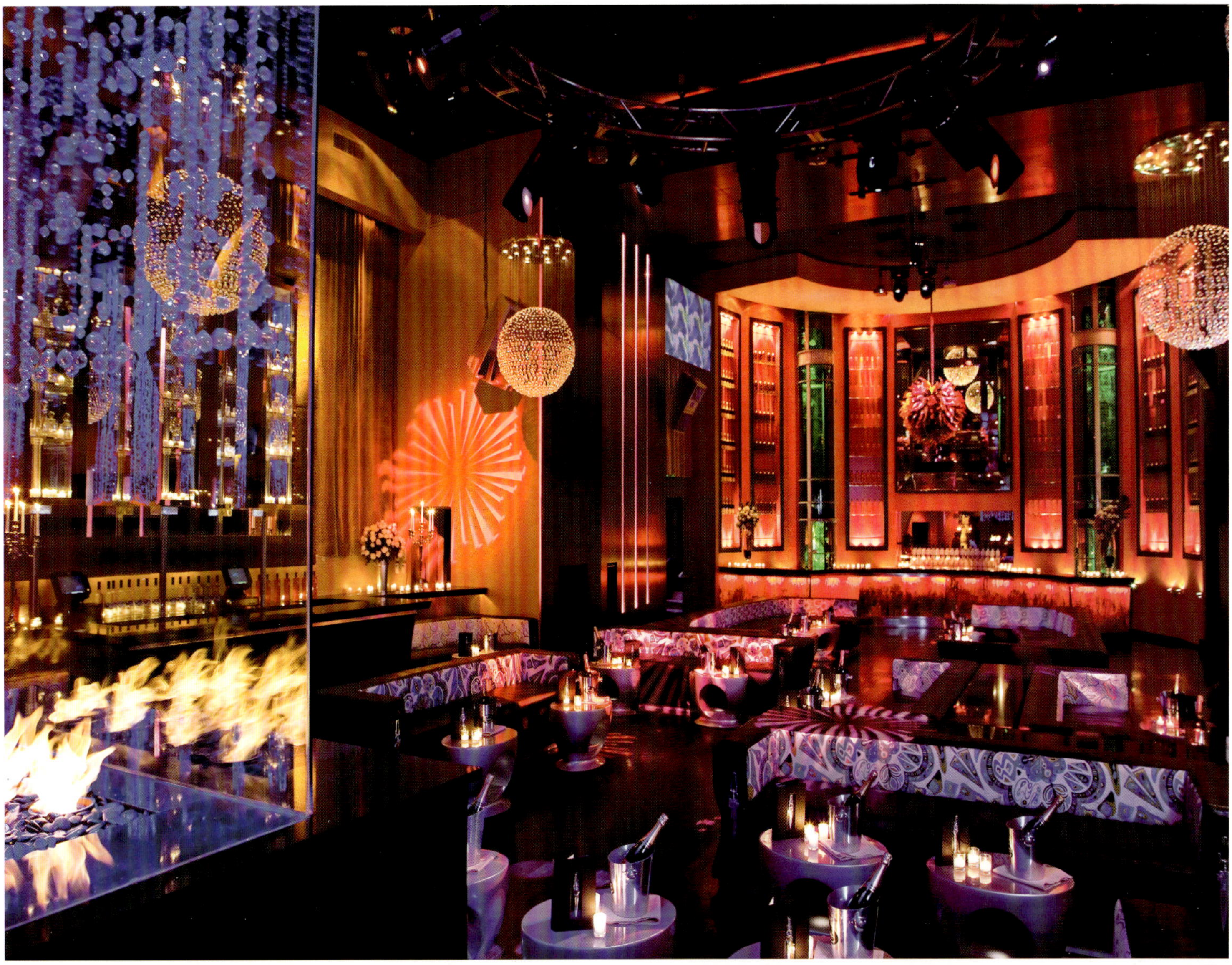

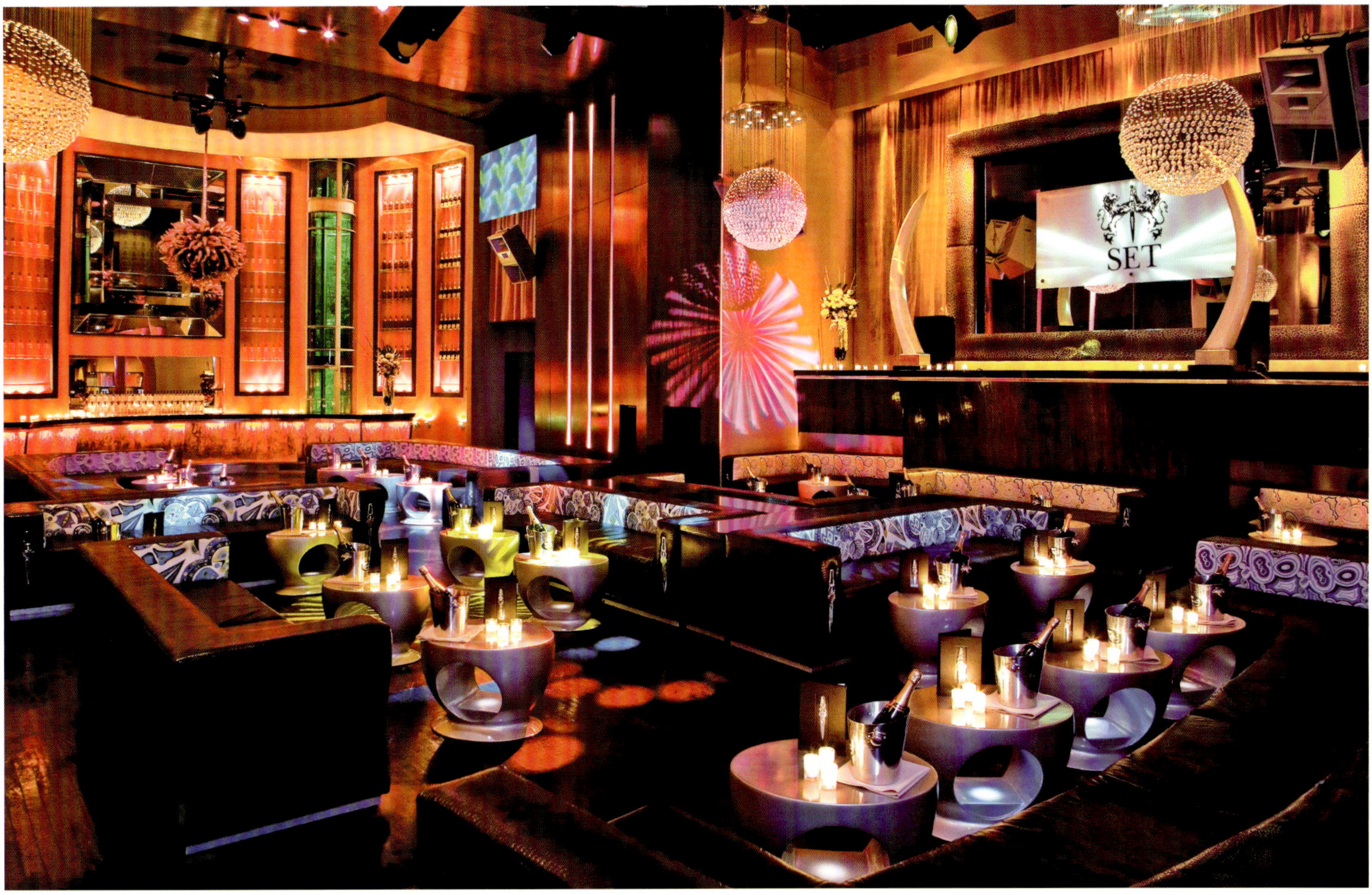
SET

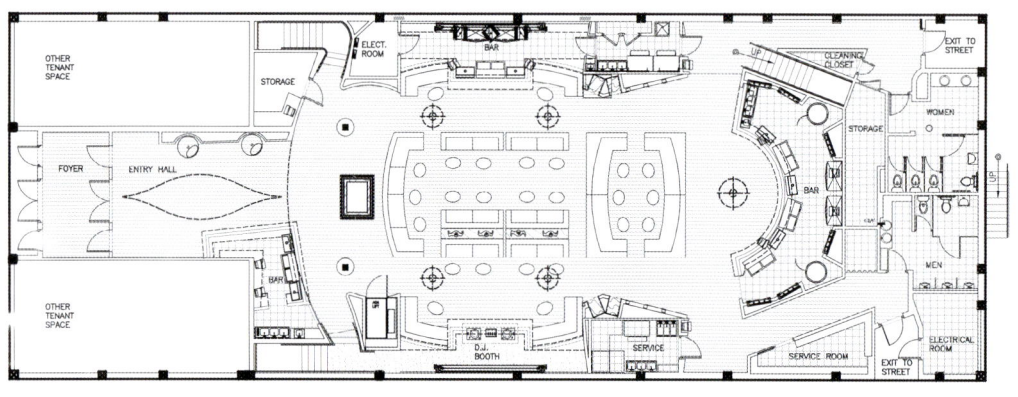

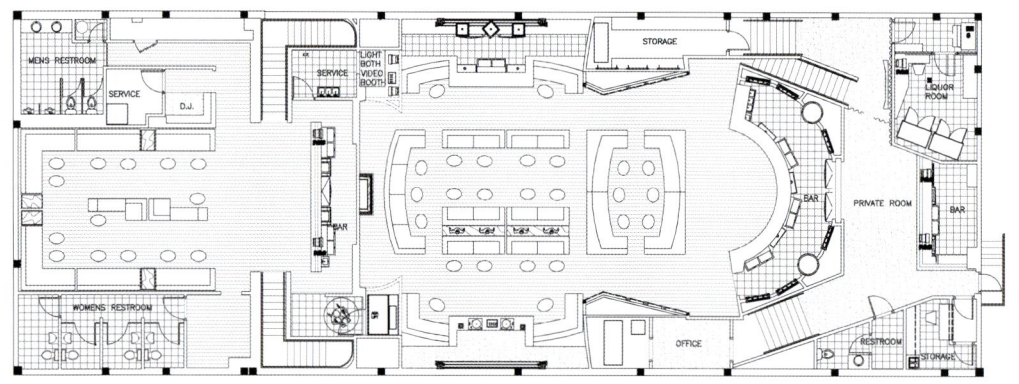

　　"SET"属于迈阿密海滩的夜总会项目。设计师Francois Frossard赋予整个空间"绚丽迷人、高贵优雅"的格调，创造出他所谓的"20世纪40年代好莱坞豪宅氛围"。

　　小舞池空间以壁炉、装饰象牙、织物等物品布置、装饰。挑高的空间配有气动升降机，在舞池周边墙上为舞者打造出专门的移动式舞台。升降舞台由金属架和玻璃面打造，两边的墙上是整齐的置酒架，红色的墙面在灯光的辉映与玻璃的反光作用下，增添了空间魅力。普遍使用的迪斯科球被四个闪闪发光的球状施华洛世奇水晶吊灯所取代，由此产生的层次感丰富了空间色彩和形式，加上欧式风格与现代元素的结合，空间显得豪华而优雅，充满吸引力。

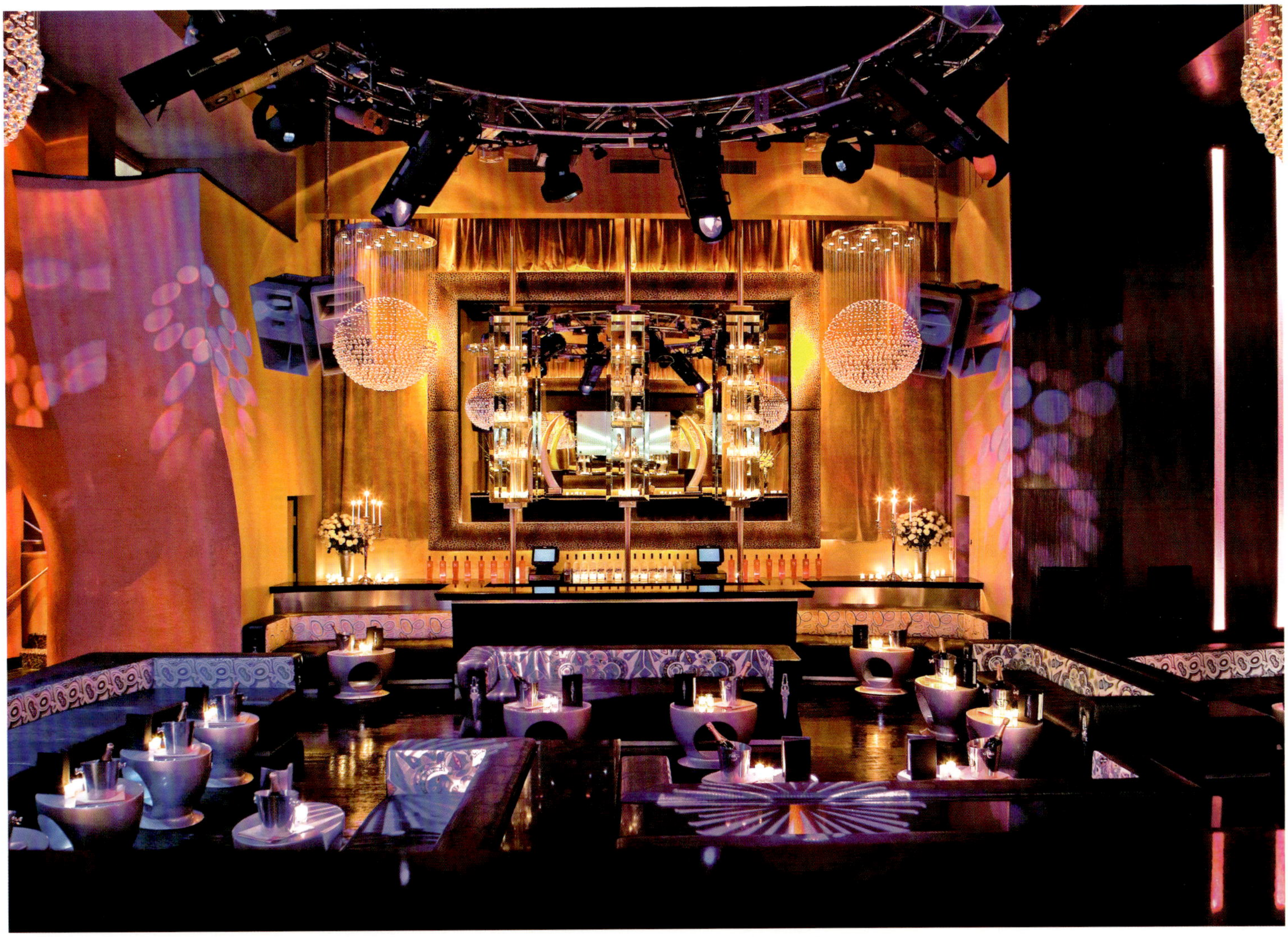

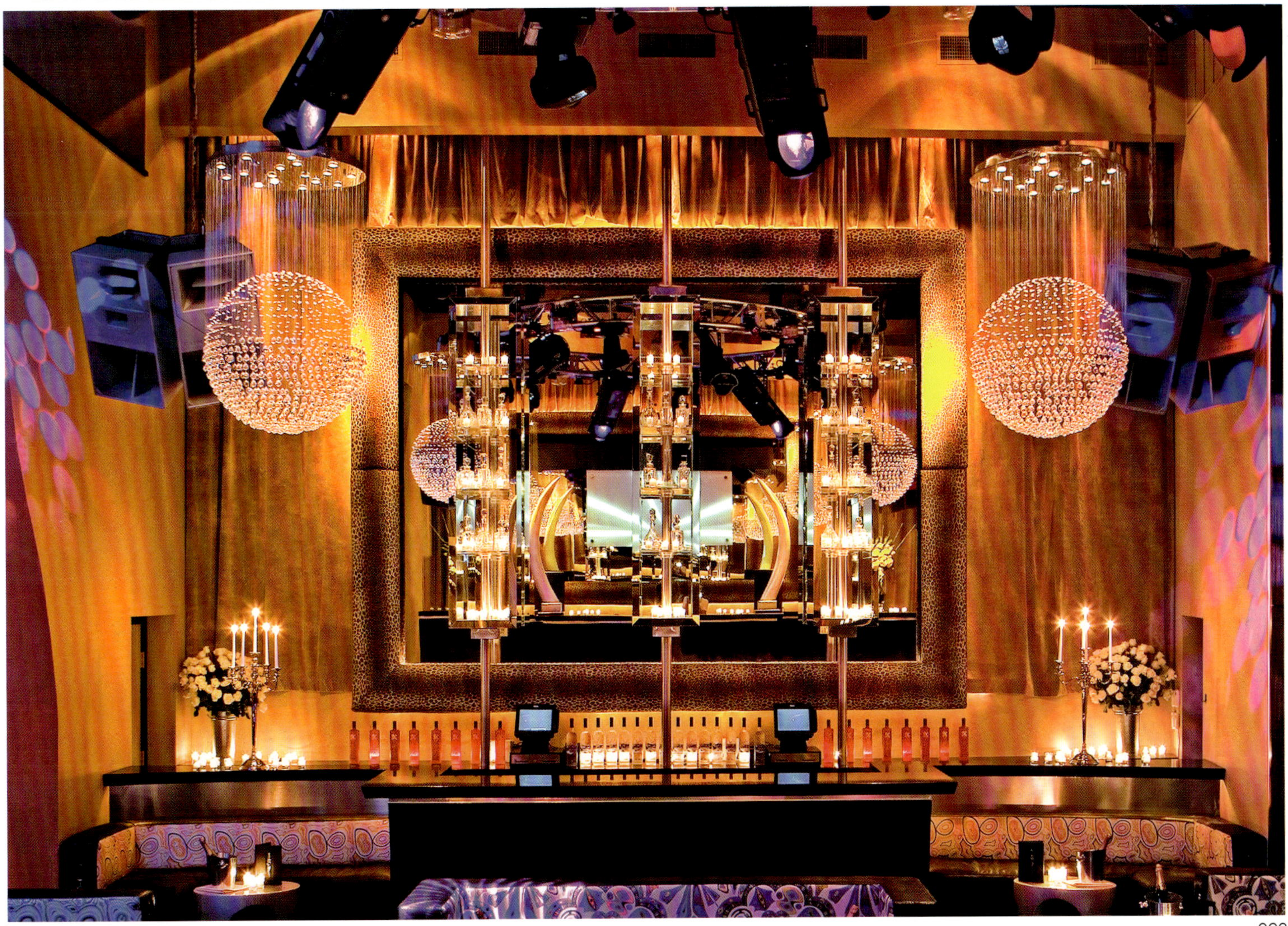

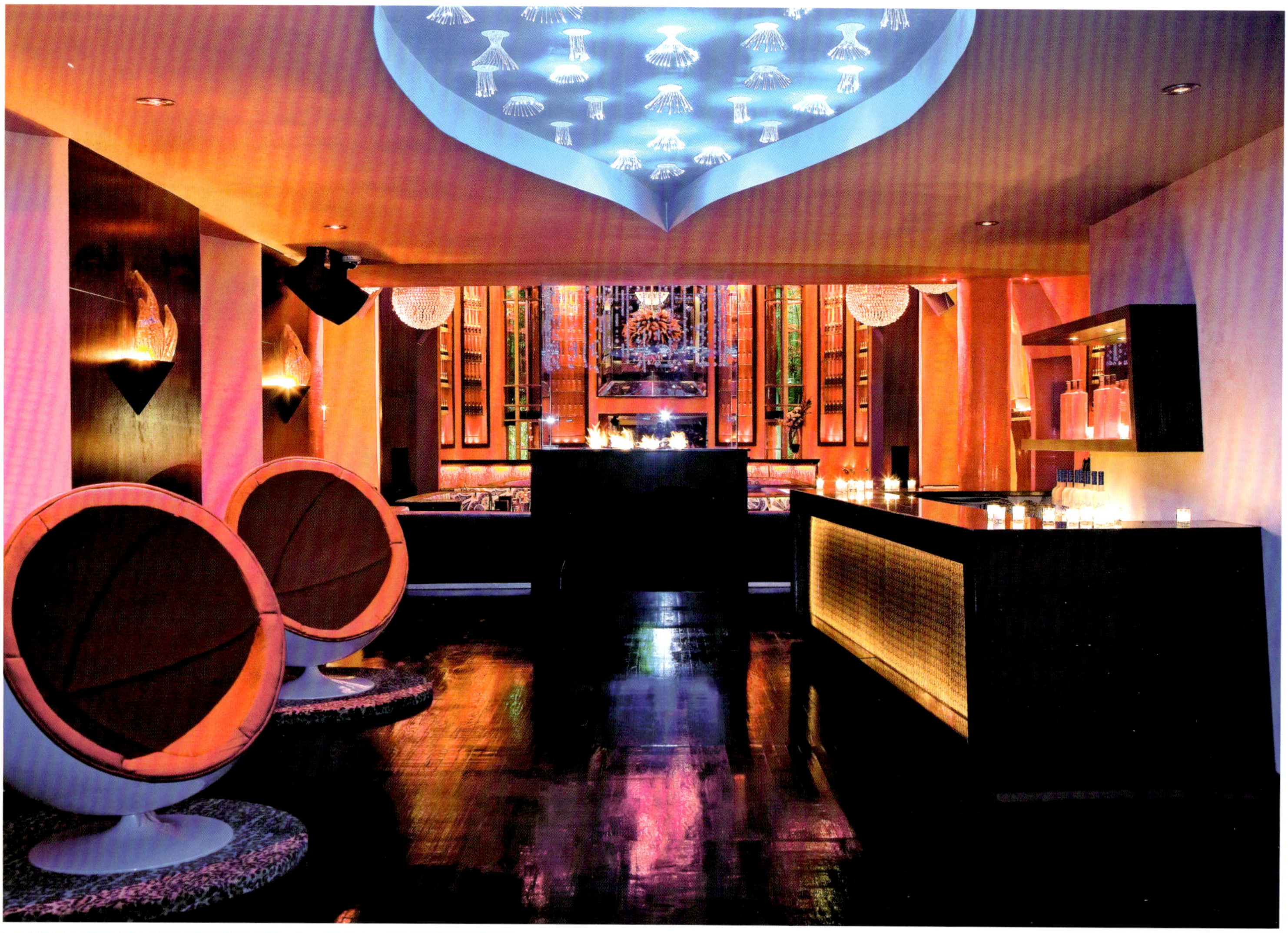

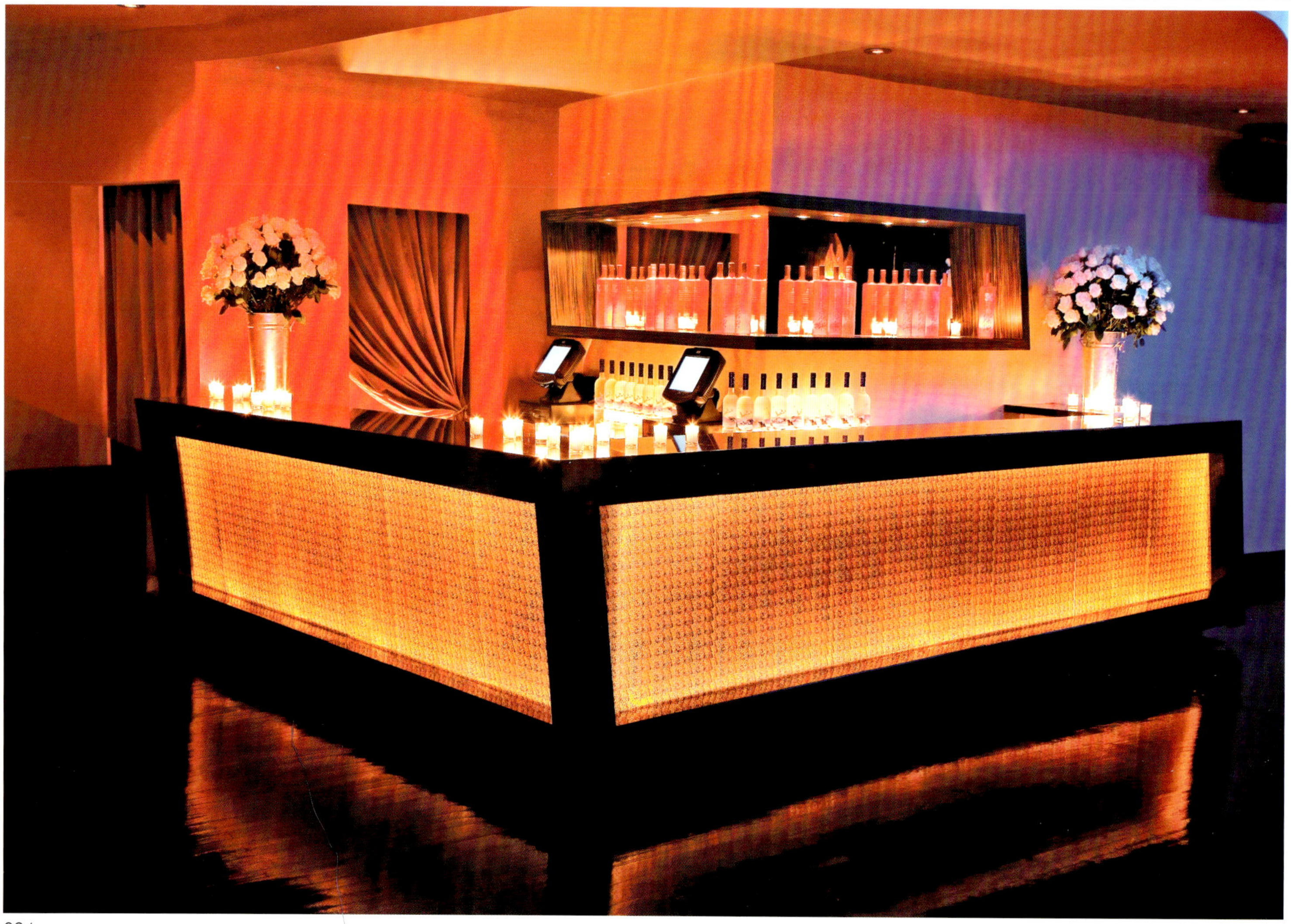

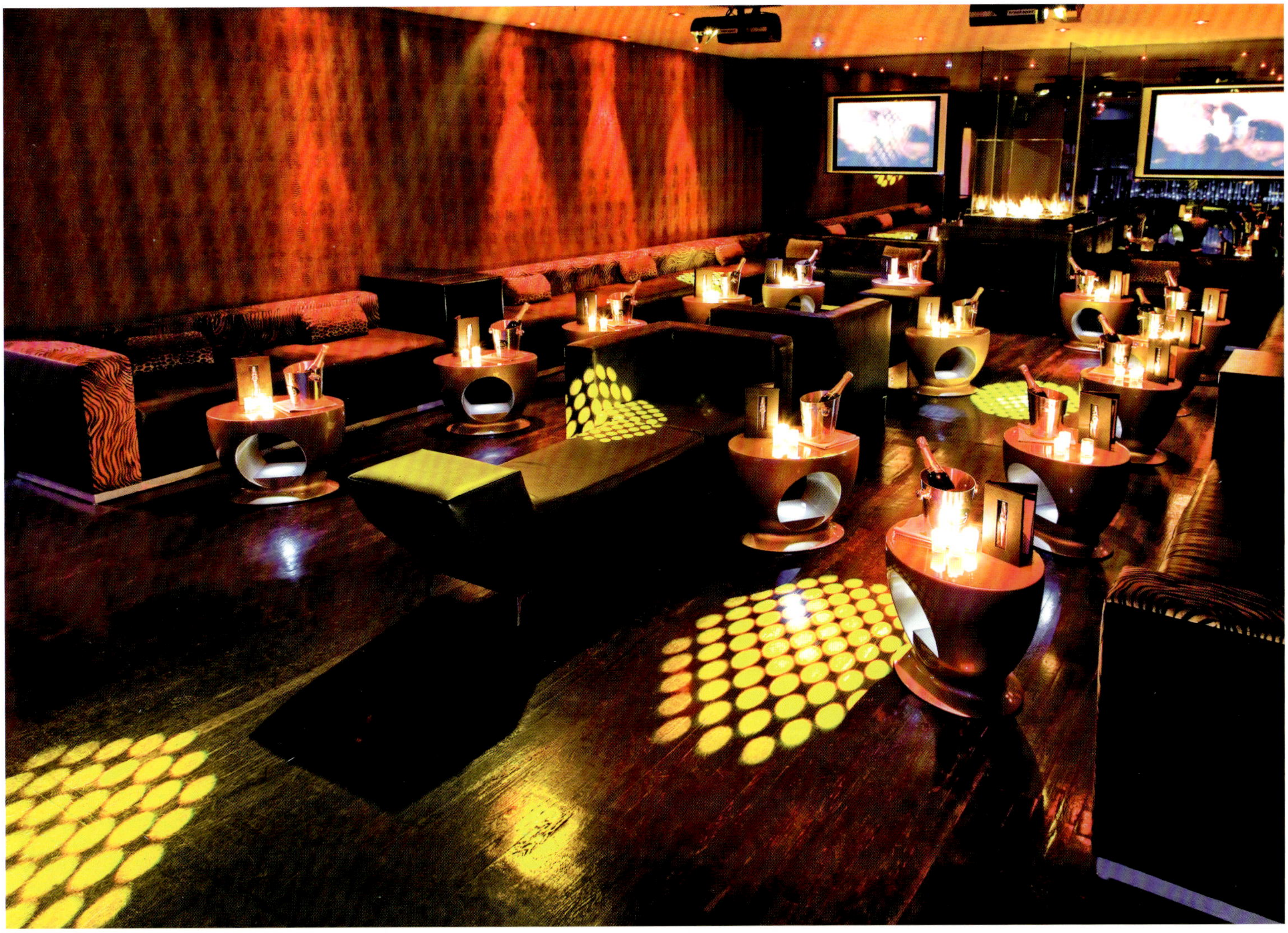

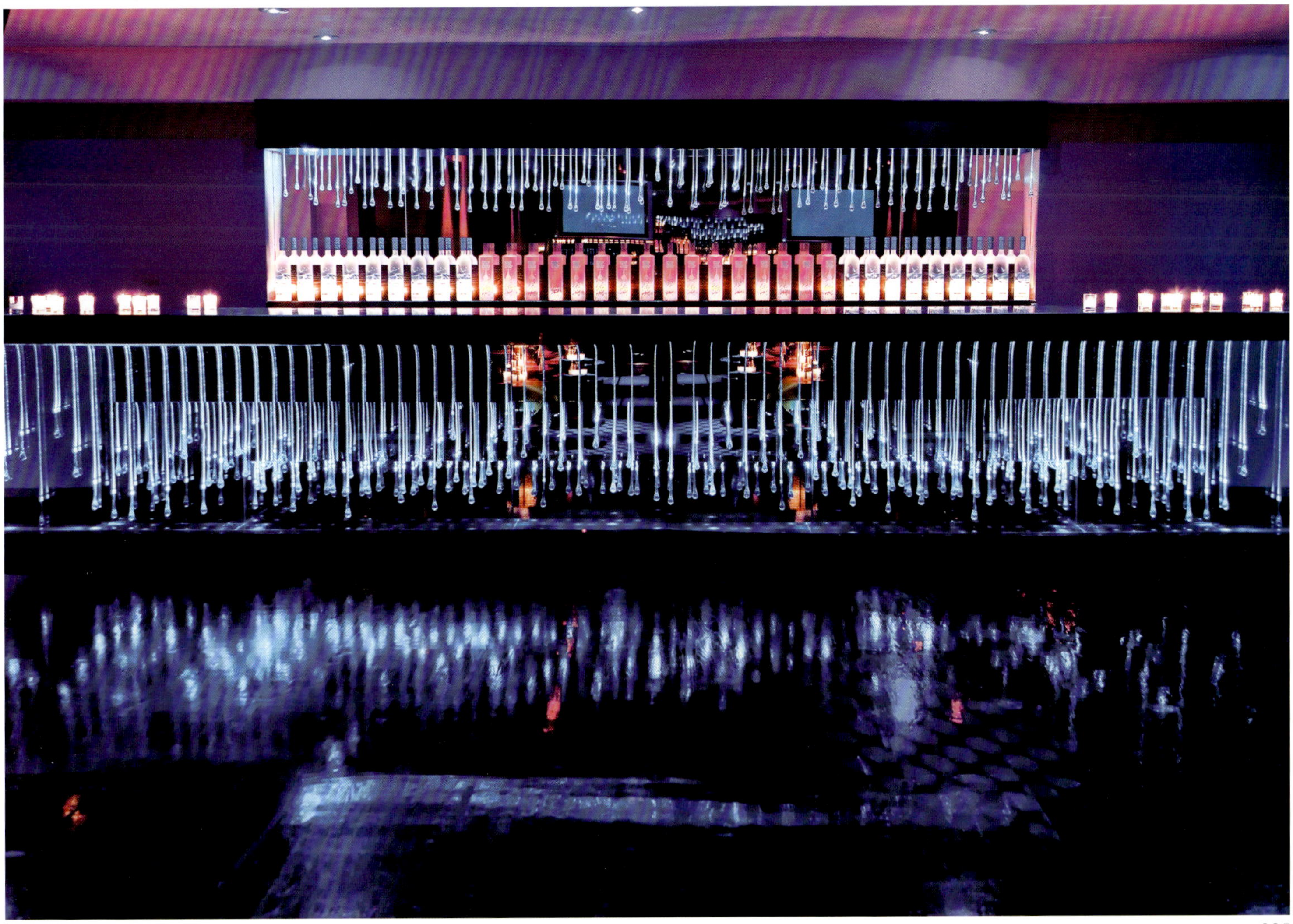

"LOUIS" Gansevoort Hotel

—— 甘西沃特酒店的"LOUIS"

It is a top nightclub project operated by Opium Group. Louis Nightclub of Gansevoort Hotel is one of the fashionable nightclubs under Opium Group on Miami Beach. Opened since the end of 2008, LOUIS has become a hot spot within a very short time, eye-catching like Russell Simmons, Marilyn Manson and Lindsay Lohan.

The interior space of the nightclub is mainly classical French style, where the modern black decorations and the theme full of classical art flavor are integrated together, and the conflict of contradictories is converted into the charm of art. With the background of crystal lighting and brilliant light, the space atmosphere is more attractive.

The VIP lounge of about 2000 square meters and the secret hotel entrance are not lack of European-style decoration of the eighteenth century. In the seemingly simple space, unique decorative elements are presented everywhere, which makes people feel fresh and shocking. It is the charm of art, and also the unique personality created by Opium Group.

Behold, my love, behold all
that I simultaneously do:
scandal, seduction, bad
example, incest, adultery,
sodomy! Oh, Satan! one and
unique God of my soul,
inspire thou in me something
yet more, present futher
perversions to my smoking
heart, and then shall thou see
how I shall plunge myself
into them all !
Marquis De Sade

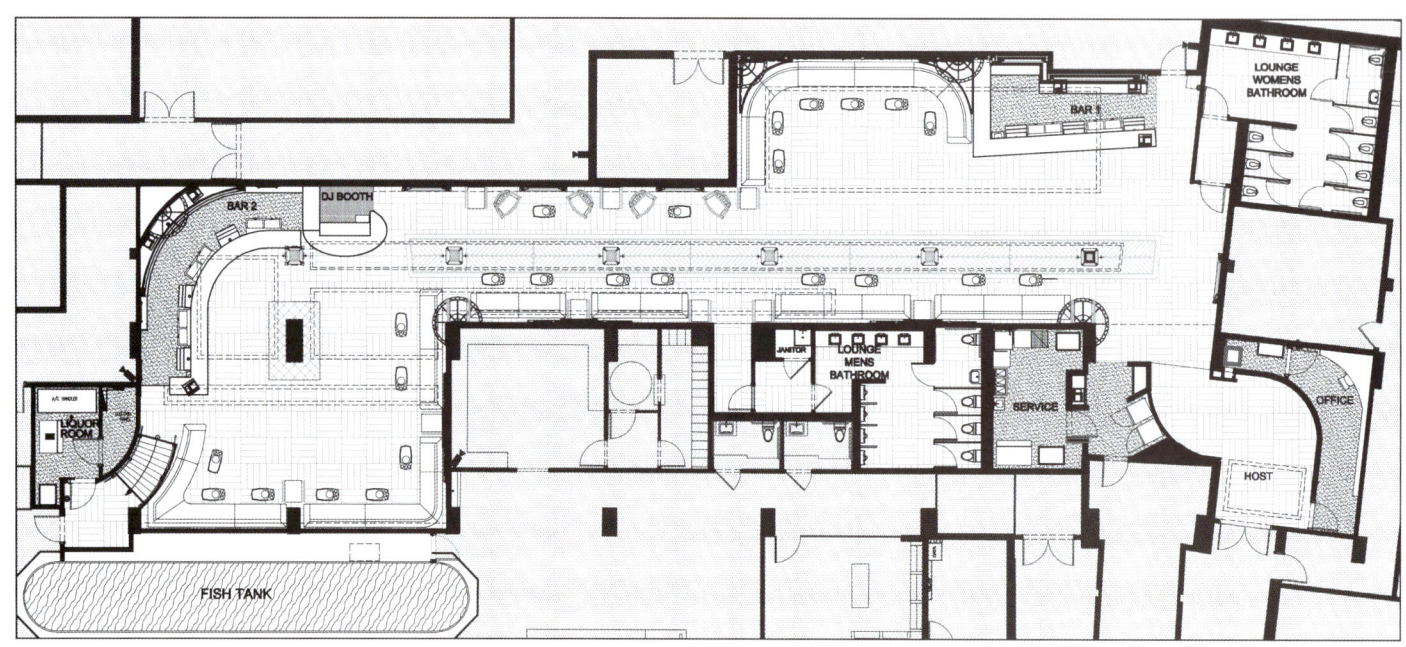

本案属于 Opium Group 经营的顶级夜店项目。在迈阿密海滩诸多 Opium Group 经营的时尚夜店中，甘西沃特酒店的 Louis 夜总会便是其中之一。自 2008 年底开业以来，LOUIS 在很短的时间内就成为一个热点，如同罗素·西蒙斯、玛丽莲·曼森和琳赛·洛翰一样引人注目。

夜总会内部以法国古典风格为基调，黑色的现代装饰与极具古典艺术气息的主题墙融合在一起，矛盾的冲突在这里被转化为艺术的魅力，加上水晶灯饰和绚丽灯光的映衬，空间氛围更具魅力。

在近 2000 平方米的贵宾休息室和秘密的酒店入口处，也不乏 18 世纪欧式风格的装饰。在看似简约的空间里随处可见独特的装饰元素，带给你新奇和震撼。这是艺术的魅力，也是 Opium Group 要营造的独特个性。

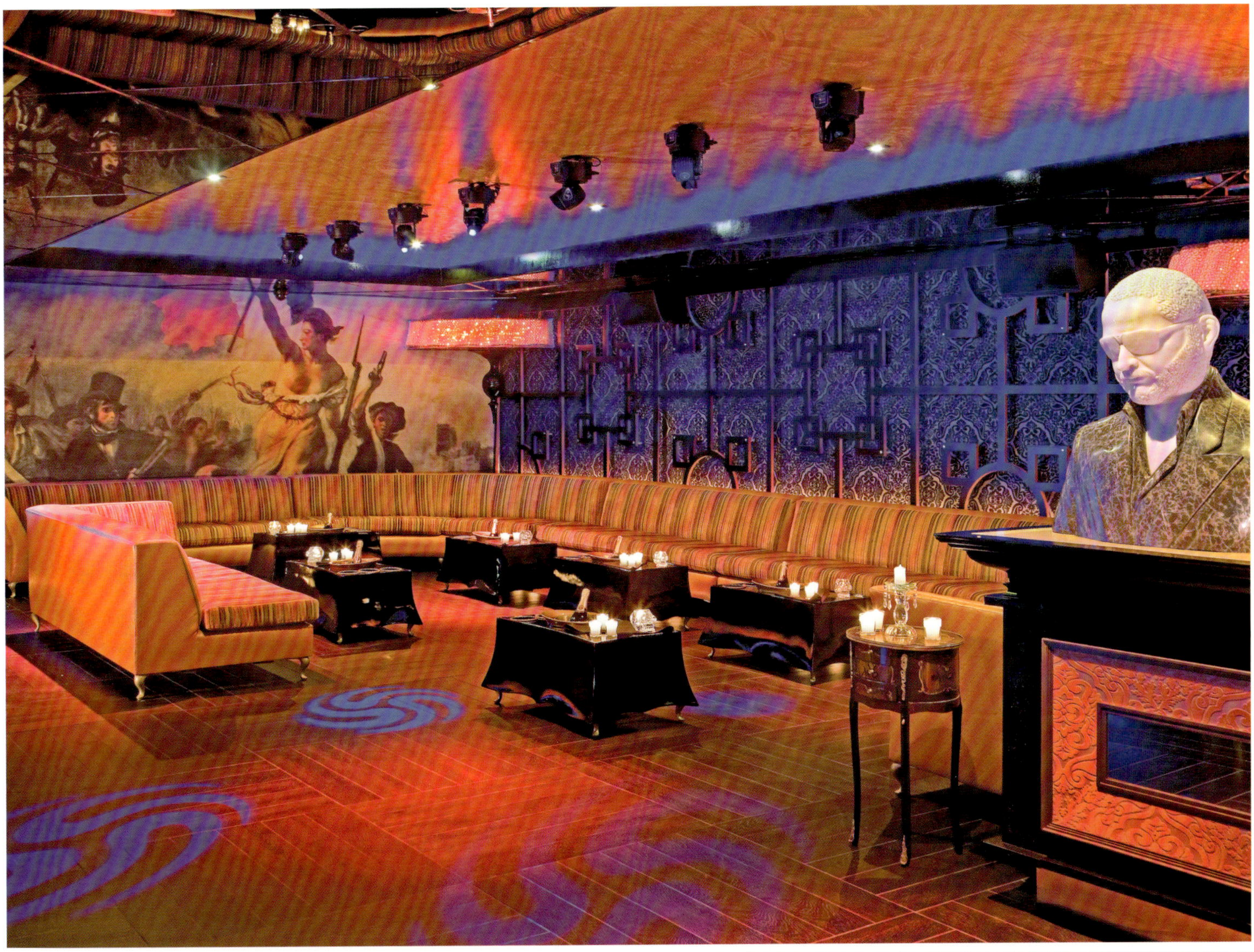

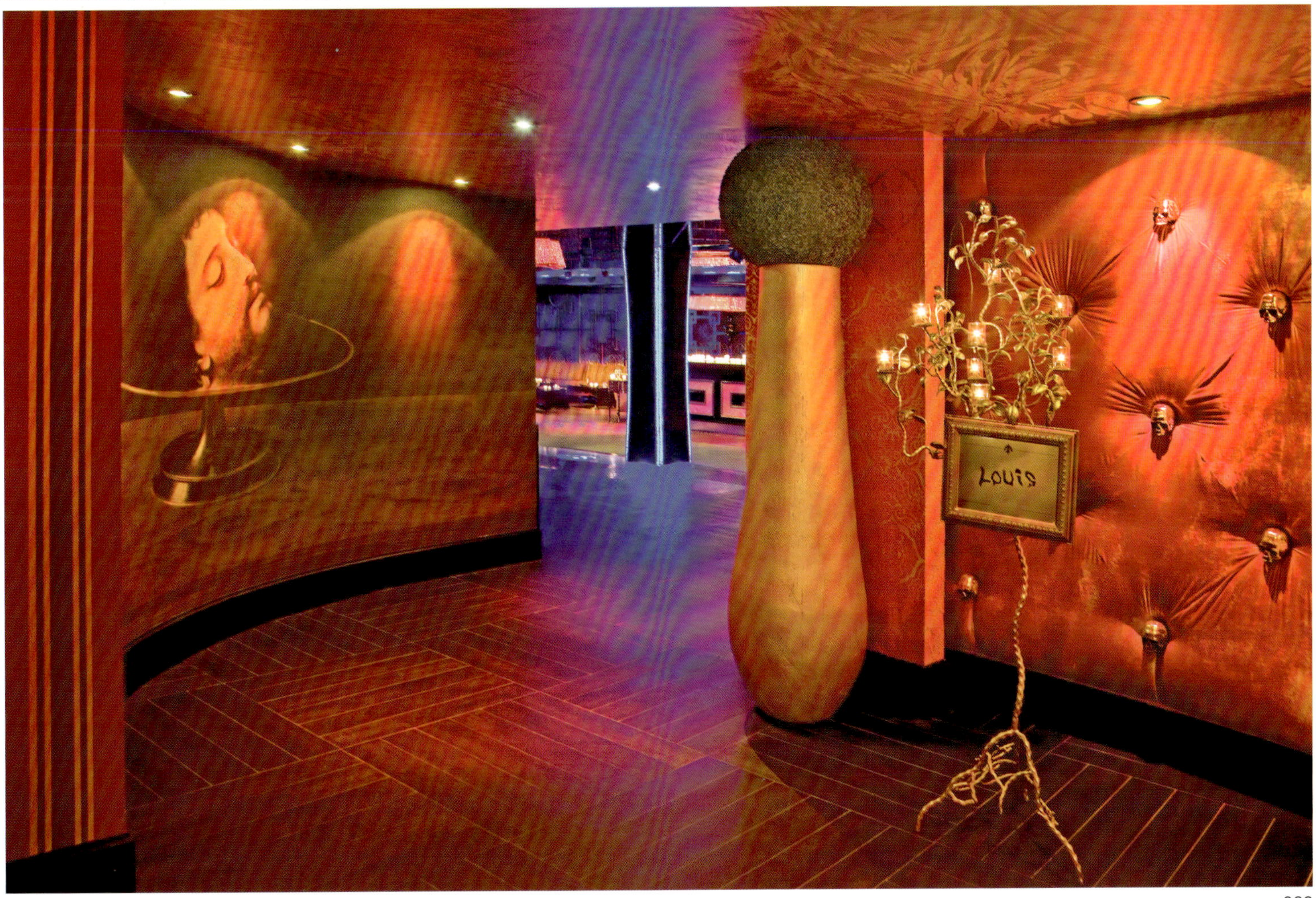
Louis

地点	面积	设计师	设计公司	主要材料
/ 上海	/ 760m²	/ 甘泰来	/ 齐物设计	/ 米色洞石、挪威森林石材、铁刀木石材、实木染色、马赛克、铜镜、铁件、皮革、织品

Official Residence

—— 官邸

The whole project of the duplex-style dining space takes use of the existing structural columns coated with hollowed carving board and built-in LED lights, and by the relationship between litera and space, the Chinese character "Guan" is chosen as a prototype to produce geometric graphics as the shape totem of the hollowed board.

The project is designed with two entrances, respectively linking the moving lines with the heightened area on the first floor, mezzanine and private rooms. The heightened area on the first floor is an open dining space, planned with bar, booths area and stage area, and an oval-shaped Jacuzzi pool is arranged beside the stage. The semi-open private rooms are designed on sides, and the velvet drapes separate the vision from the inside seats, adjusting the privacy of the rooms. The mezzanine is decorated like a floor gallery, and the private rooms are planned with guide channel with consideration of the privacy of customers and together with sofa area, audio-visual equipment and pool tables, where the solid wood parquet floor, plush and other elements show out a low-key and warm atmosphere.

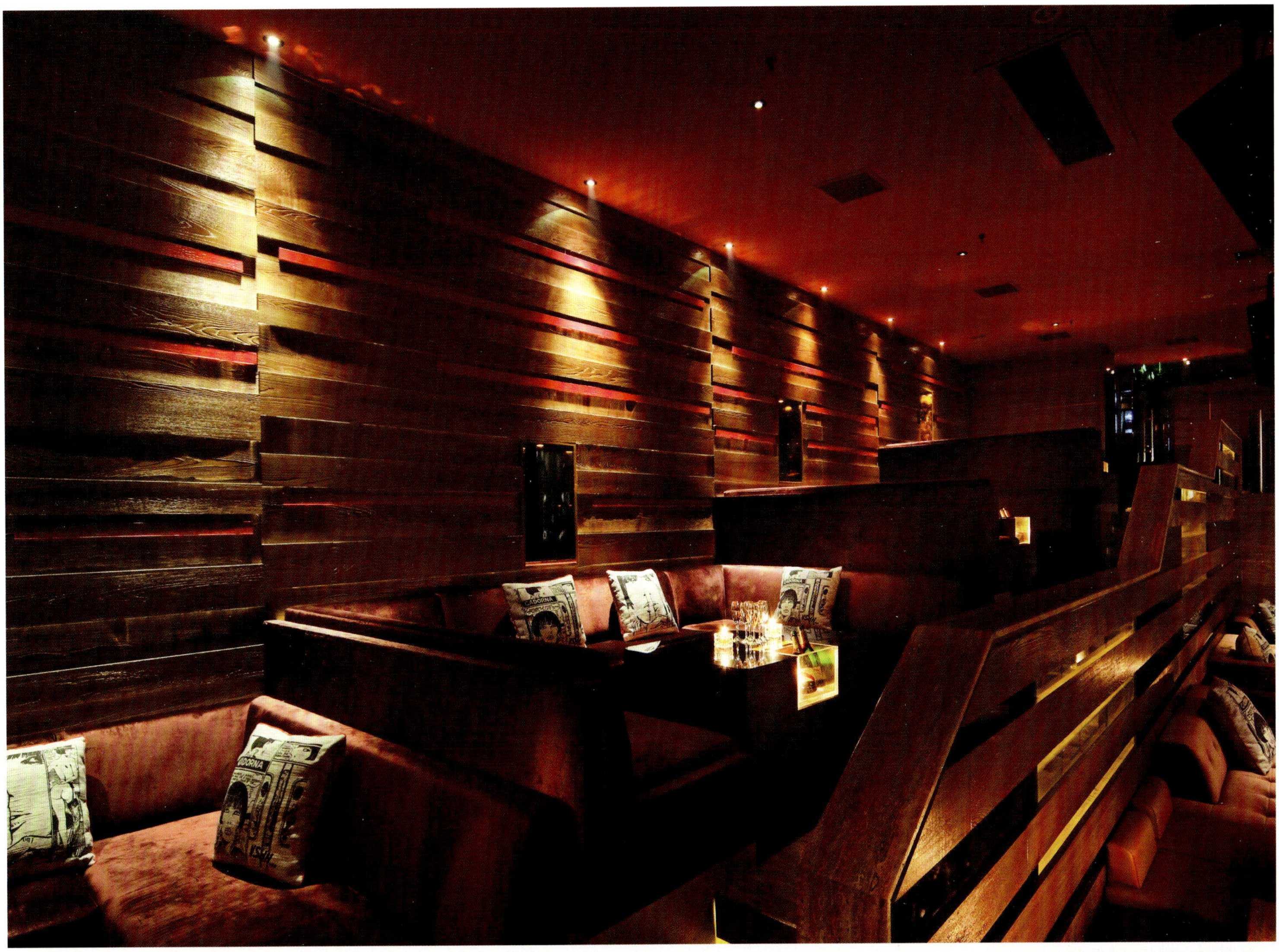

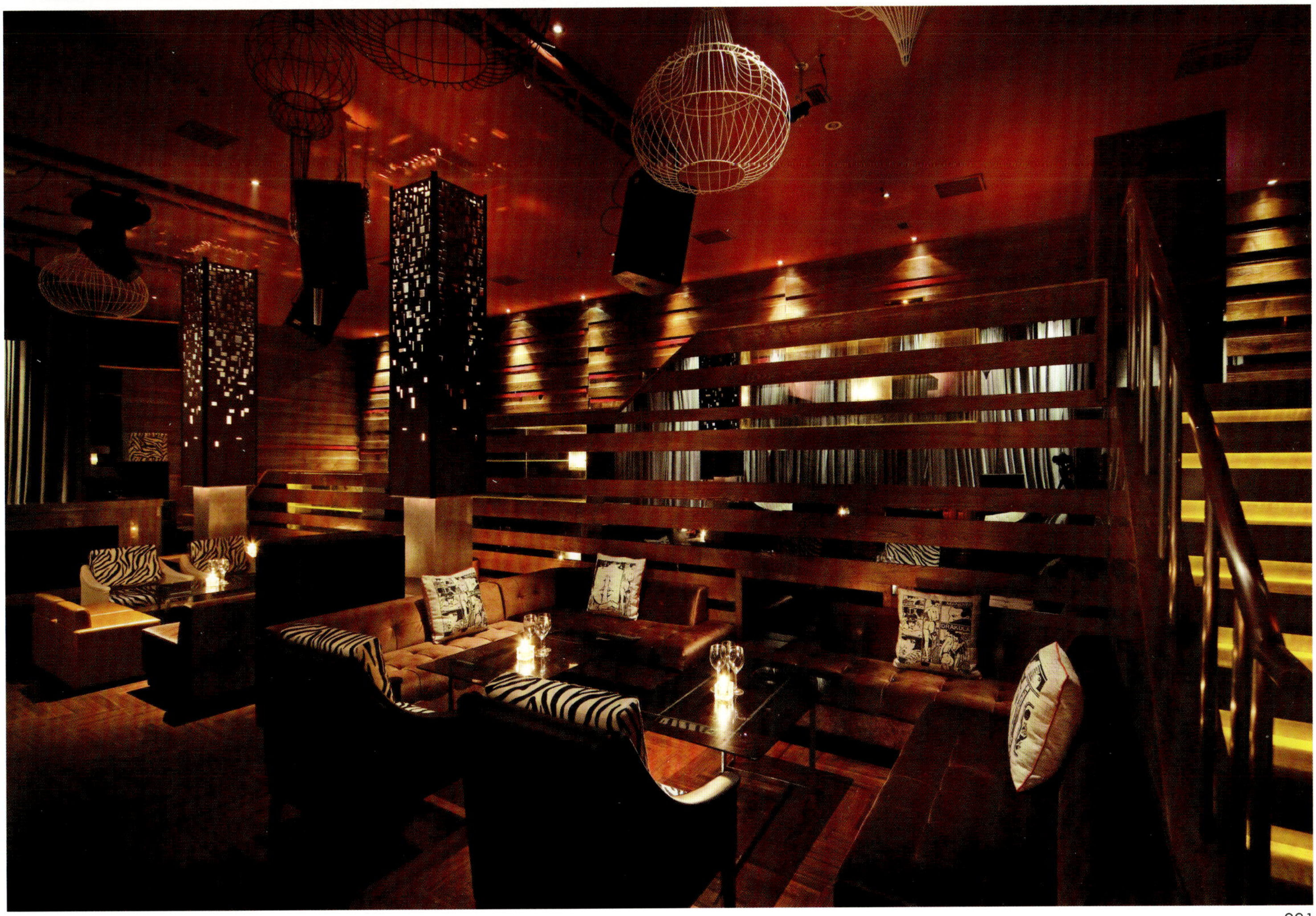

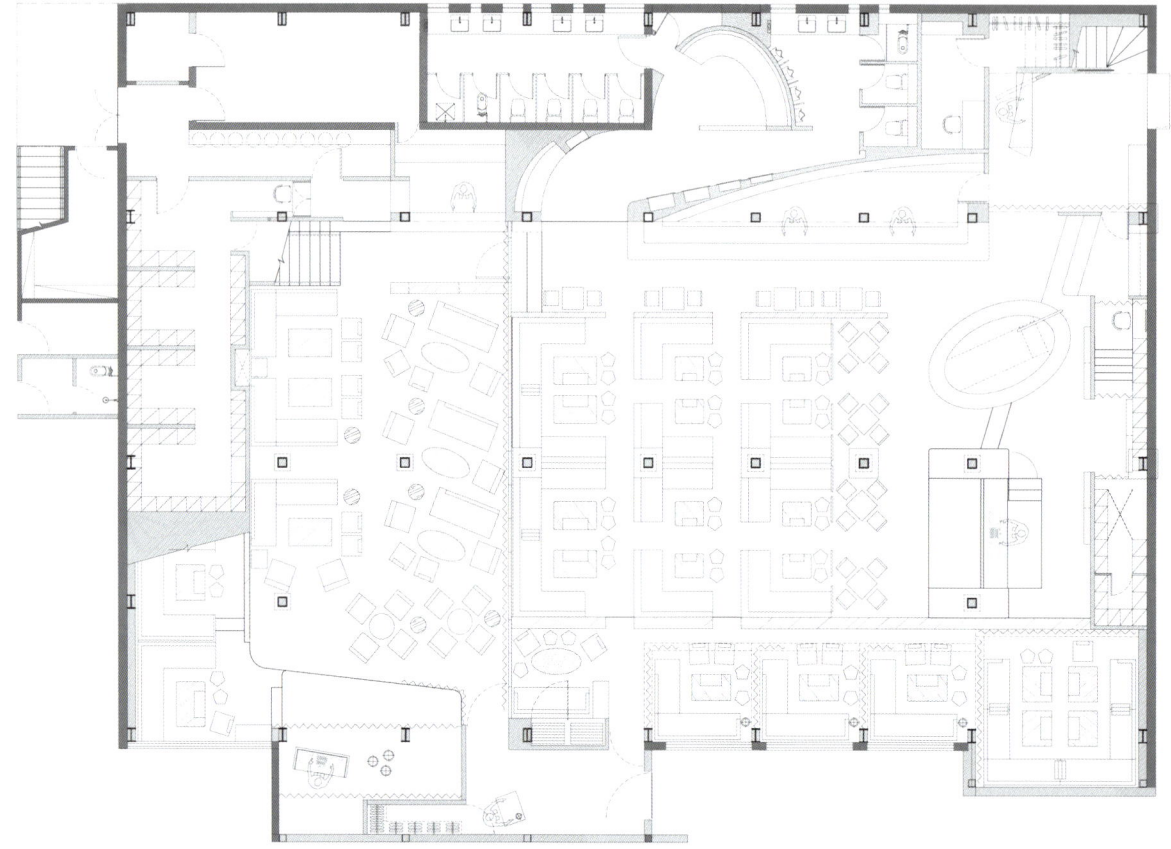

复合式餐饮空间"官邸"全案利用既有的结构柱，在柱体上包覆镂空雕刻板，内置LED灯，藉由文字与空间的关系，以"官"字作为原型，衍生出几何图形，并将其作为镂空板的造型图腾。

此案规划了两个入口，分别将动线连接至一层挑空区与夹层、包厢区。一层挑空区为开放式餐饮空间，规划吧台、卡座区和舞台区，舞台旁安置了一个椭圆形的Jacuzzi水池。侧边规划为半开放式的包厢，拉上丝绒帘幔即可区隔座位区与视景，可调节包厢内的隐密性。夹层区则采用楼廊式规划，包厢区规划引导通道并照顾到宾客的隐私，同时，规划沙发区、影音设备和撞球桌台，用实木拼花地板、丝绒等元素营造出低敛、温暖的氛围。

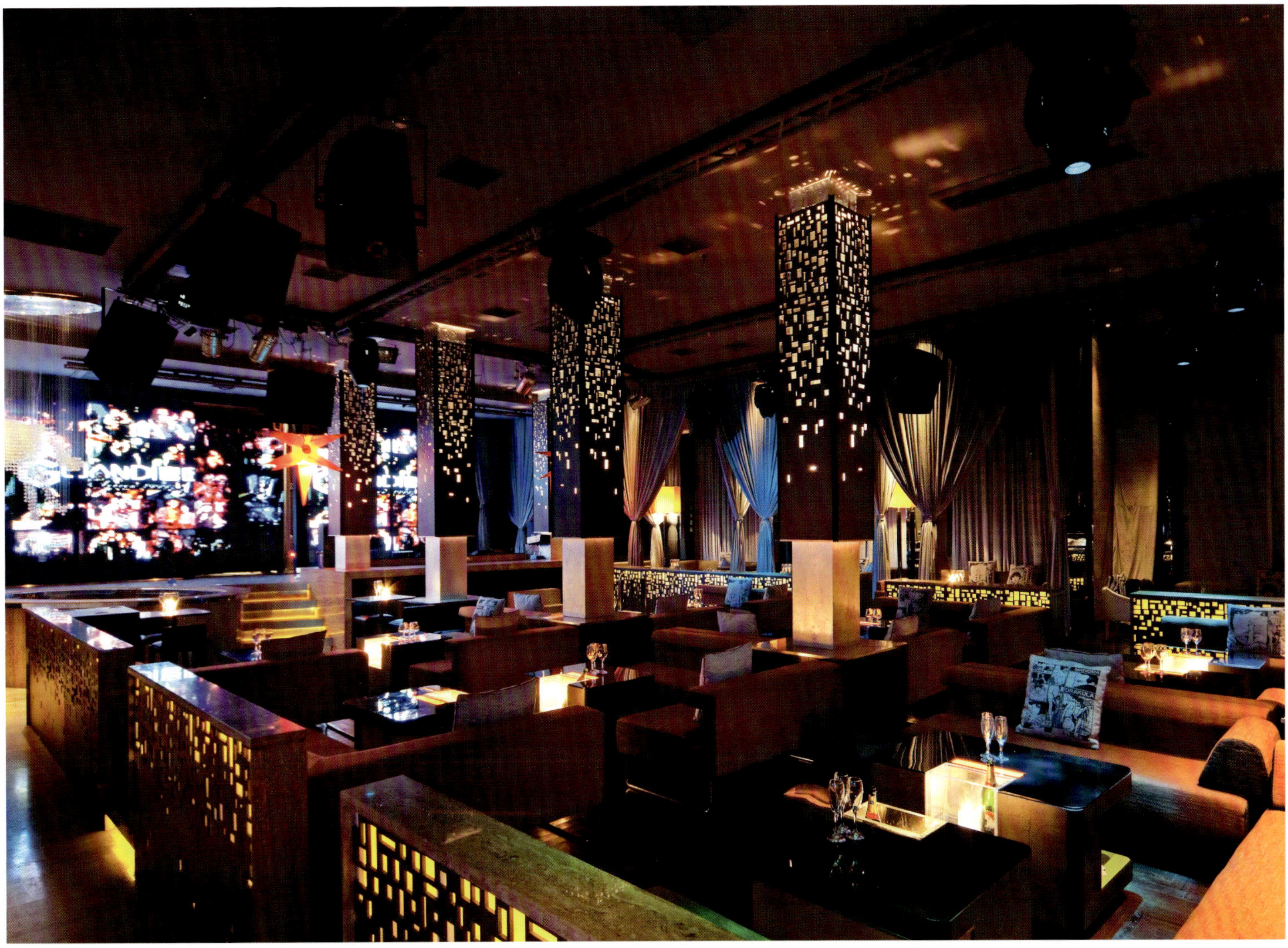

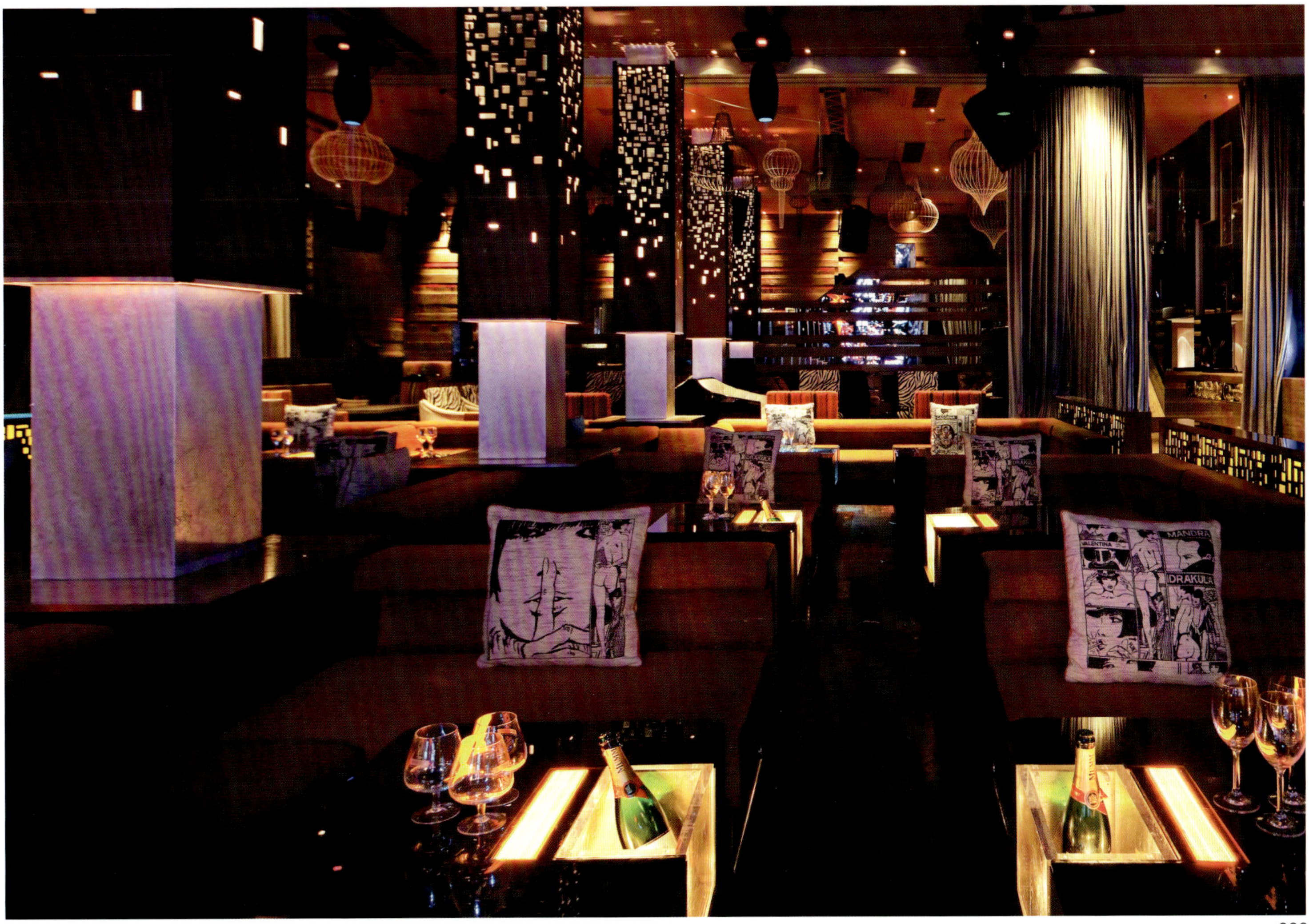

CADORNA
VSHH...
DRAKULA

DN
DN

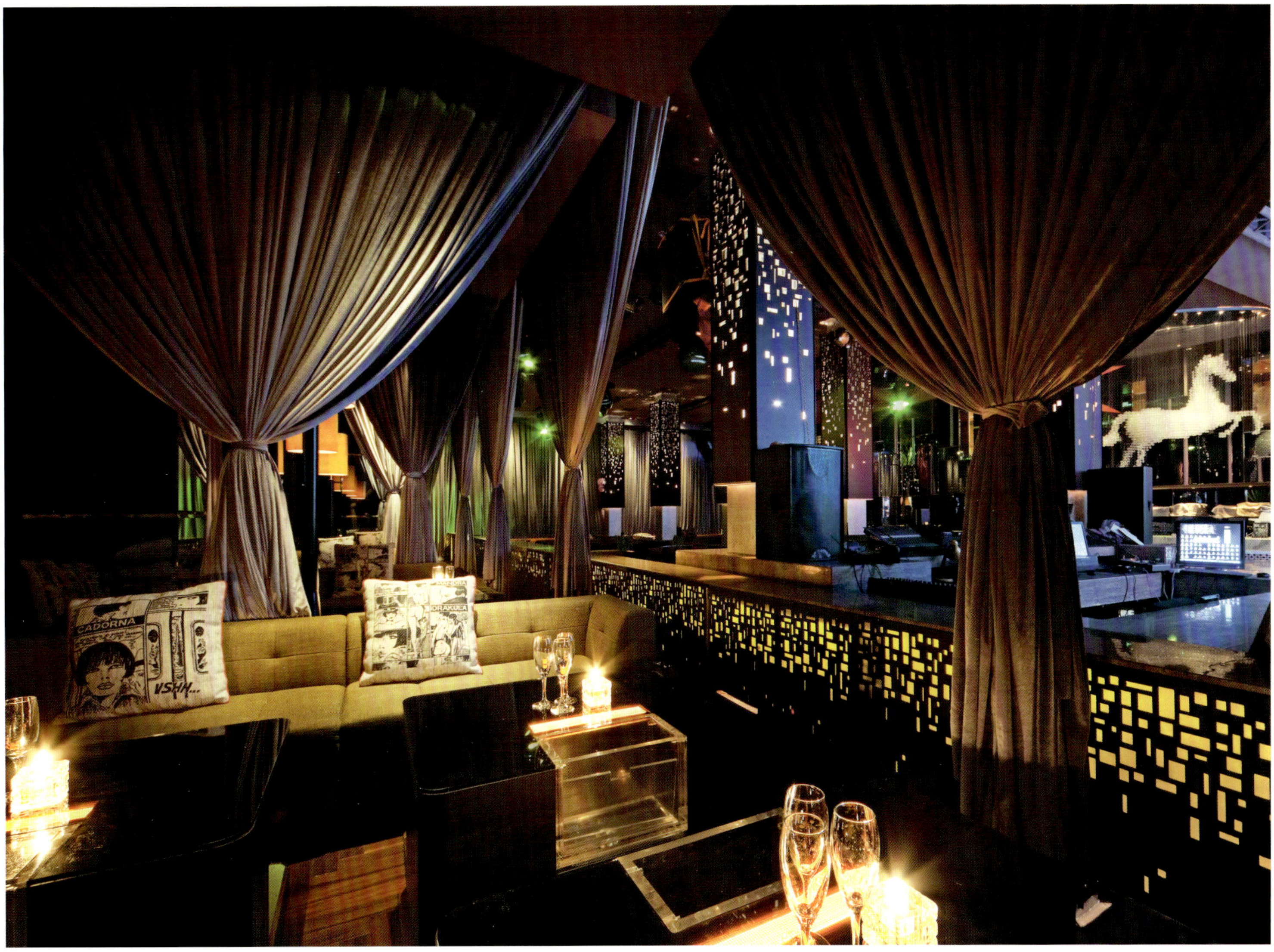

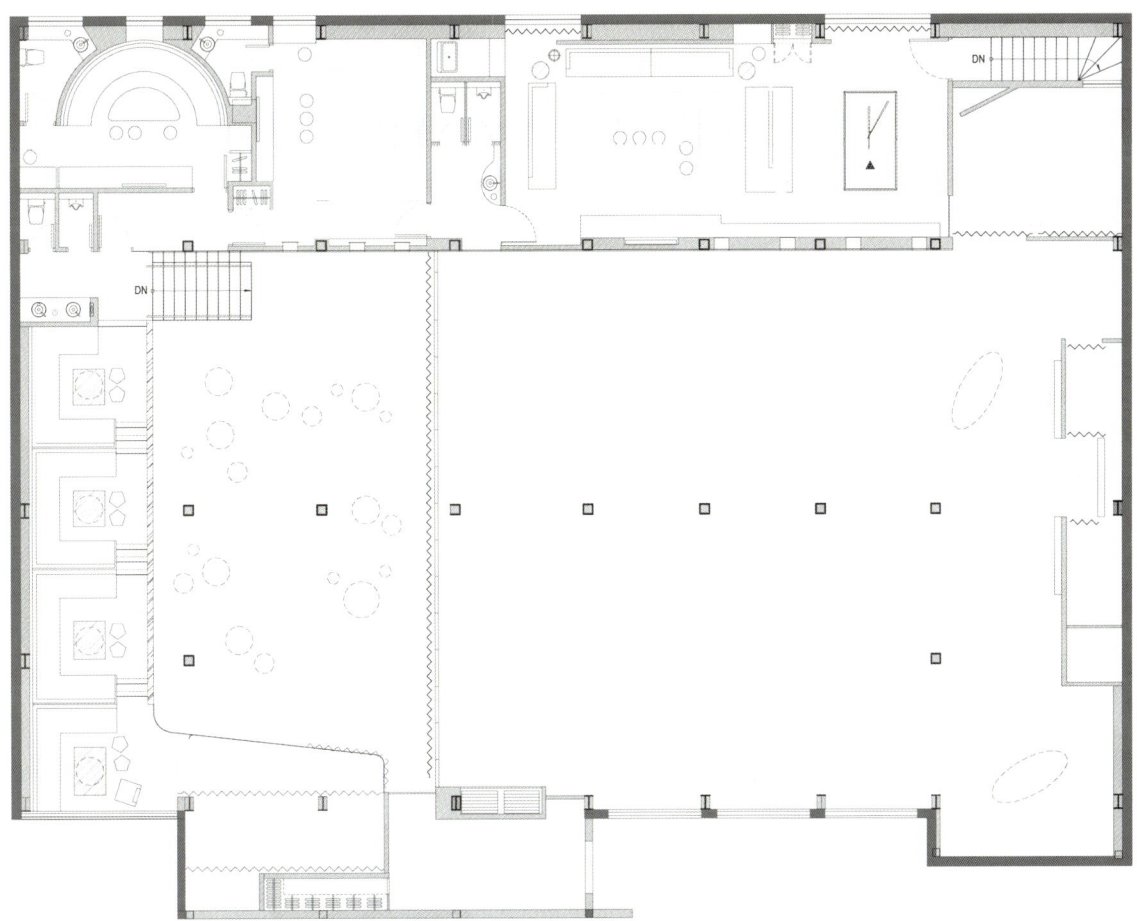

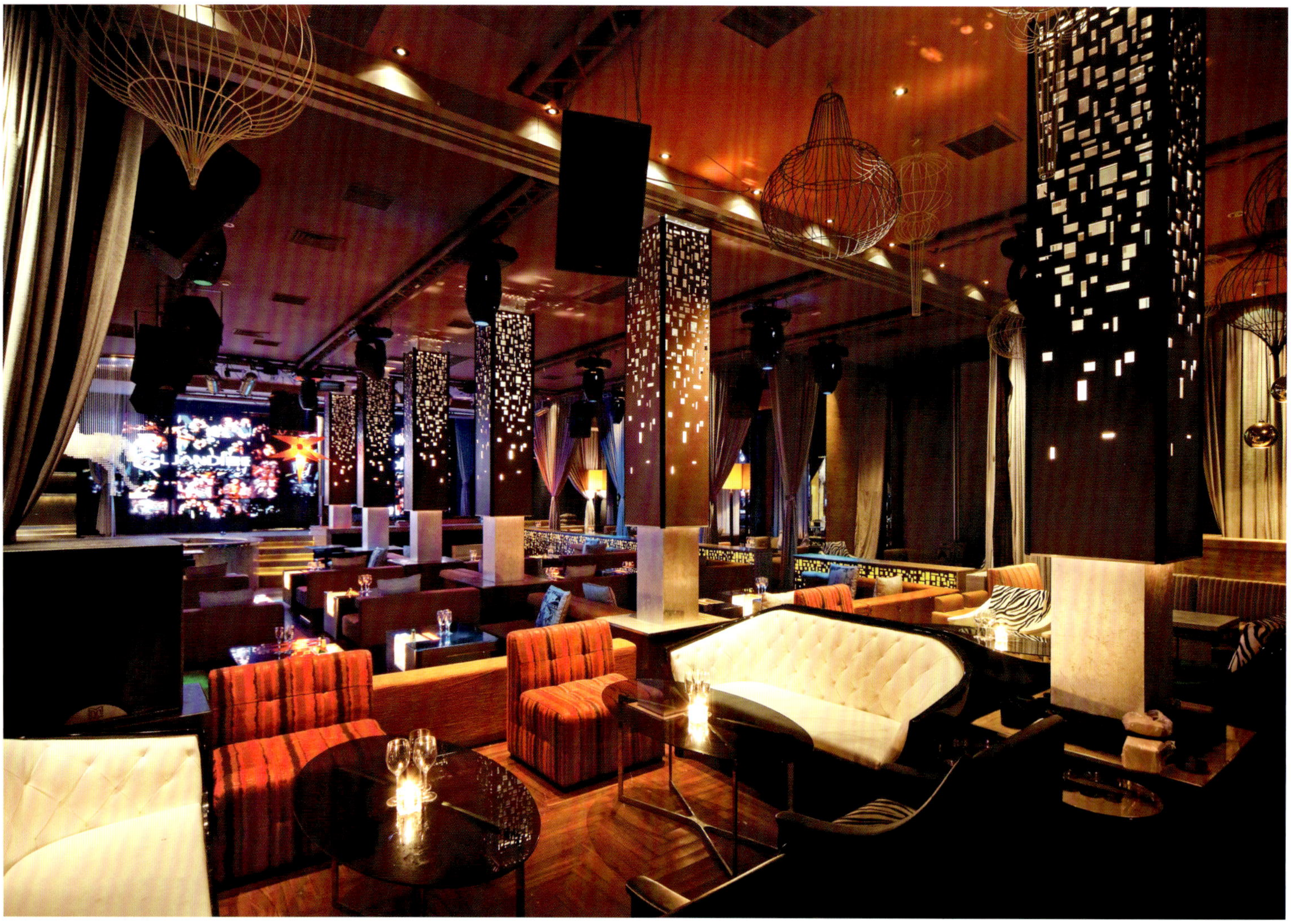

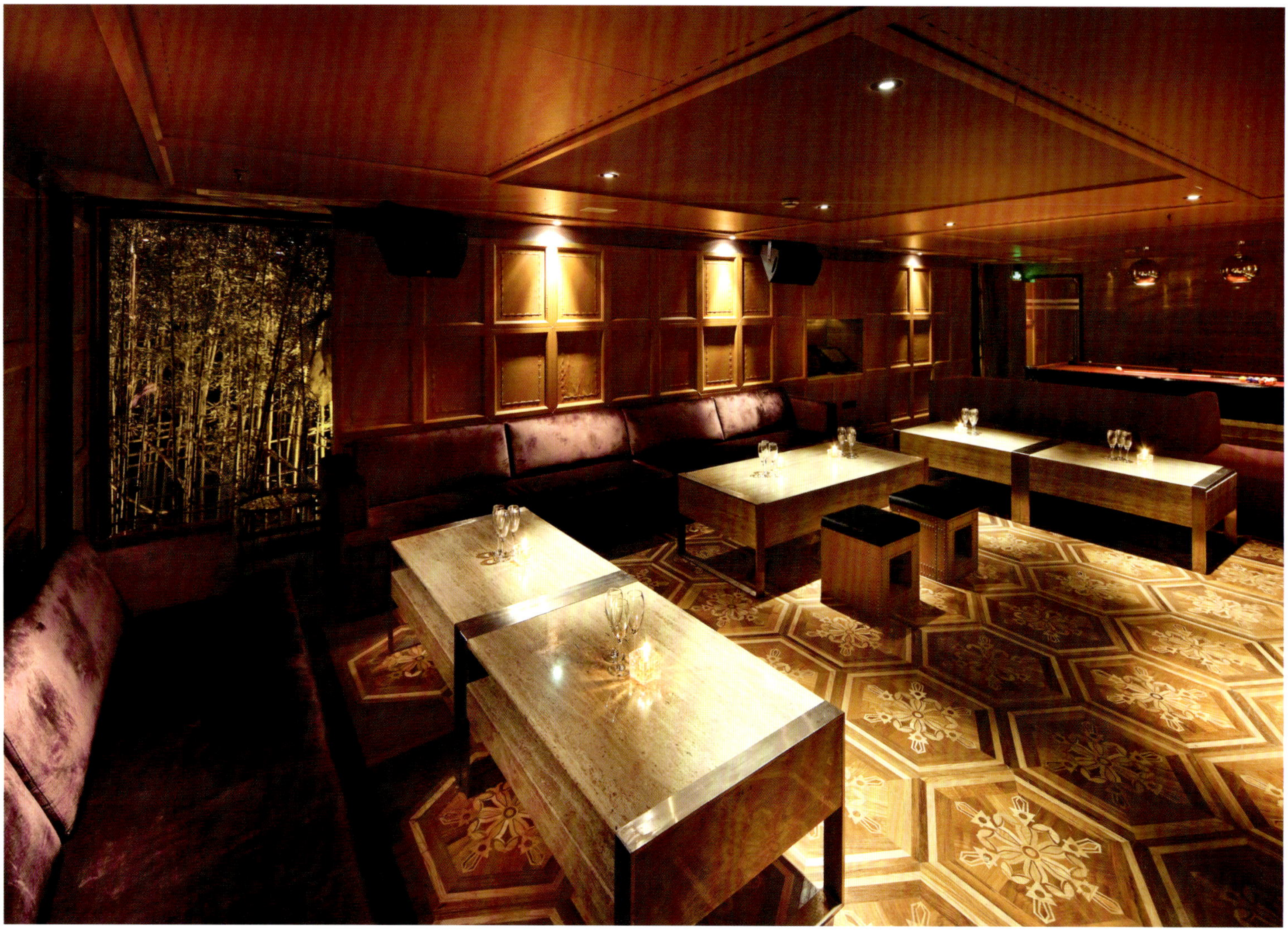

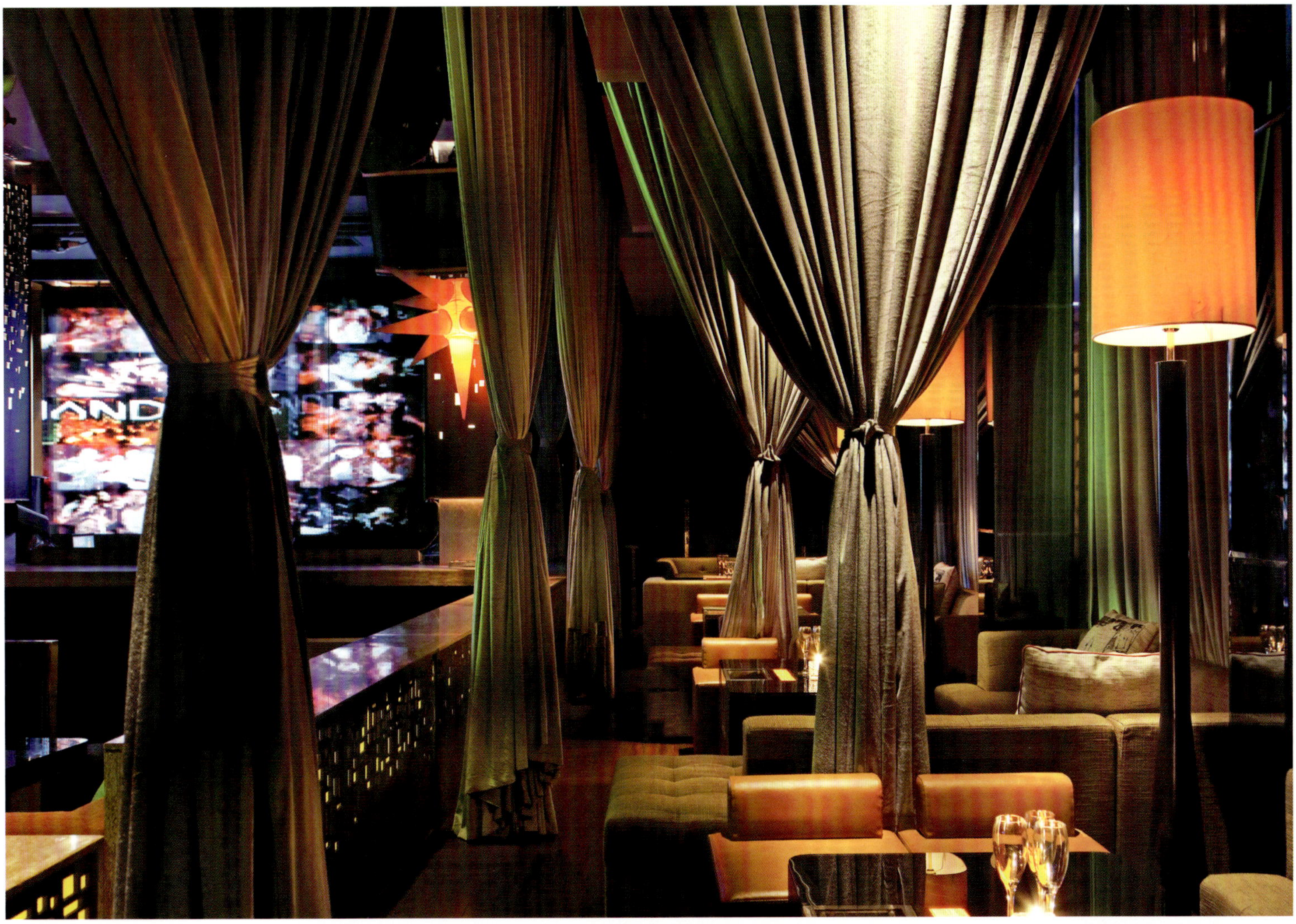

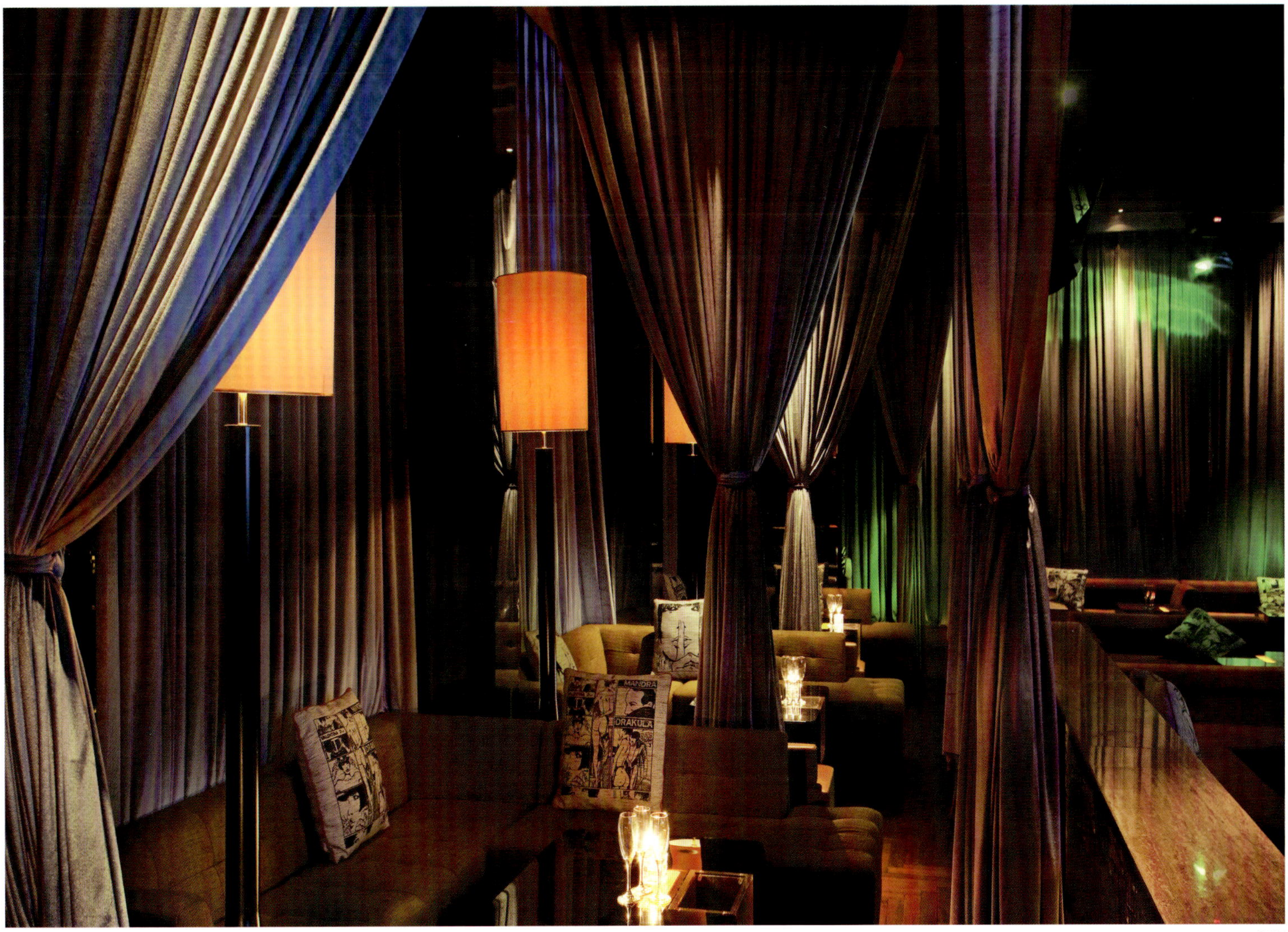
MANDRA
DRAKULA

地点	面积	设计师	设计公司	主要材料
/ 广州	/ 1500m²	/ 陈武	/ 深圳市新冶组设计顾问有限公司	/ 水泥、麻石、水晶、钢、镜子

Guangzhou True Color Bar

___广州本色酒吧

The club is located on Yanjiang Road in Guangzhou, and inspired by the historic buildings of the project itself, the designer regards the space as a large masquerade, introducing music performances and sexy and gorgeous enjoyment atmosphere into a new model, to create a unique elegant flavor in city. It becomes a landmark of night fashion in Guangzhou.

A large area of bare cement and modern materials are used to create a lobby at the entrance, turning the sense of fashion and art into the important gene of the space to create a looming publicity and quiet stunning. The designer chooses "the form of performing events" as an imagination, through people, the carrier of showing the beauty of music, to divide performing ways into "passive performances without sense" and "active direct performances", which provides a strong feeling of new experience at any time and infects everyone in True Color Bar, putting the "fashionable, artistic, good-quality and friendly" design definition of True Color into full play.

Fashion
True
House Club
MUSIC

THEME
MUSIC
Live
Fashion
TRUE
Fashion
MUSIC
Live
Live

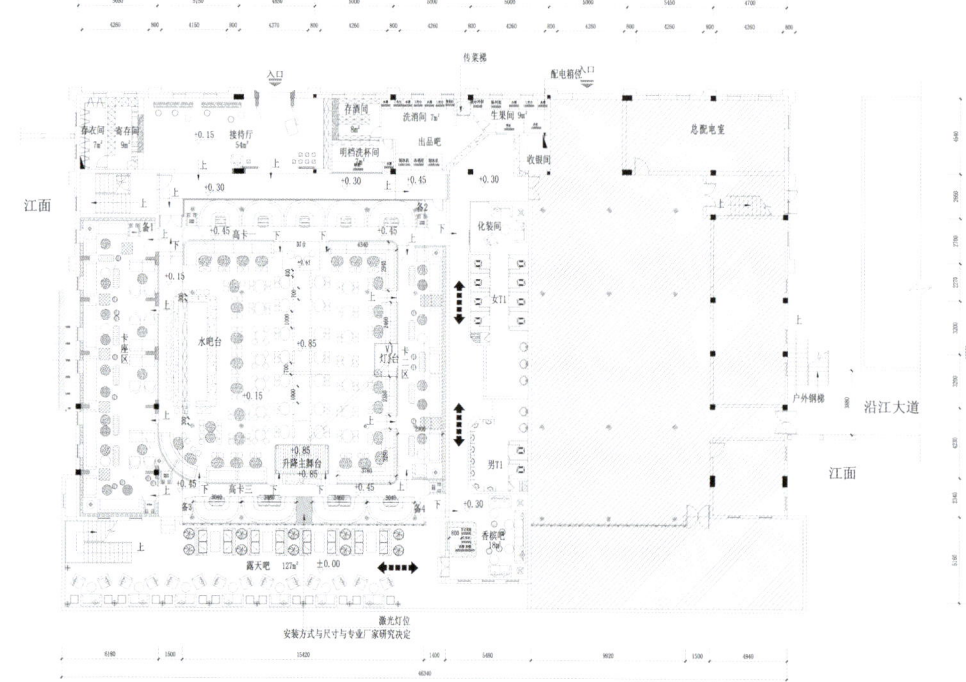

本店位于广州沿江路，带着项目自身历史建筑所激发的设计灵感，设计师将空间视为一场大型化装舞会，将音乐表演与性感华丽的享乐氛围引入新的模式，营造出独特的都会雅士风情，成为广州的夜时尚地标。

设计师用大面积裸露的水泥和现代素材来打造入口前厅，将时尚感与艺术性转为空间的重要基因，营造出一种若隐若现的张扬和不动声色的惊艳。设计师以"表演事件的形态"作为想象，透过人这一展现音乐之美的载体，将表演方式划分为"被动式无感展演"及"主动式直接演出"，给人一种时时都有新体验的强烈感受，感染着来本色酒吧的每一个人，将本色酒吧的"时尚、艺术、品质、亲切"的设计定义发挥得淋漓尽致。

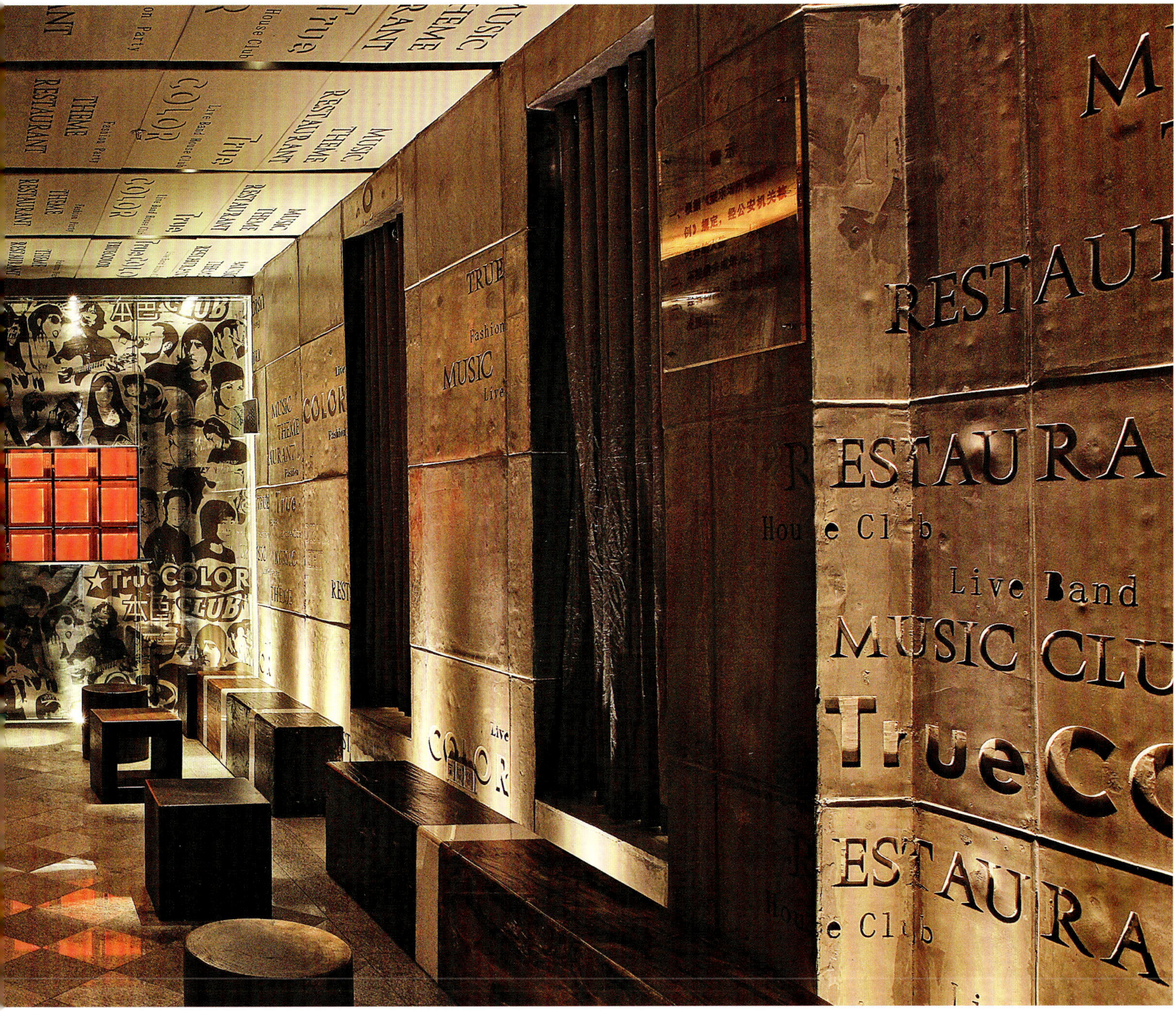
MUSIC
THEME
RESTAURANT
True
COLOR
House Club
TRUE
Fashion
MUSIC
Live
Live Band
MUSIC CLUB

RESTAURANT
House Club
Live Band
MUSIC CLUB
TrueCOLOR

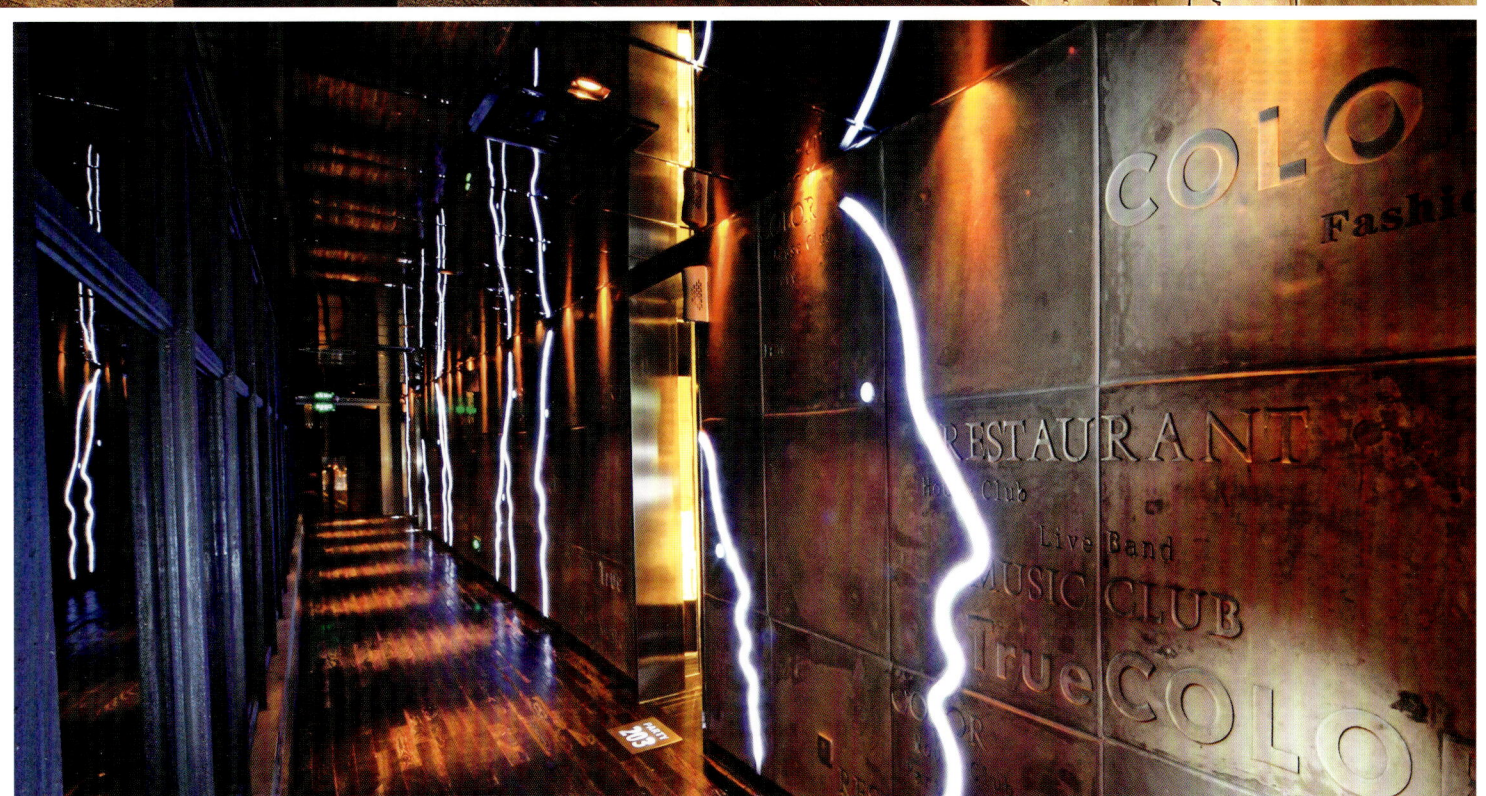

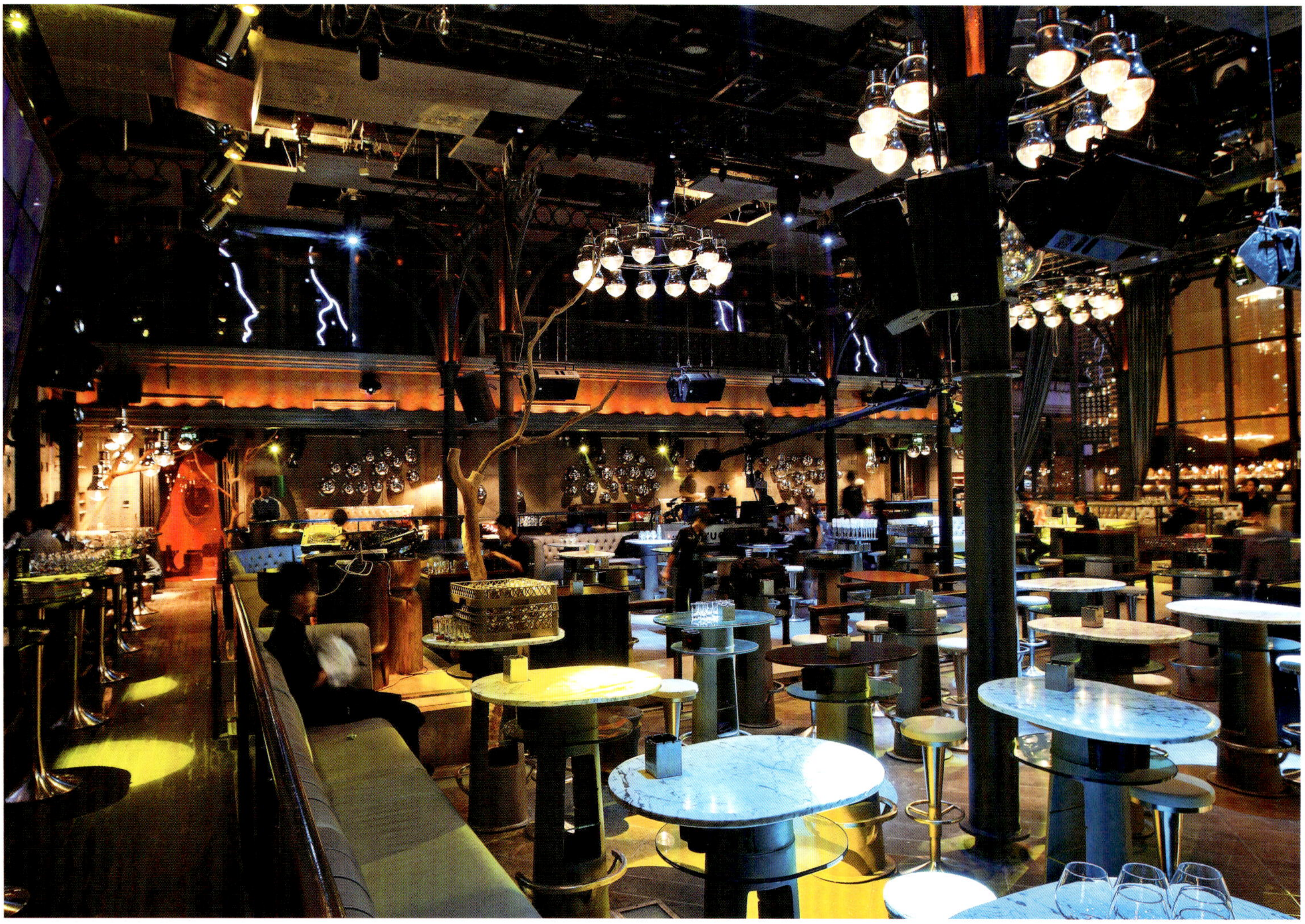

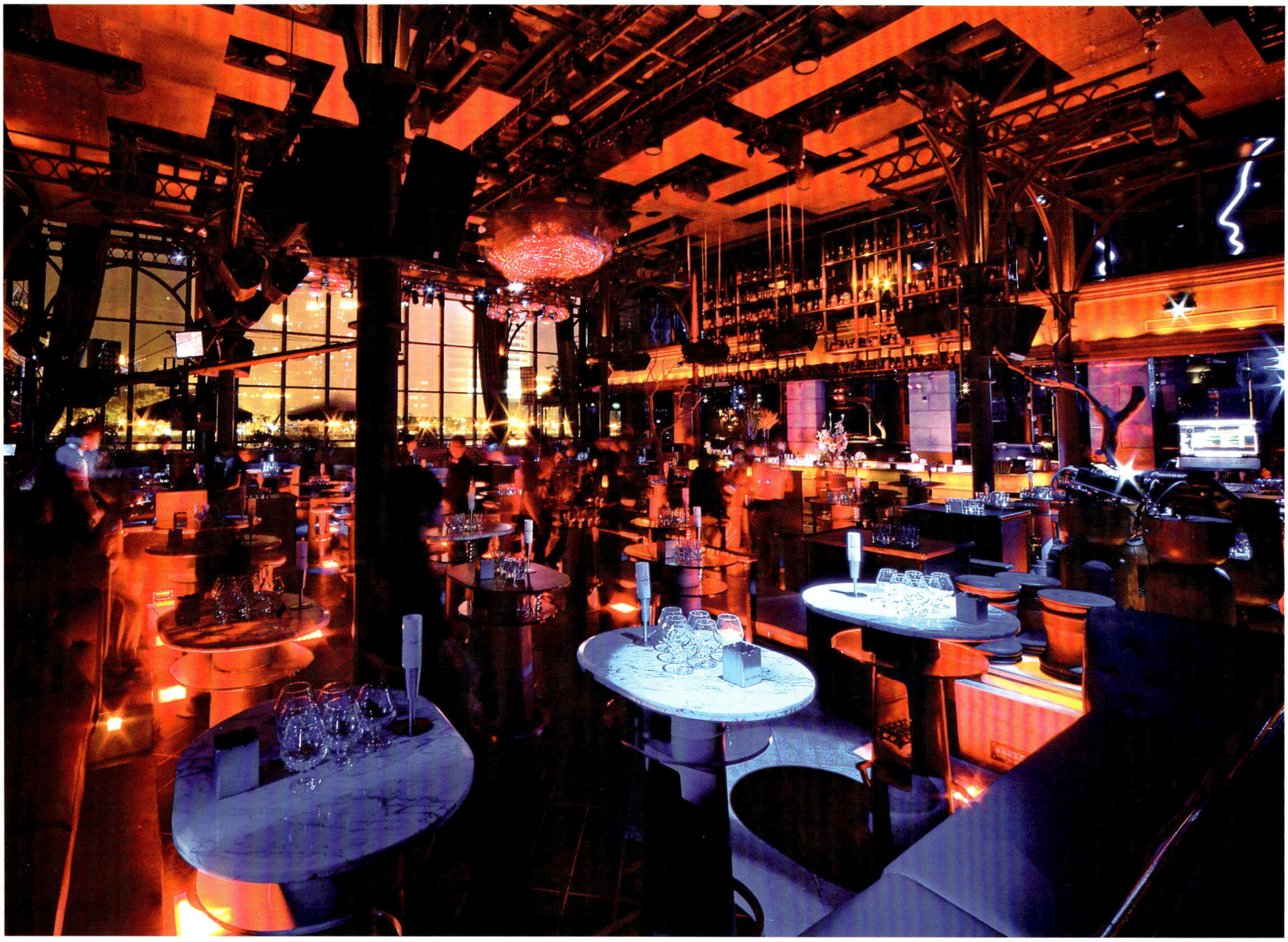

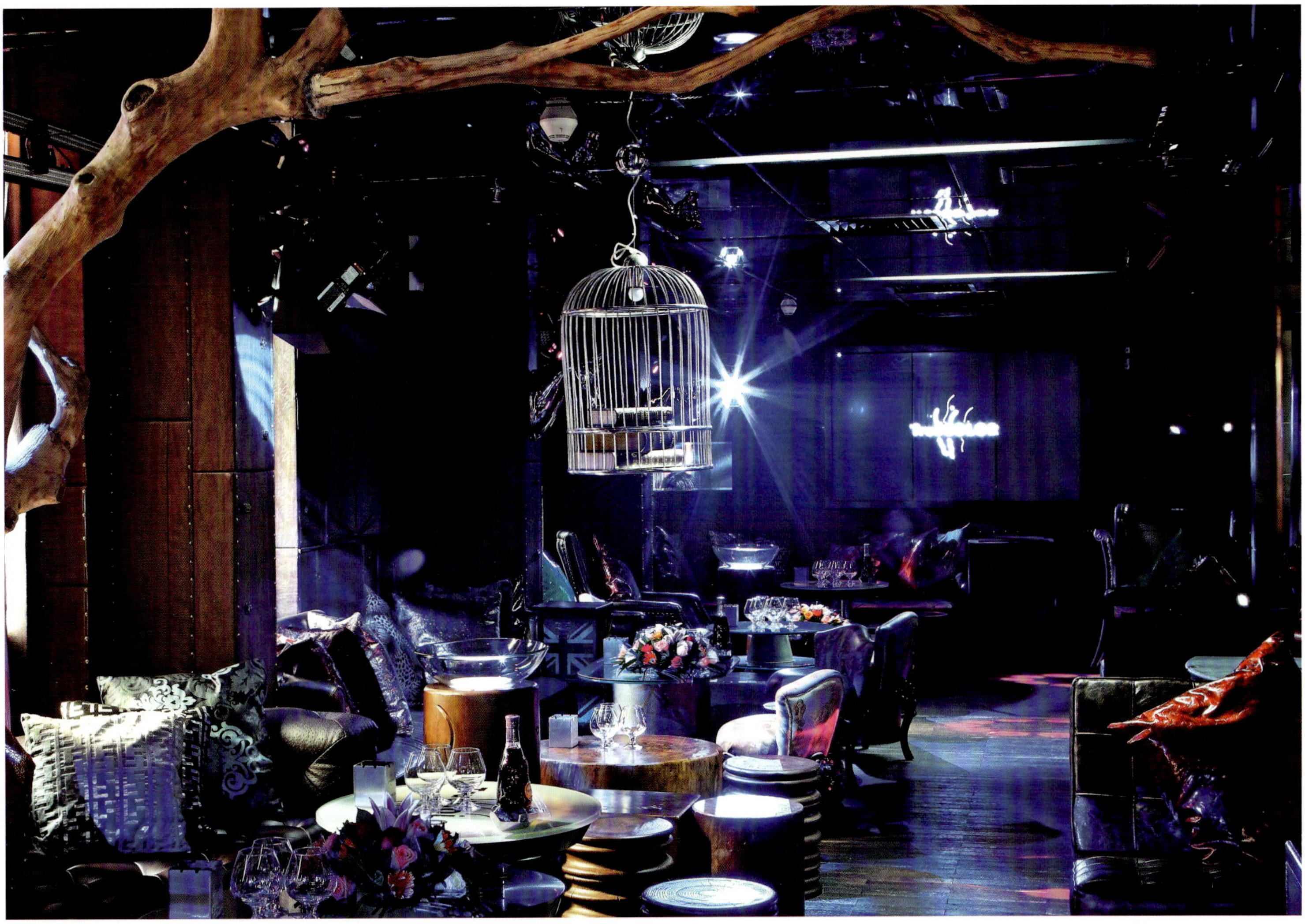

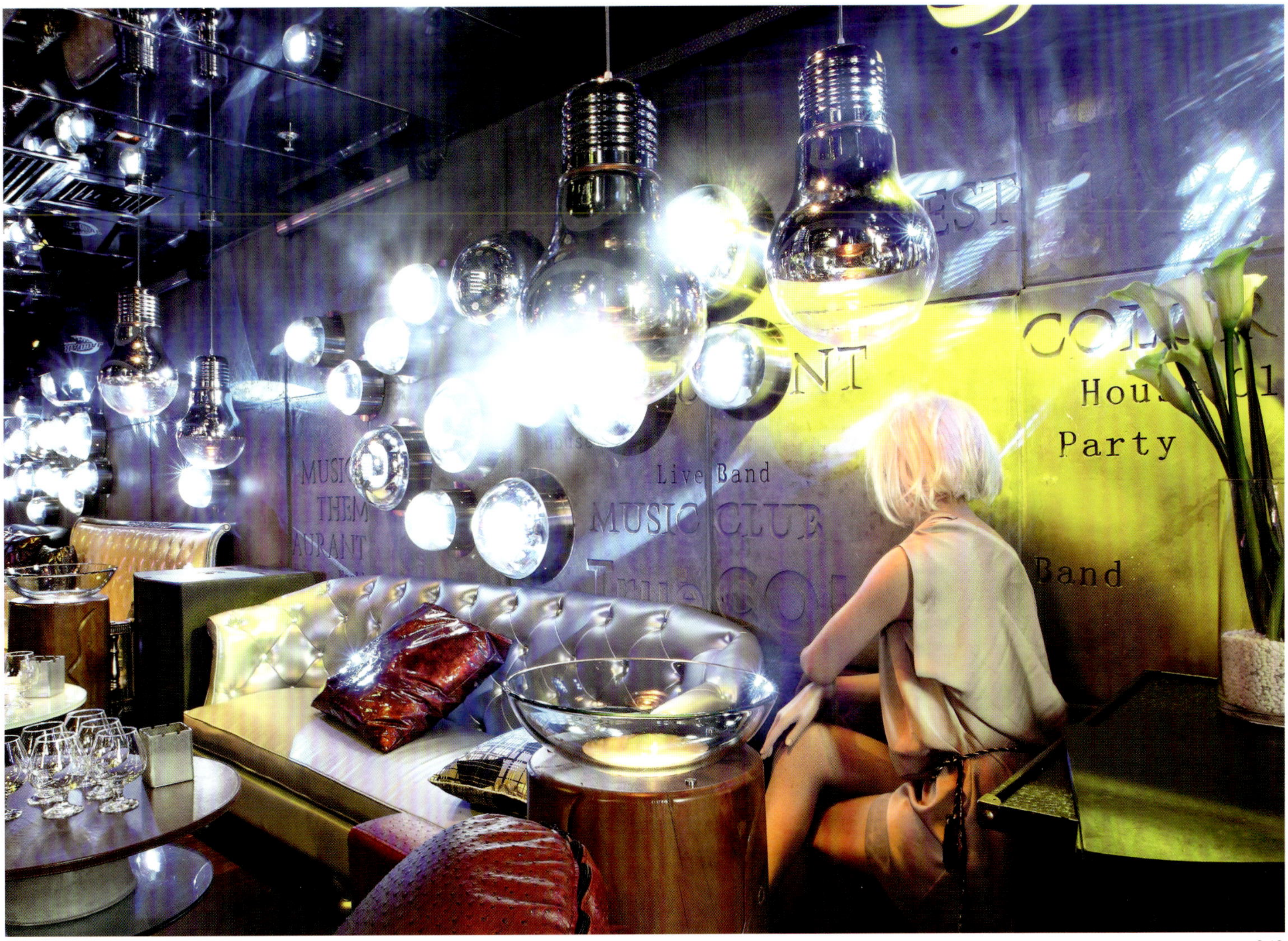
Live Band
MUSIC CLUB
Party
Band

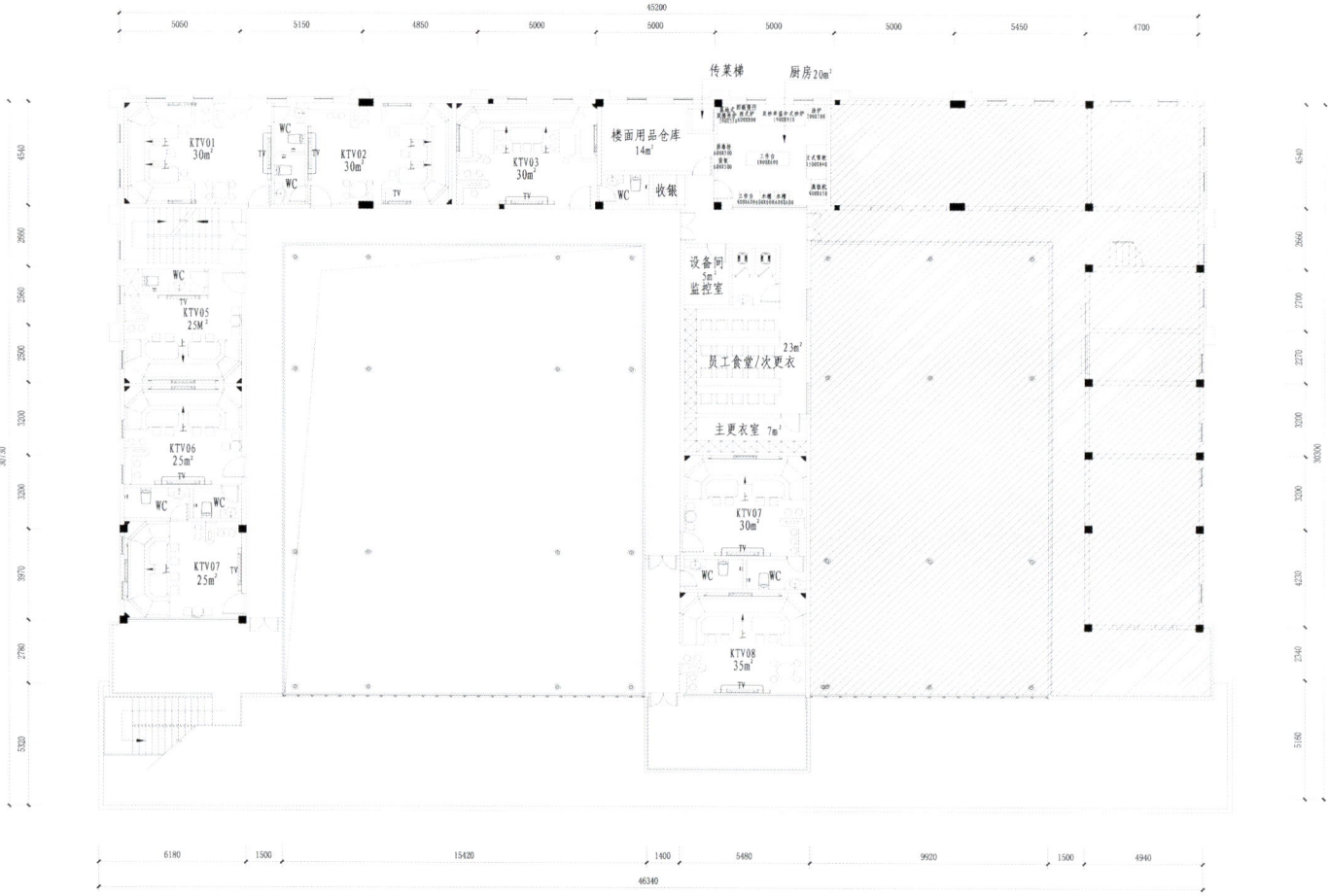

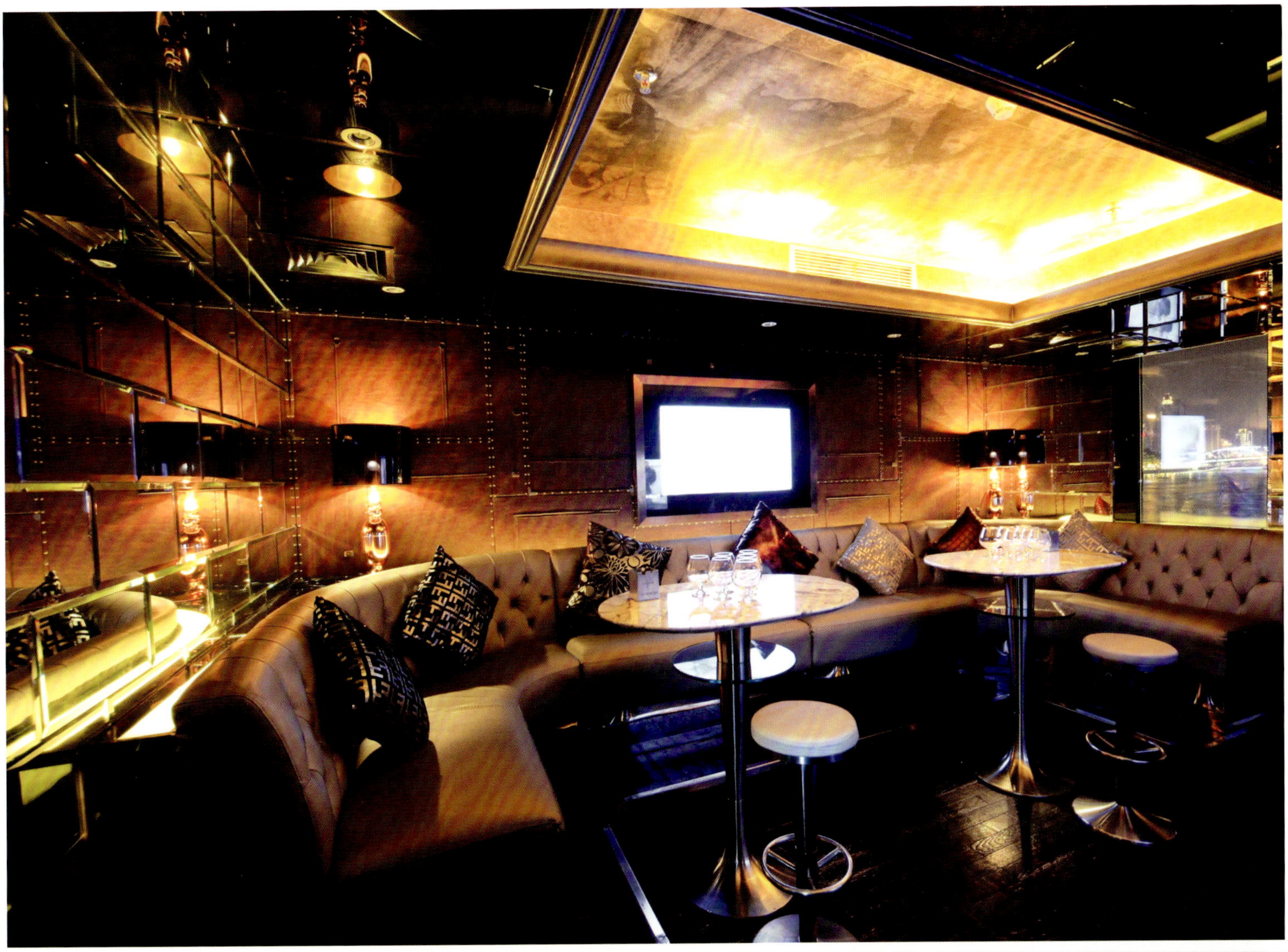

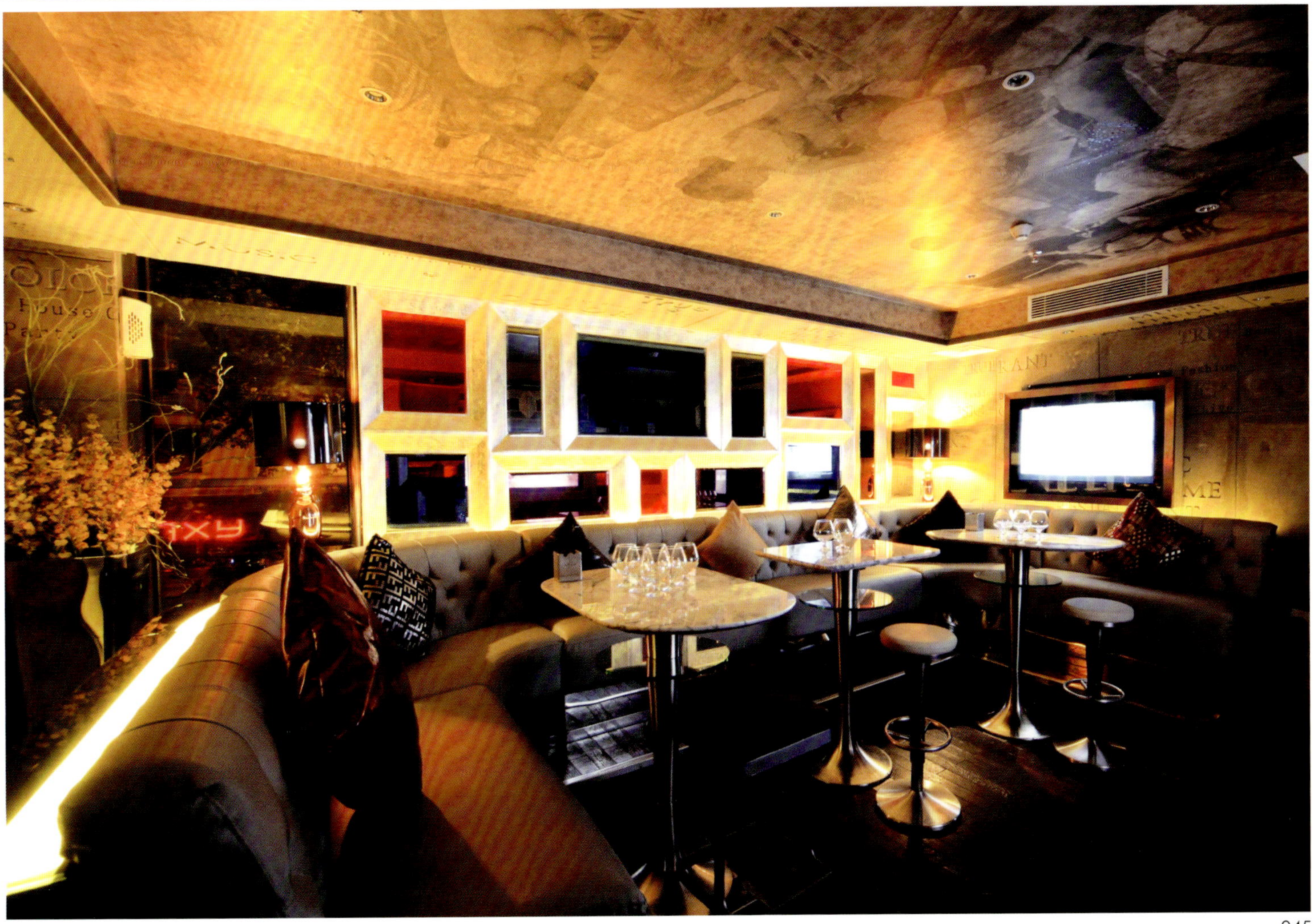

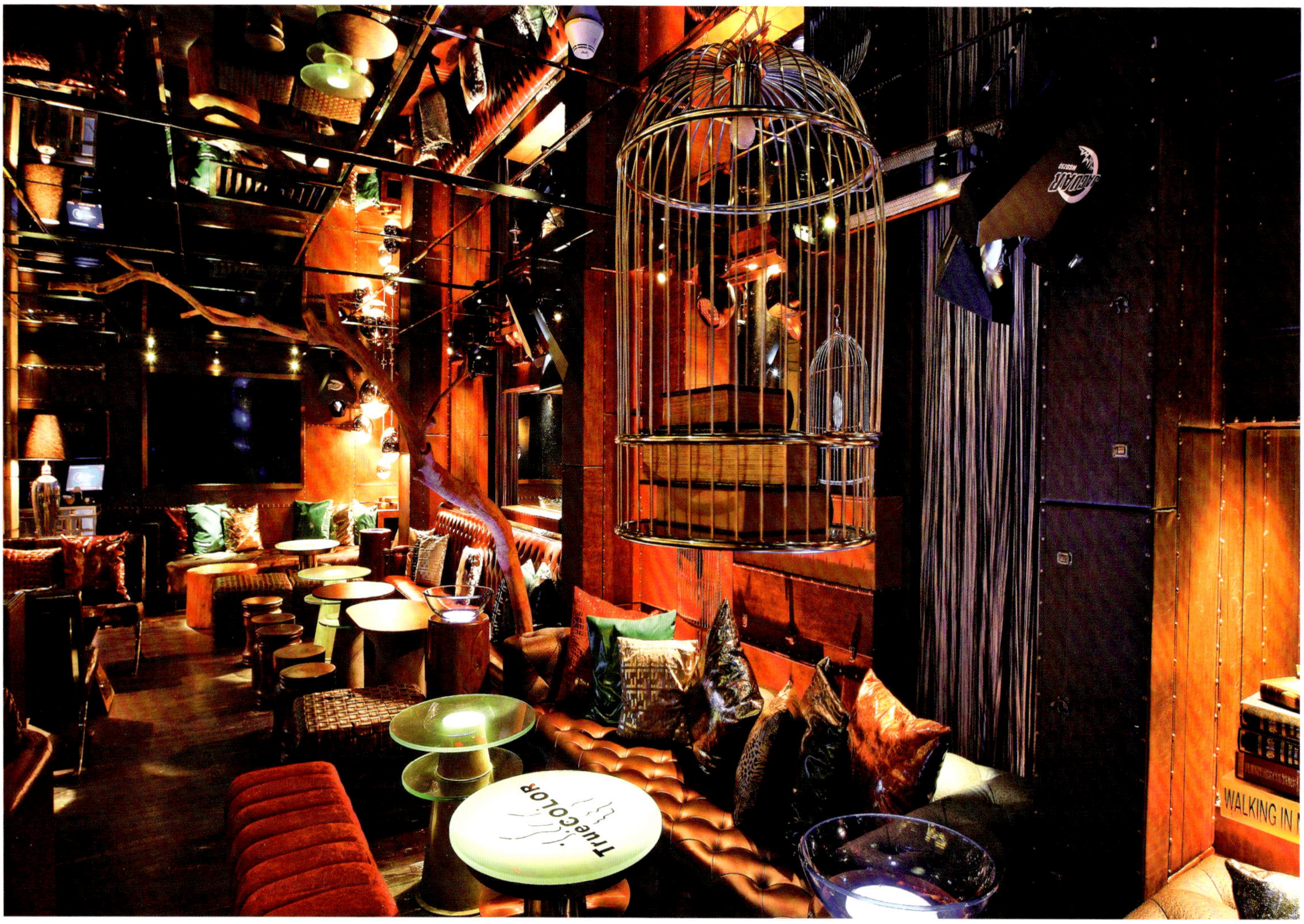
TrueCOLOR
WALKING IN

RESTAURANT
Live Band House Club
TrueCOLOR
House Club
MUSIC

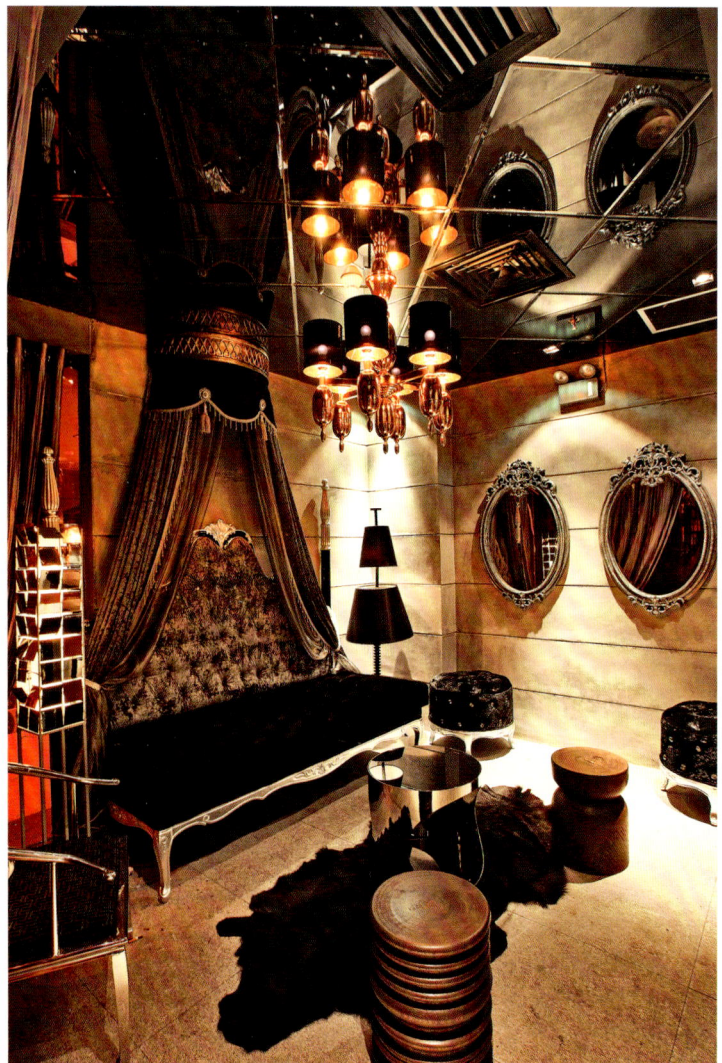

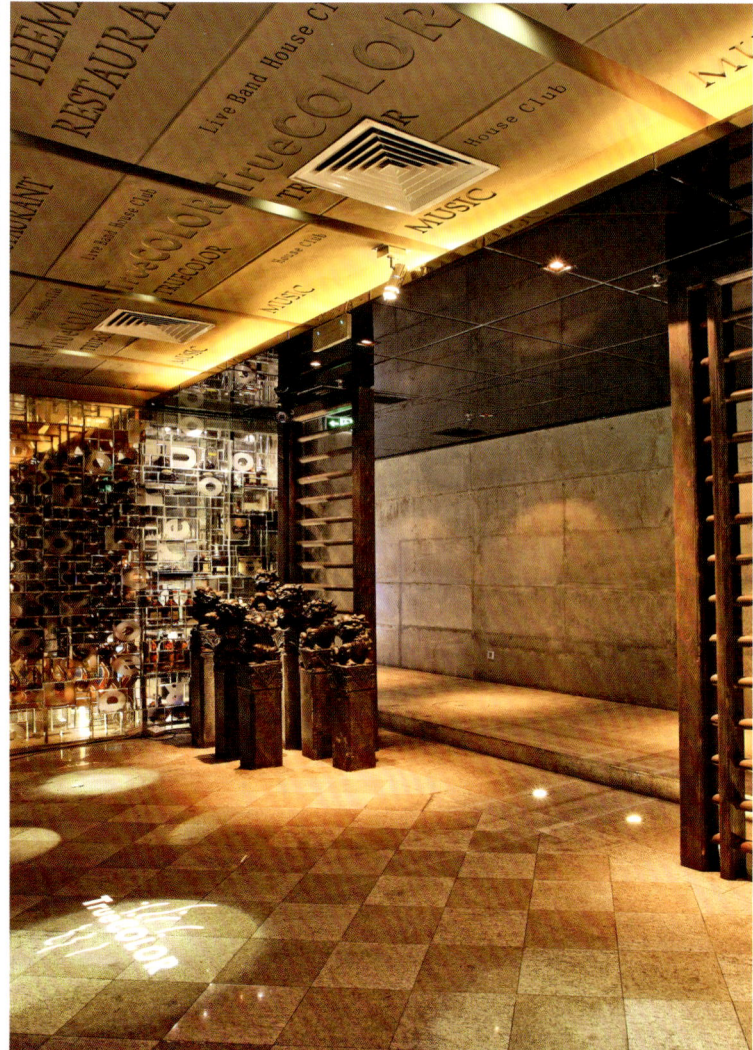

地点	面积	设计师	设计公司	主要材料
/ 湖南	/ 2400m²	/ 吴守峰	/ 吴守峰设计有限公司	/ 烤漆板、抛光砖、玻璃、墙纸

Kamade Wholesale KTV

—— 嘉麦迪量贩 KTV

Kamade Wholesale KTV is designed with the customer orientation of its surrounding white-collar workers and young consumers, so space with unique design is produced for this specific group of people—simple, fashionable, gorgeous and elegant, with unique taste.

In terms of the selection of materials, it is mainly environmental protection materials, reducing the use of soft decoration and increasing the use of shape plates, glass and wallpaper. In addition, the exposed-style ceiling reduces the loss of plate materials, and LED lights decrease the energy consumption. At the same time, the designer unconventionally gives up the common black, but chooses white as the main tone to change the light as the protagonist, so the whole space is dreamy and colorful. The designer introduces a way of changing the architectural modeling instead of the usual flat approach of arranging materials, so the theme of the whole space is brighter, leaving a deep impression to consumers. At the same time, the various forms of rooms, lobby added with studio, internet lounge area, DJ booth and other fresh projects are refreshing and shining.

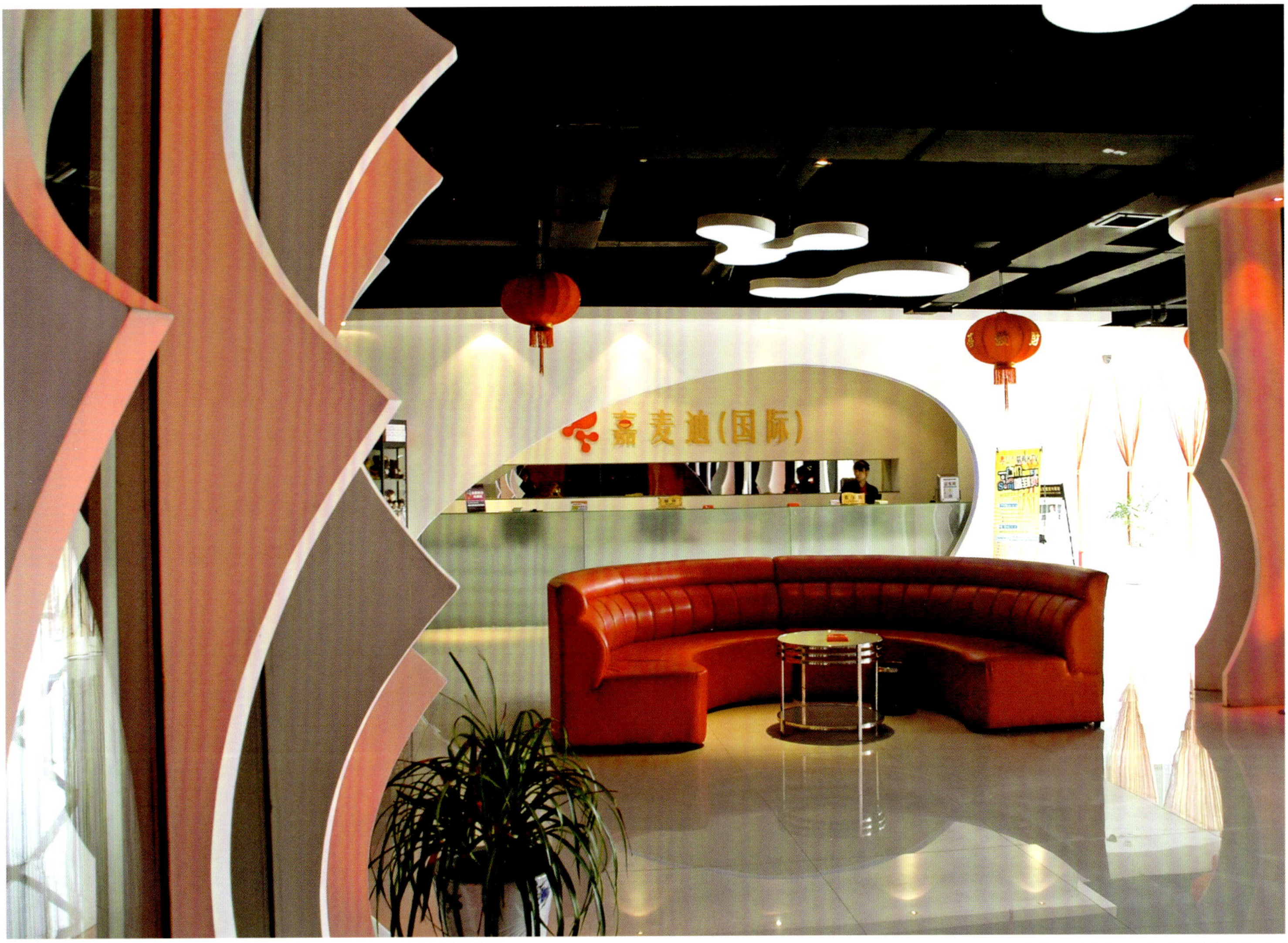
嘉麦迪(国际)

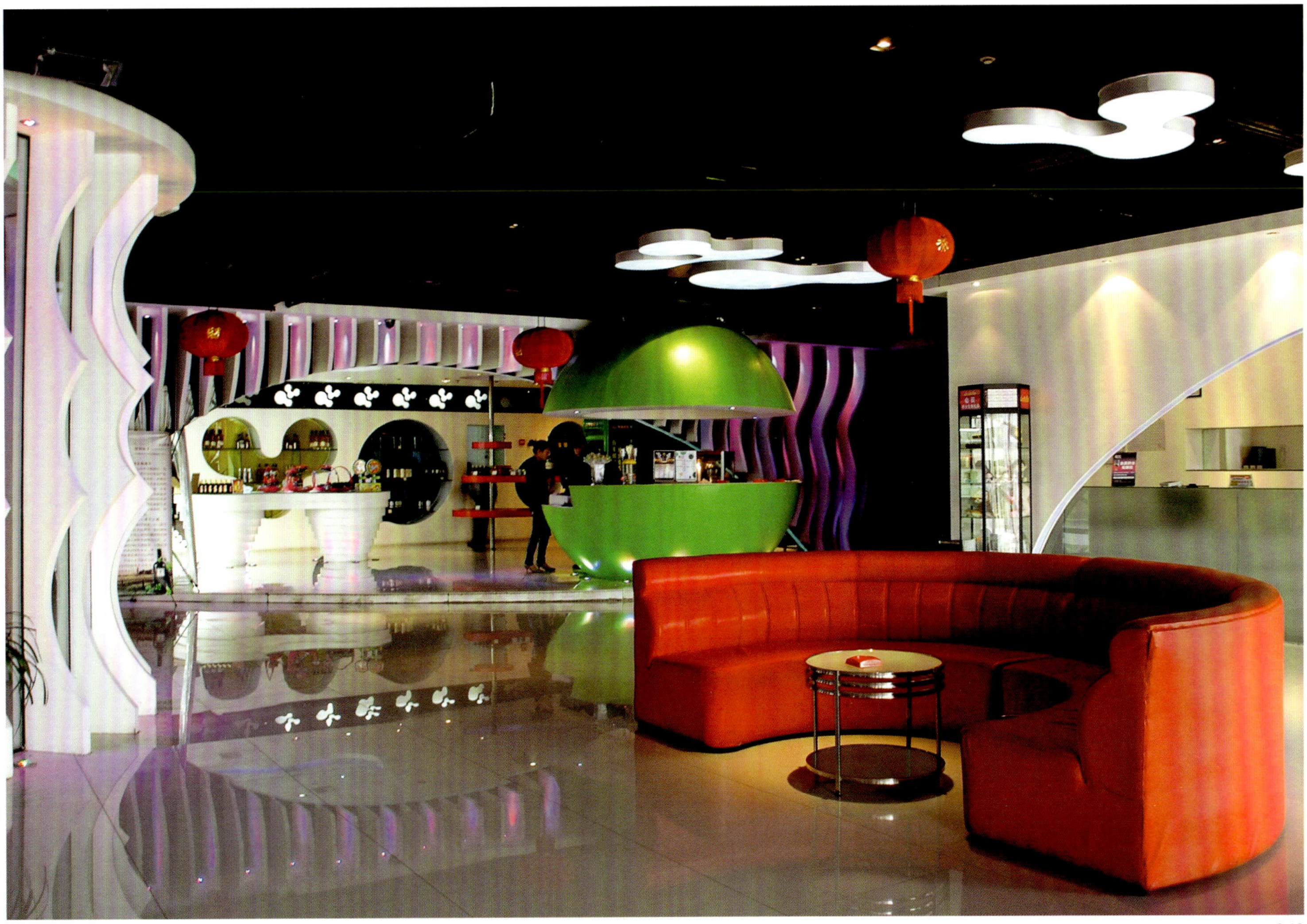

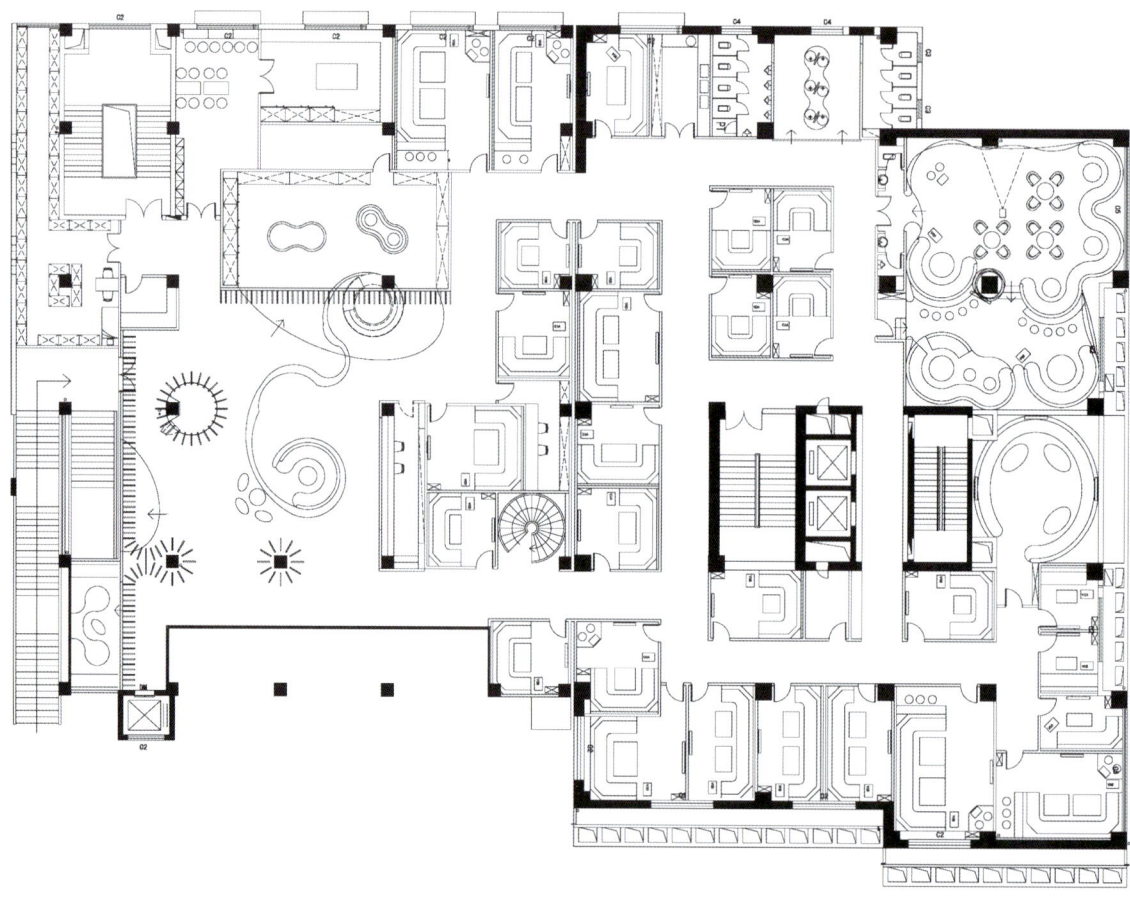

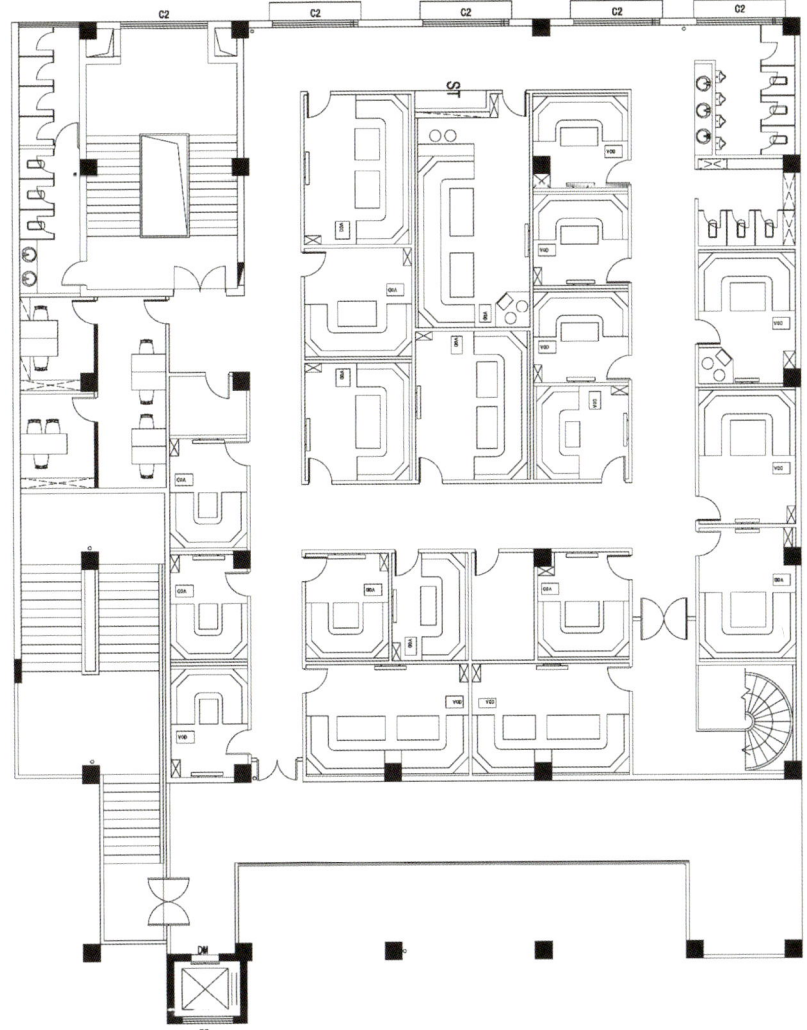

嘉麦迪量贩 KTV 是一项以其周边的白领和年轻的消费群为客户定位的设计，因此，设计师以独特的设计策划出一个符合这一特定人群品位的空间——简约、时尚、绚丽且优雅，独具品位。

在材料的选择上，以环保材料为主，减少了软包的运用，增加了造型板、玻璃和壁纸的运用。此外，外露式的天花也减少了板材的损耗，LED 灯则减少了能耗。同时，设计师一反常规，放弃以往惯用的黑色，采用白色为主调，让灯光变为主角，使整个空间变得梦幻、多彩；放弃了以往惯用的平铺材料的手法，采用改变建筑造型的手法，使整个空间的主题更为鲜明，给消费者留下深刻的印象。同时，多种形式的房型、增加了录音棚的大厅、上网休闲区、DJ 台等新鲜项目也让人耳目一新，眼前一亮。

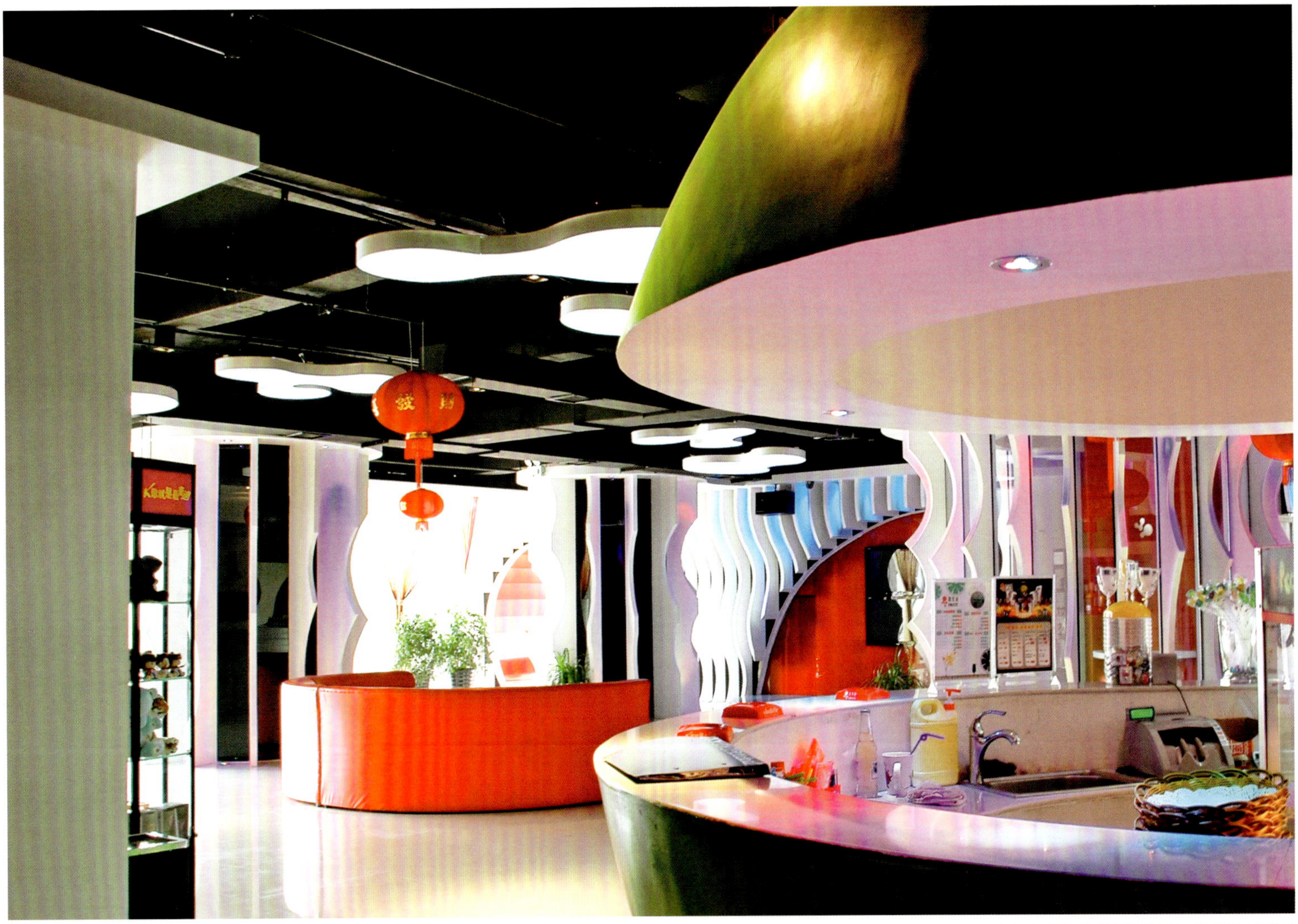

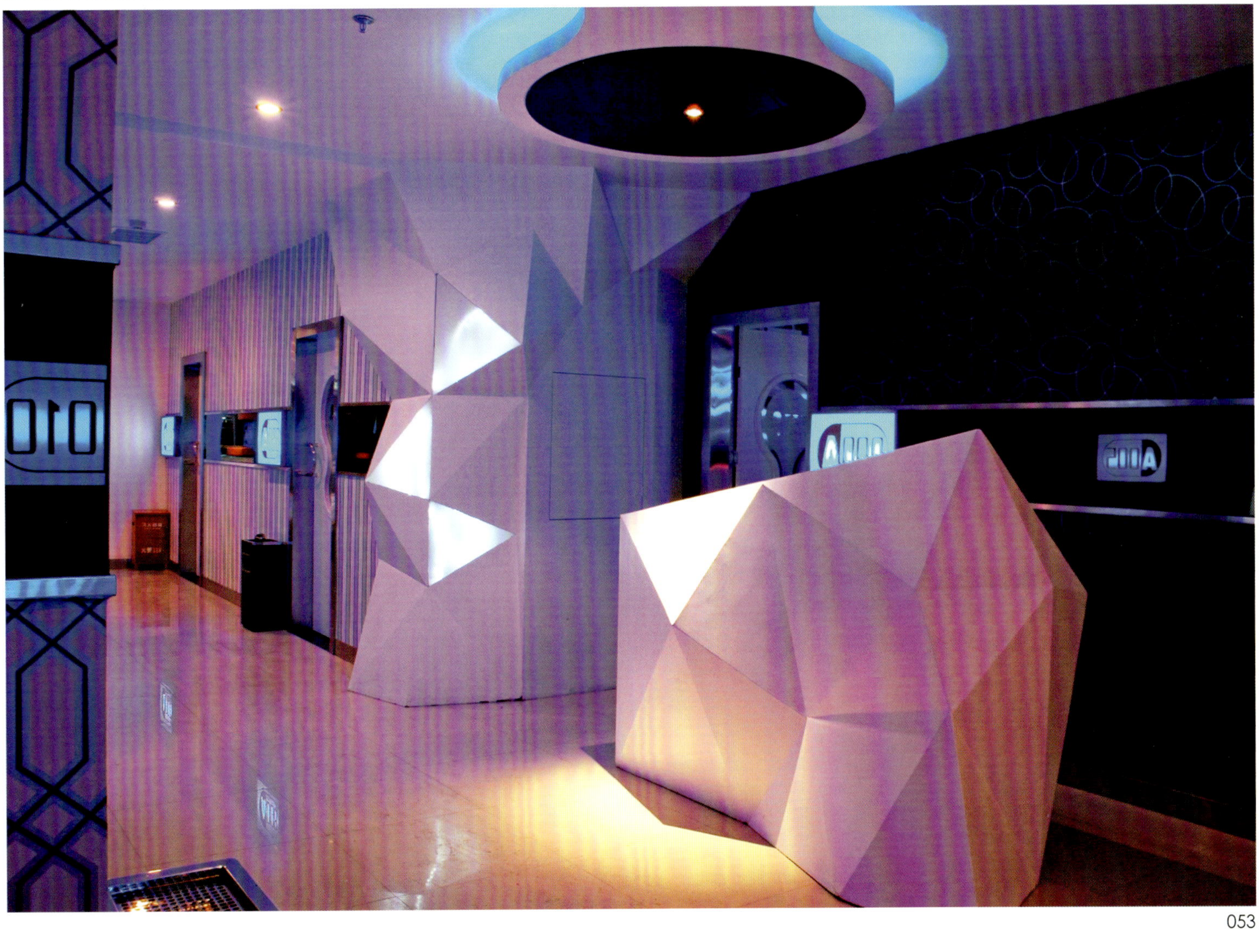

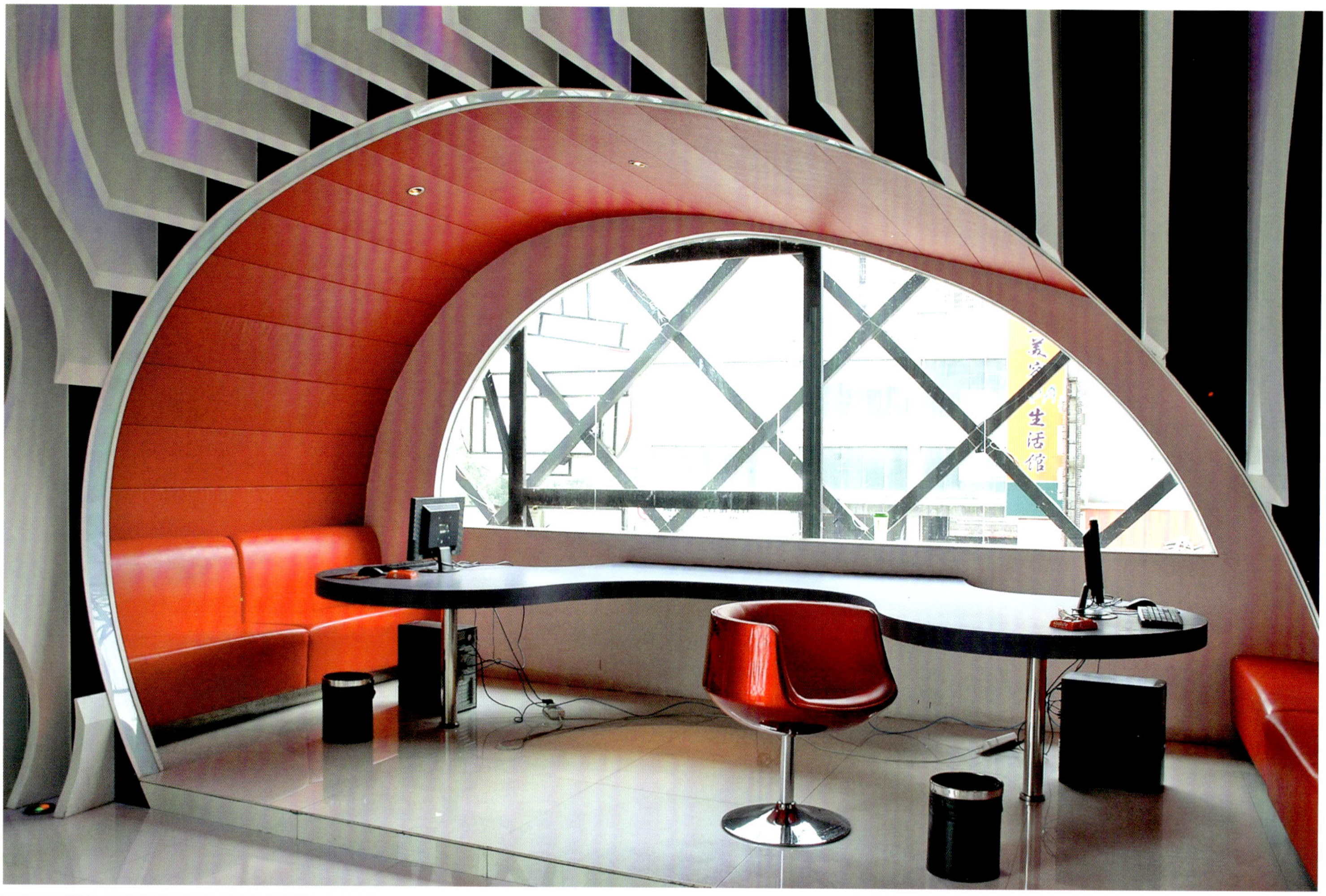
生活馆

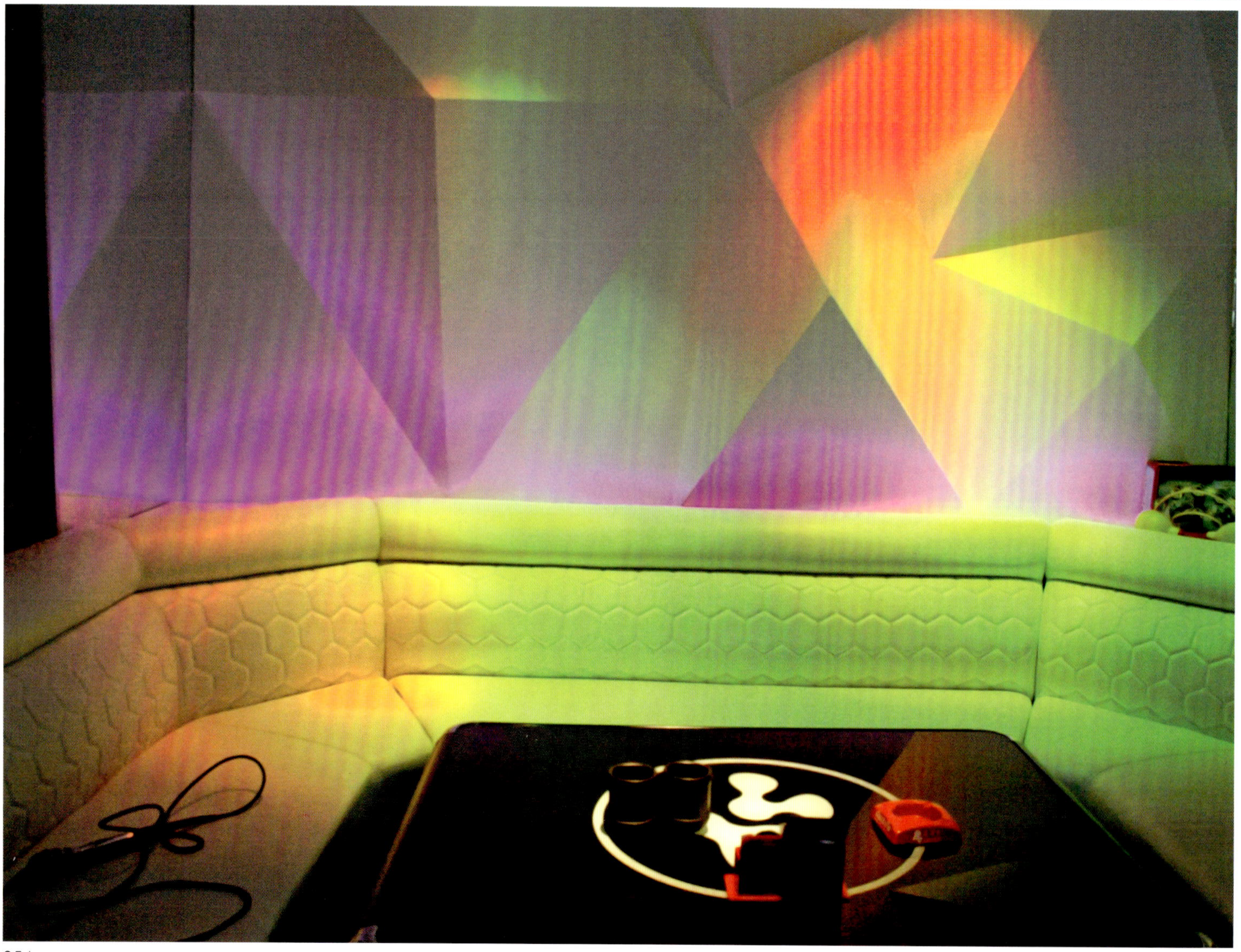

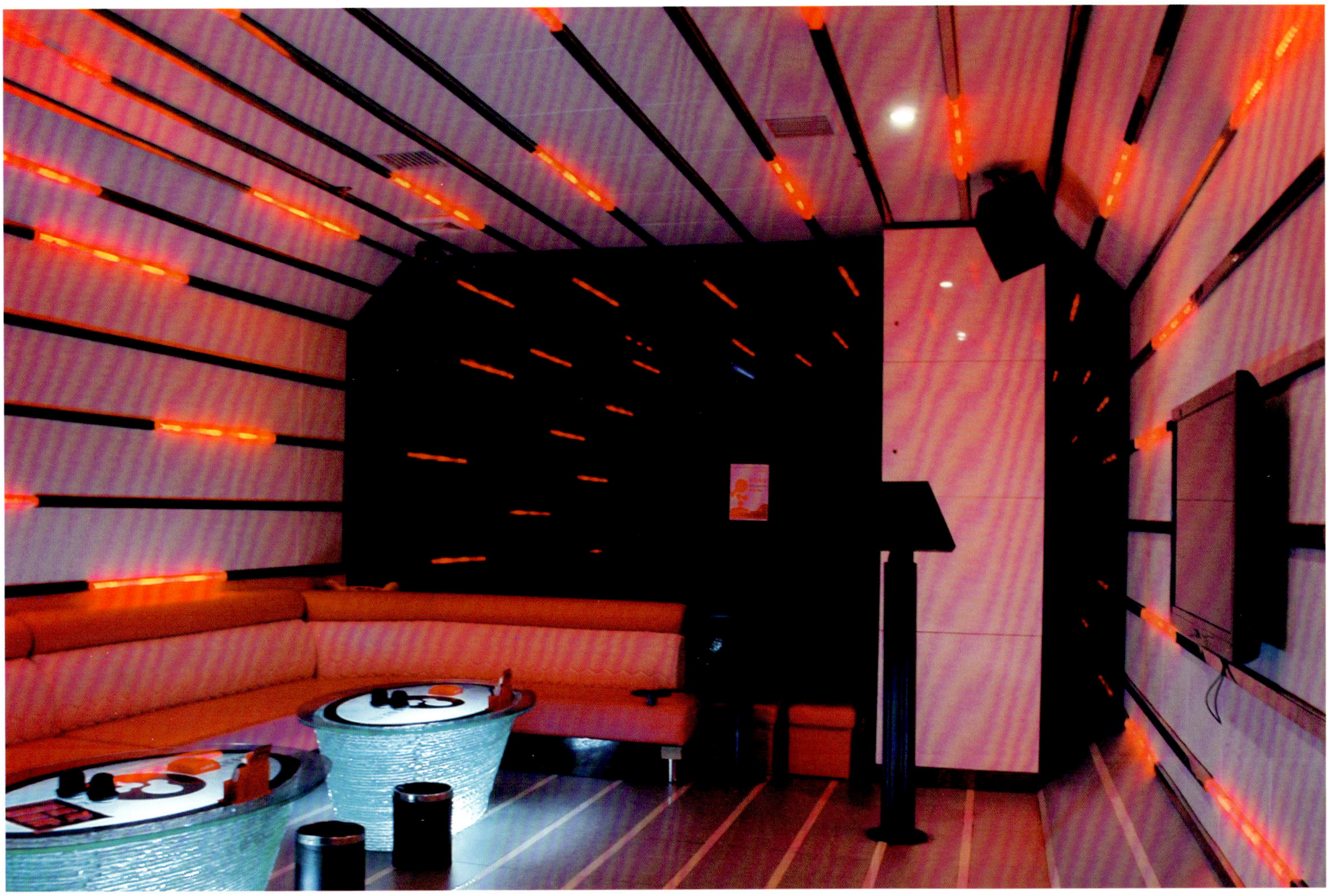

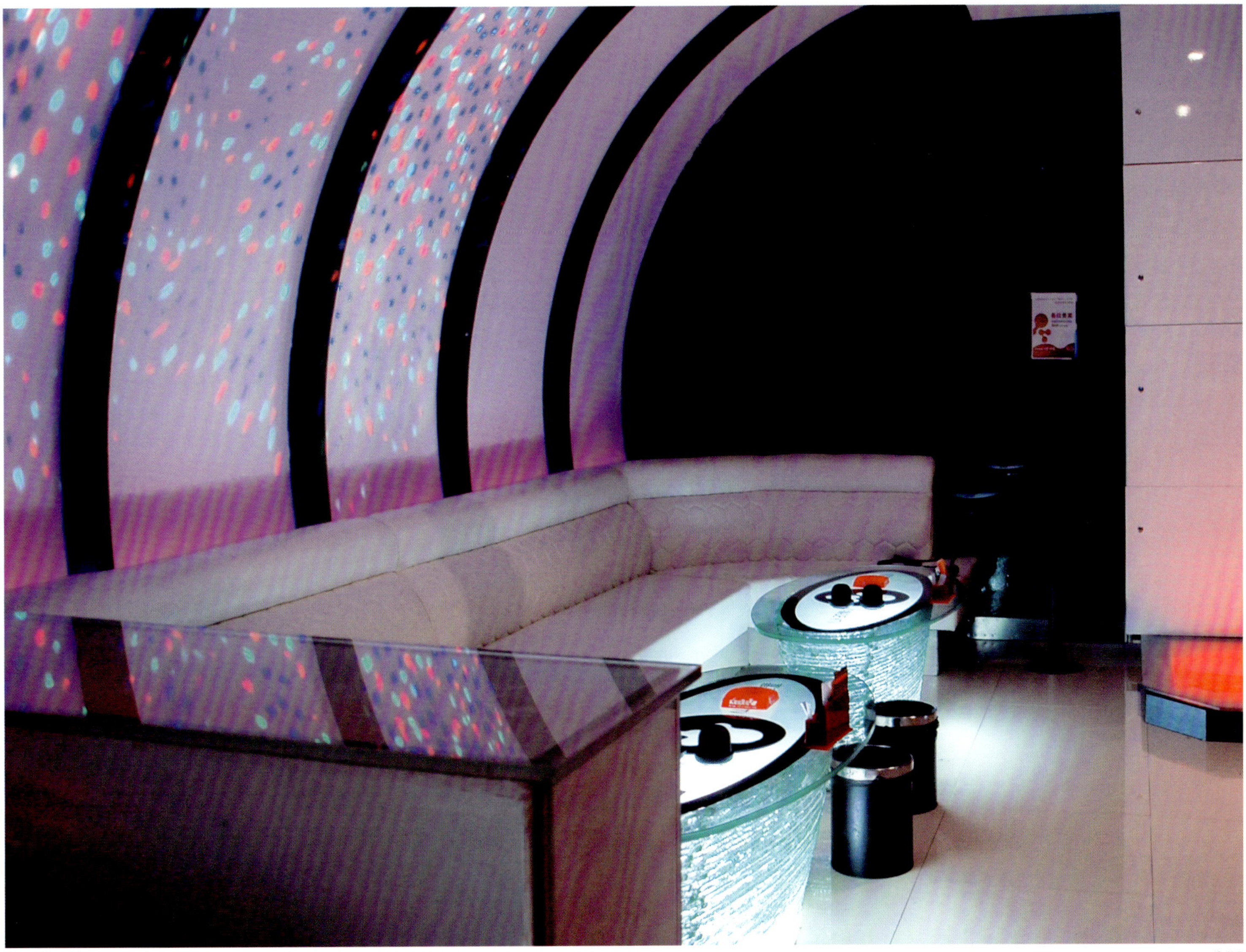

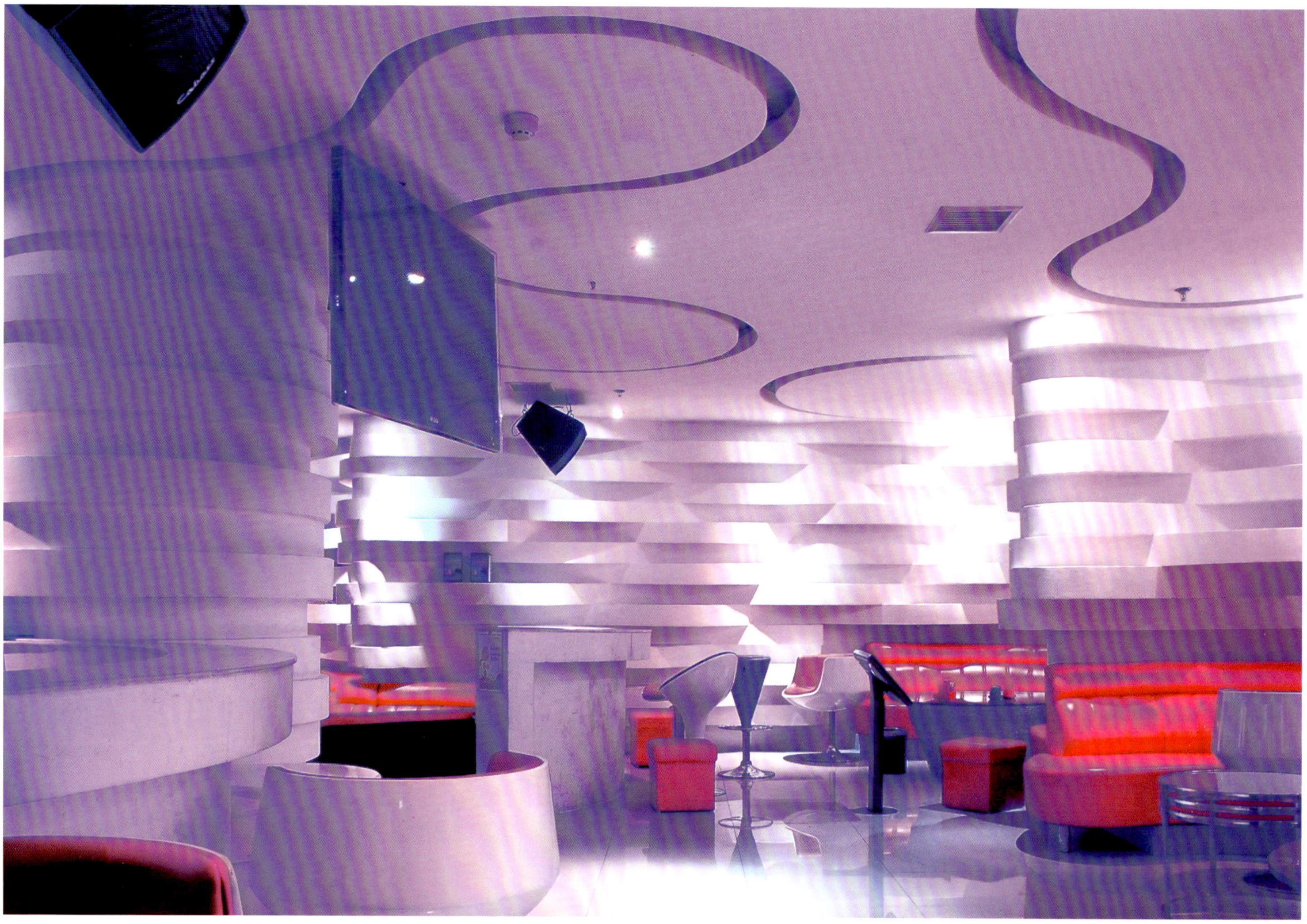

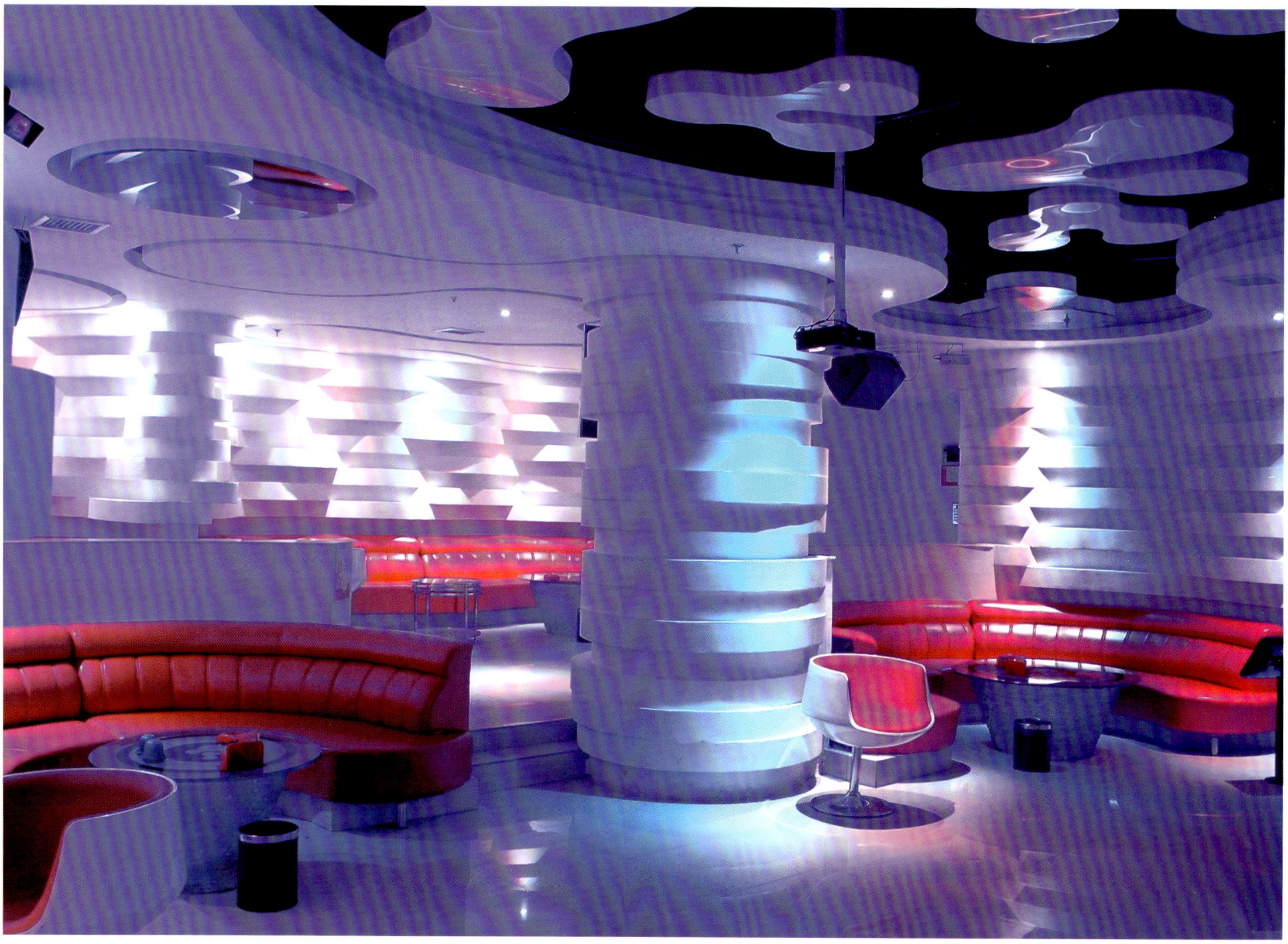

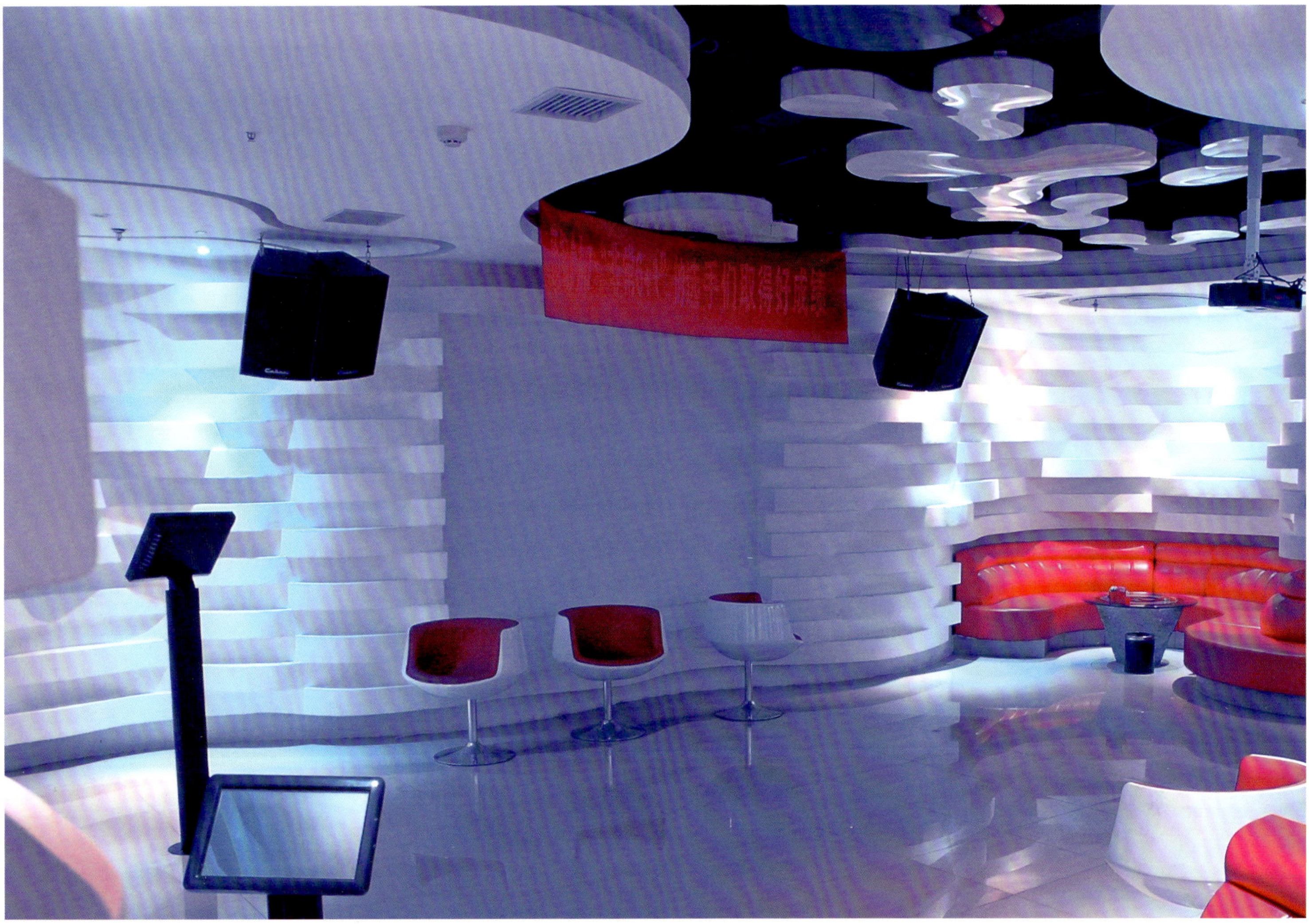

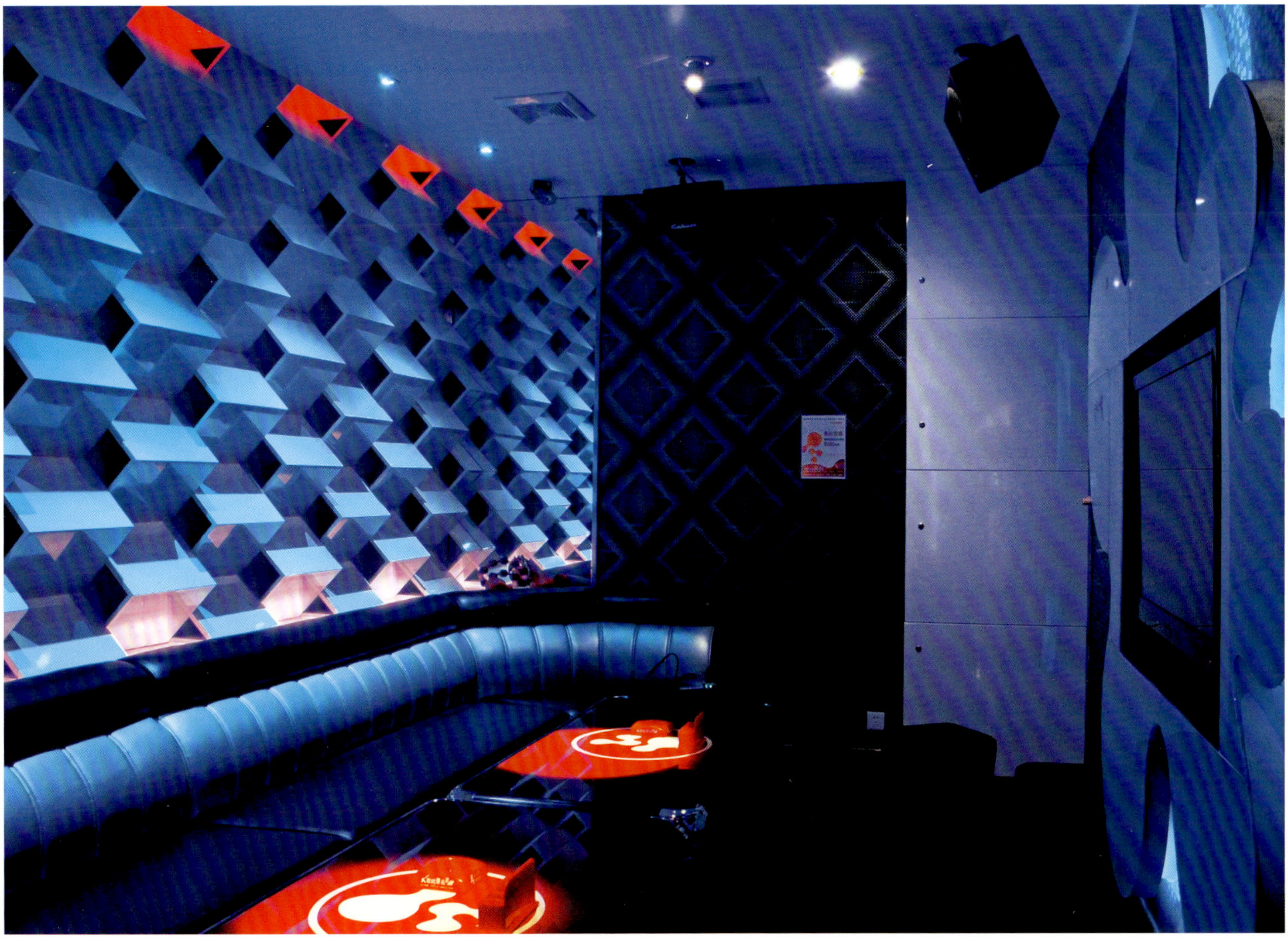

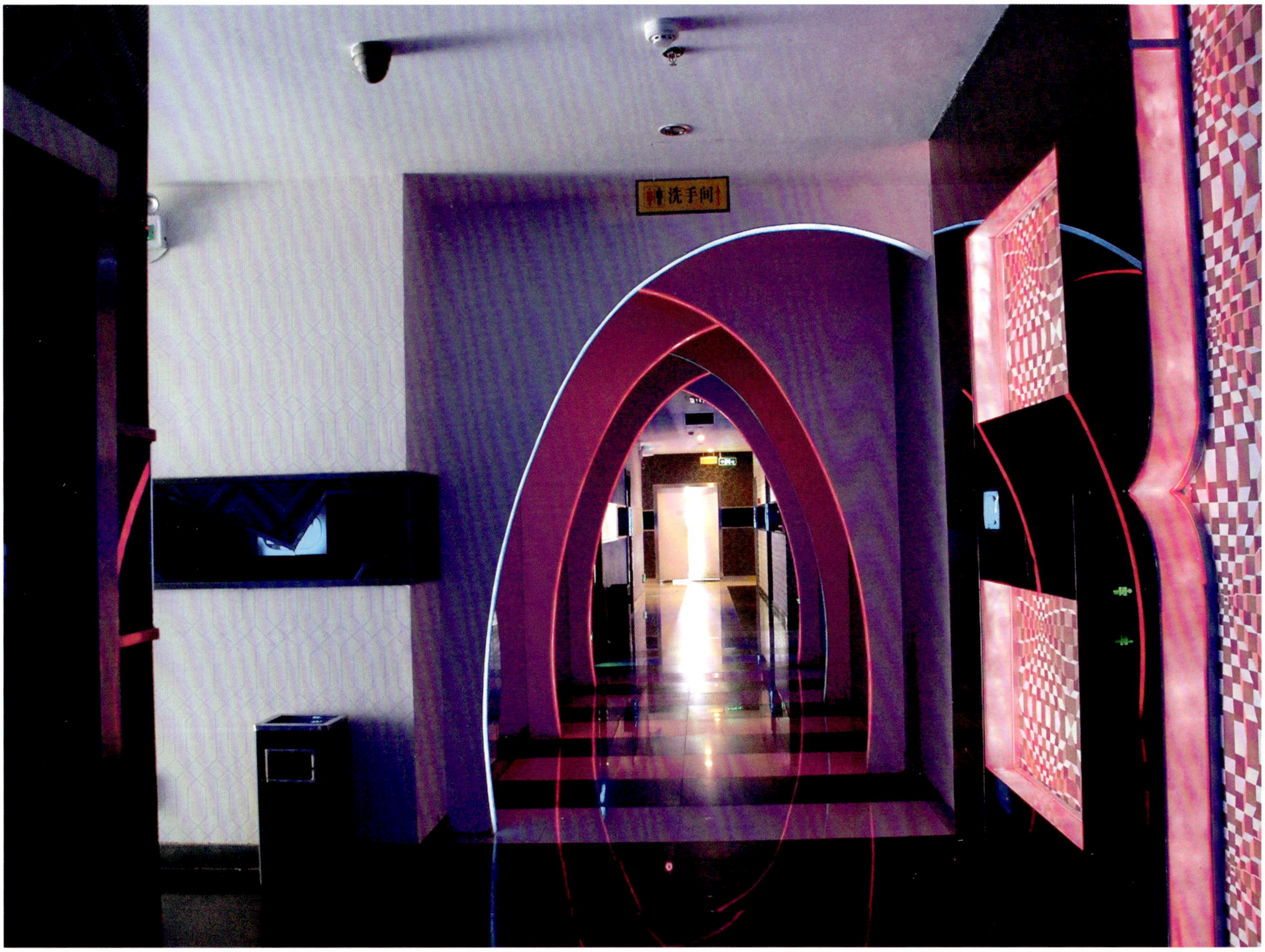
洗手间

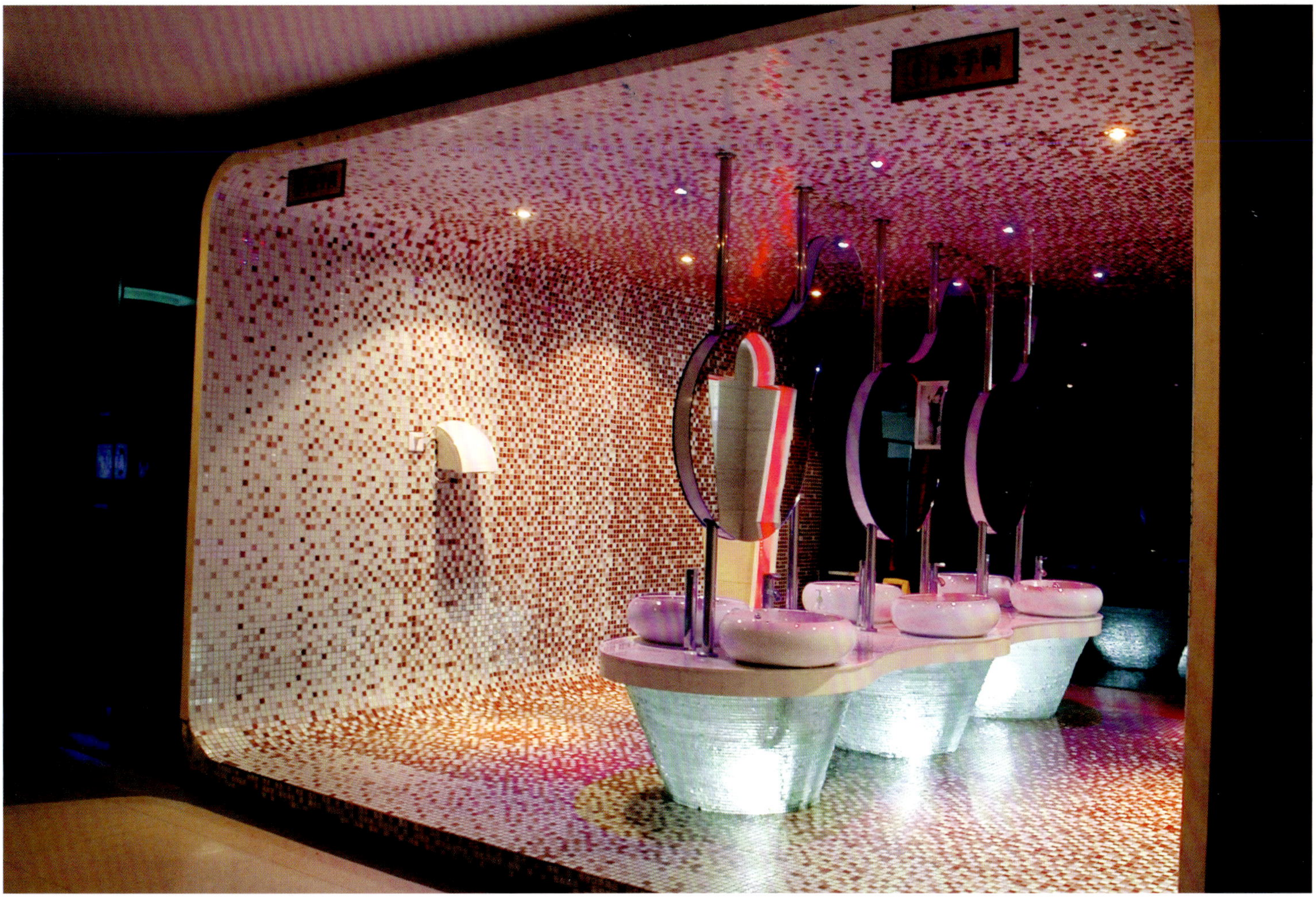

地点	面积	设计师	设计公司	主要材料	摄影
/ 杭州	/ 6000m²	/ 王建强	/ 杭州金白水清悦酒店设计有限公司	/ 大理石、透光云石、不锈钢、GRC、木饰面	/ 林德坚

Hangzhou New Royal Yongli International Entertainment Club

杭州新皇家永利国际娱乐会所

The project introduces the light and shadow match of bright and perfect amber and golden color, with the levels of light, through transmission, reflection, refraction, diffusion, absorption and other ways to create the space together. It sets off the true color of material texture, contrasts the interior atmosphere, produces a wide variety of performances within the specific space, and cleverly creates an entertainment environment integrated with charm and art.

The lighting design fully mobilizes the characteristics of light, combined with each function of environment and designing concept. It creates an ideal world of light and paradise for entertainment in order to achieve the artistic and entertaining design effect. Decoration and lighting are the most fascinating part of the entertainment club, where various lights echo with each other, and the main and auxiliary are clear to form a beautiful light contrast. The color match is graceful, making people feel like in a magnificent golden palace, or like in a fantastic maze waved with lights and shadows. The luxurious atmosphere reaches a perfect space effect, realizing the purpose of relaxing and entertaining.

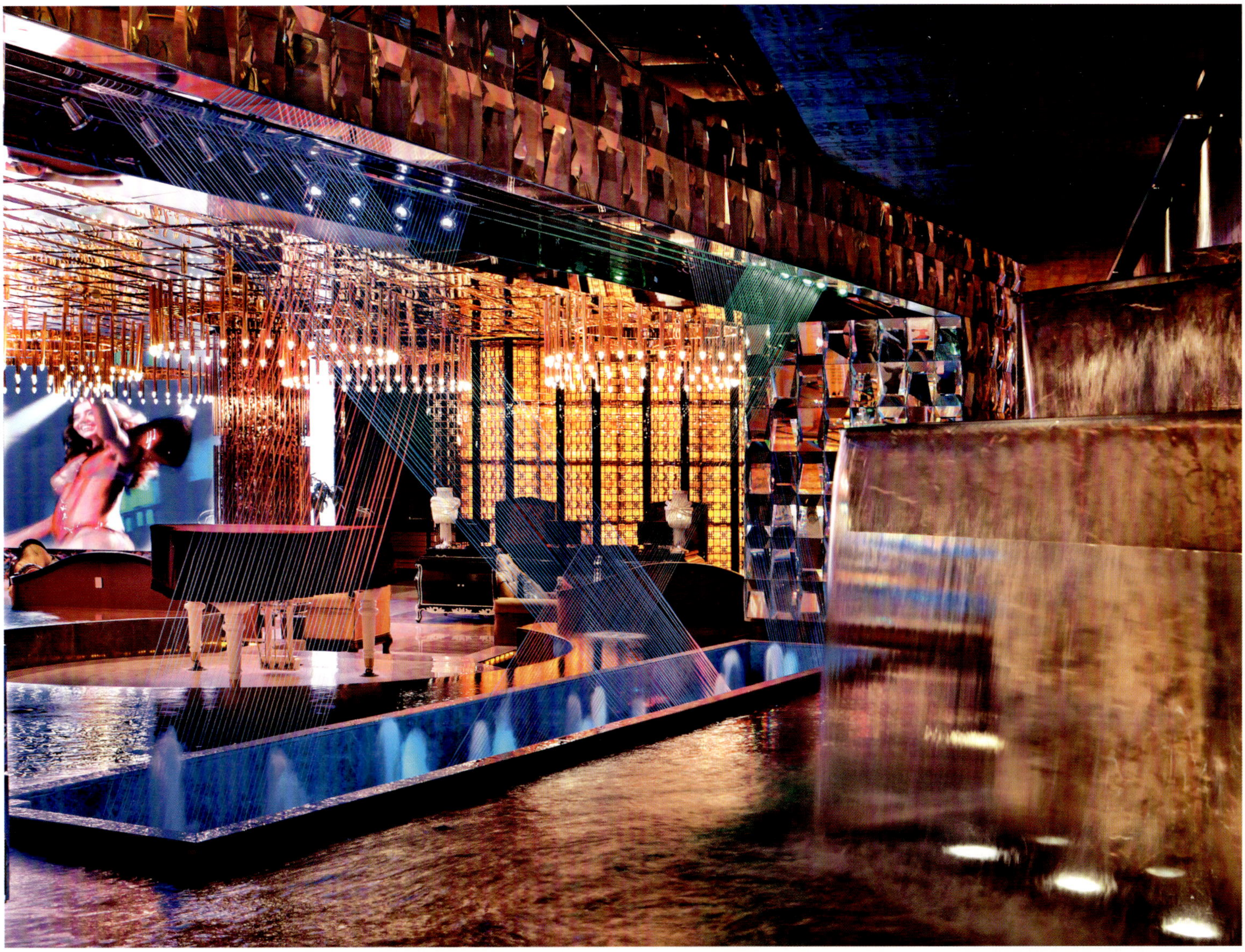

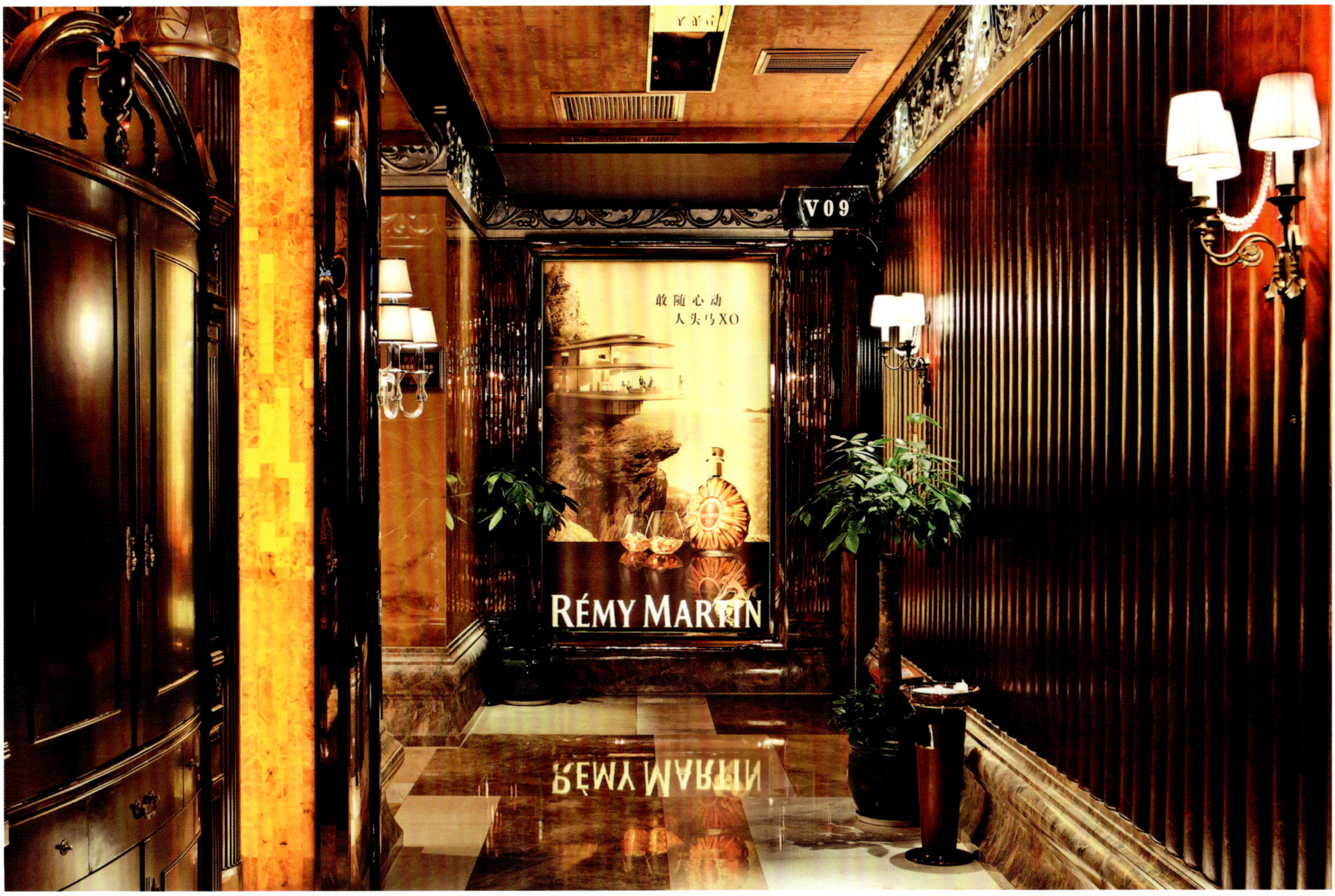
V09
敢随心动
人头马XO
RÉMY MARTIN

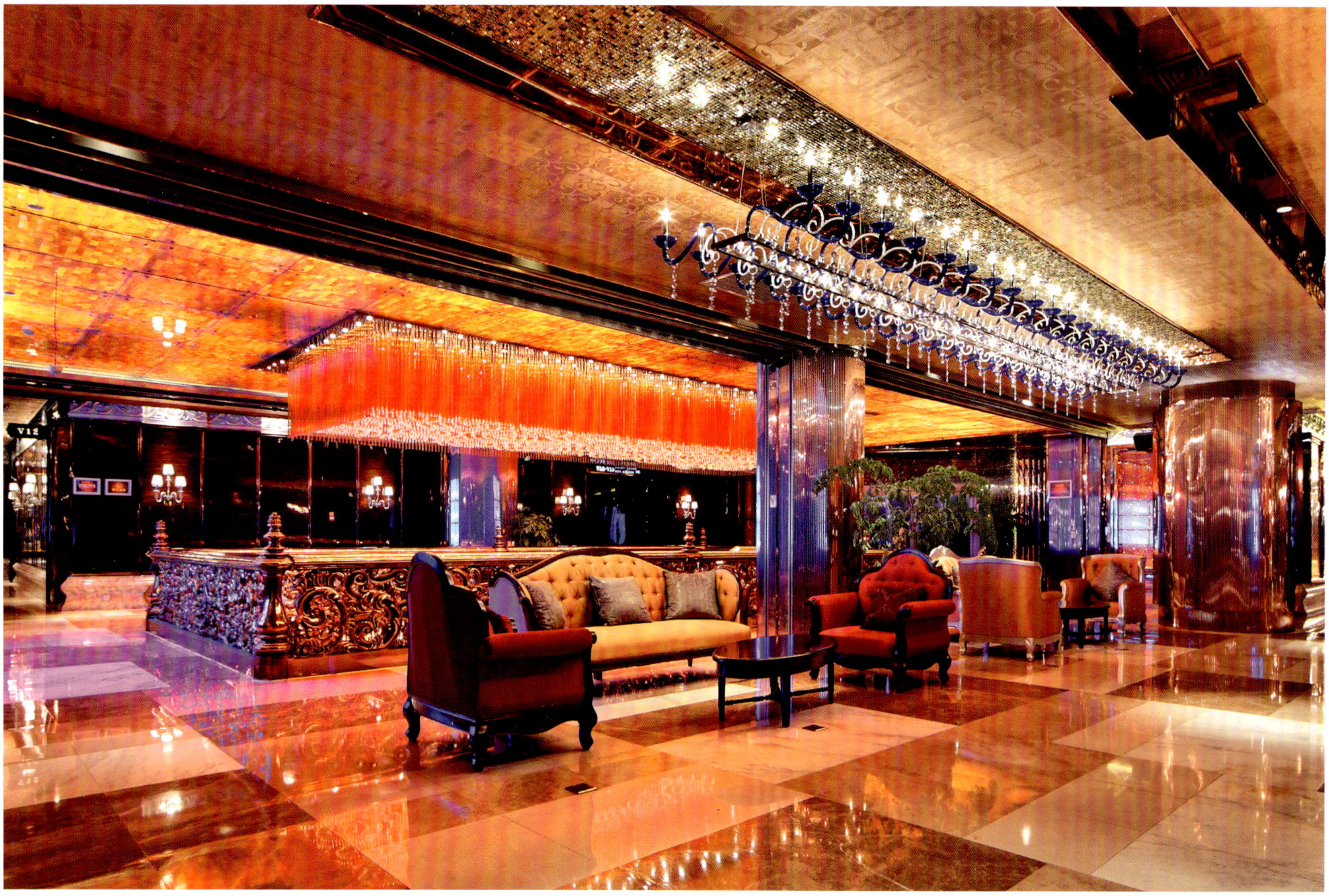

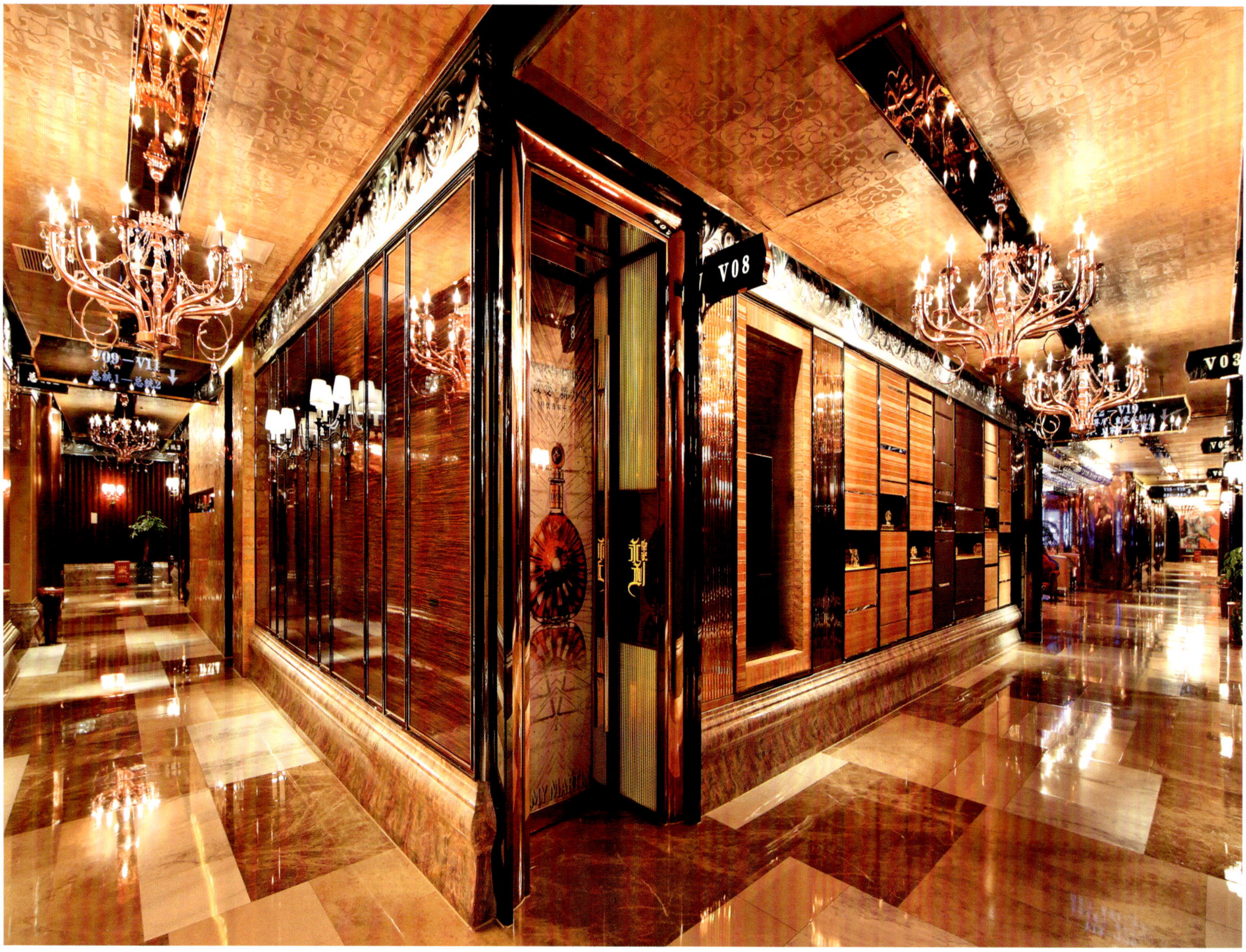
V08

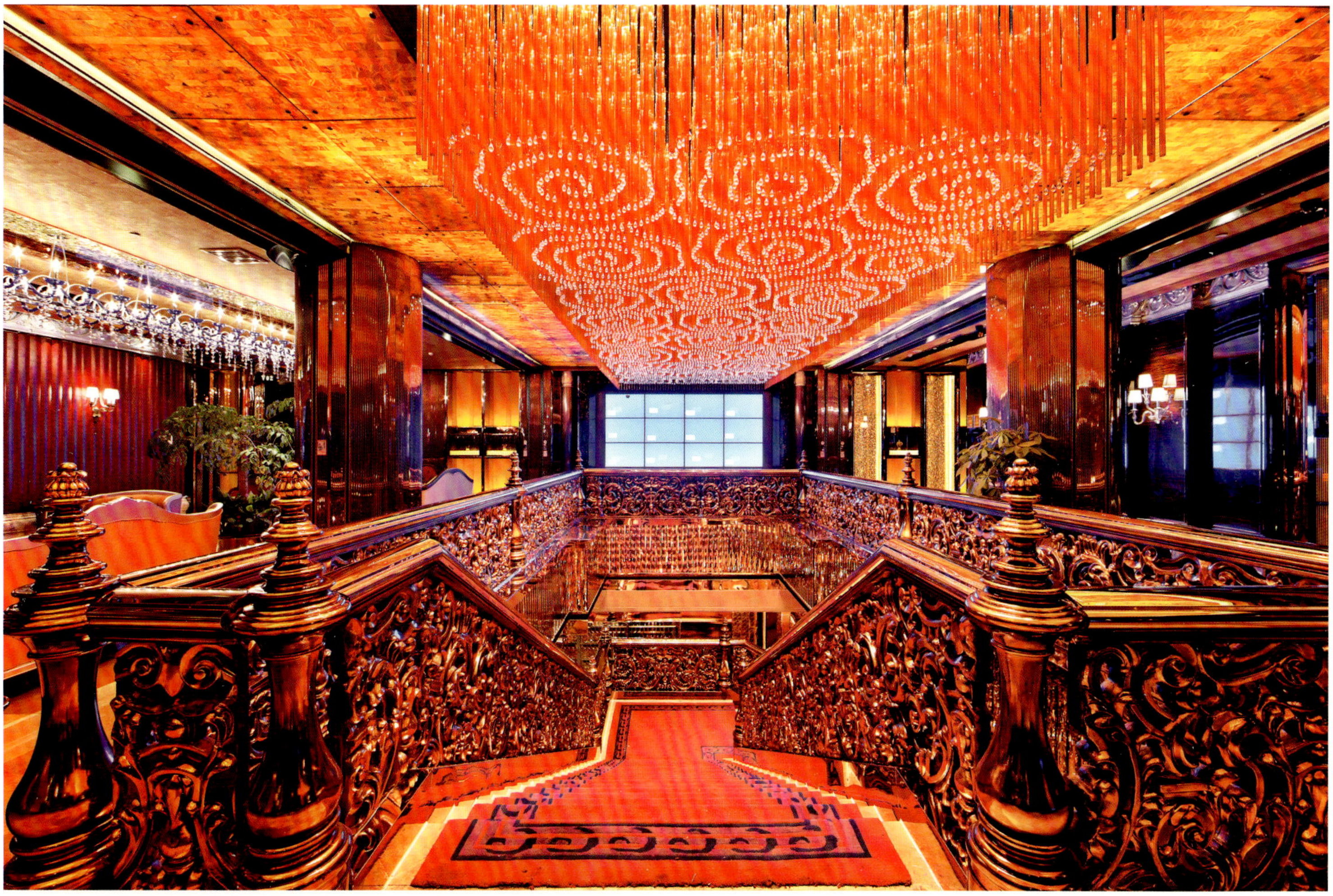

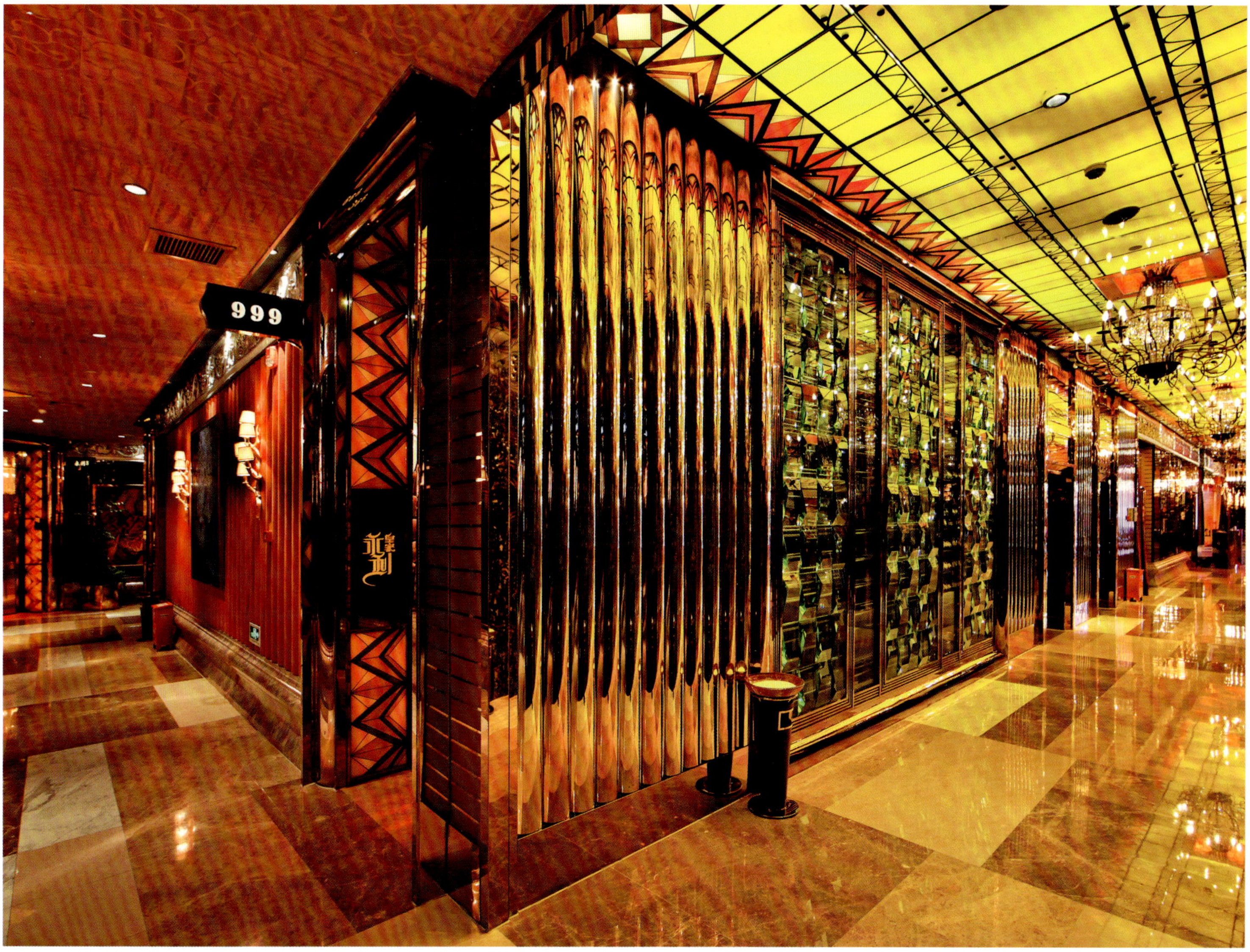
999

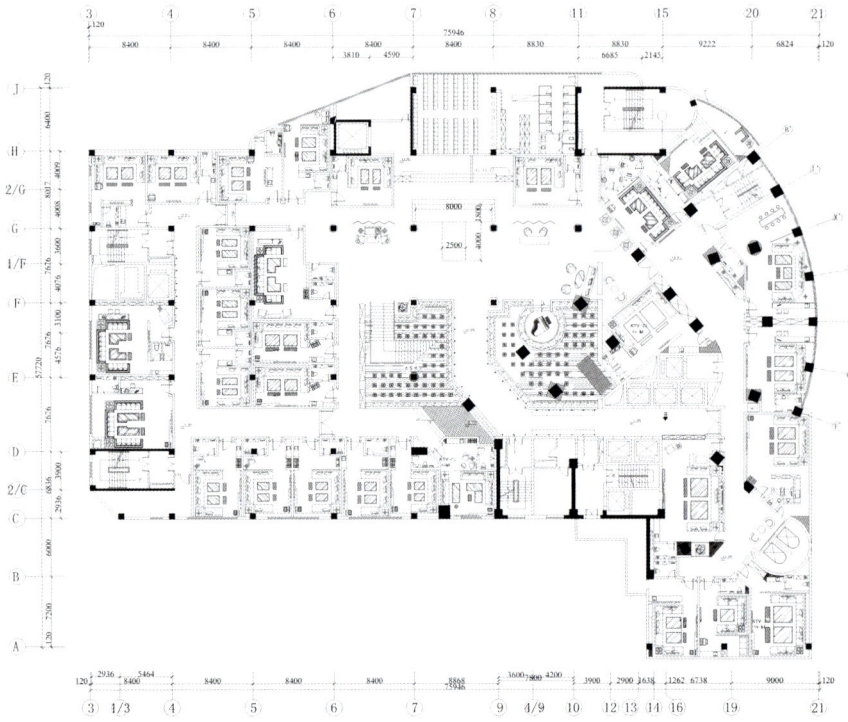

本案采用亮丽剔透的琥珀色与金黄色的光影搭配，运用光的层次，通过透射、反射、折射、扩散、吸收等方式共同打造该空间。衬托材料质感的本色，烘托室内的氛围，在特定的空间内展现多种多样的表现力，巧妙地塑造出一个魅力和艺术共生的娱乐环境。

照明设计充分调动光的特性，结合环境的各项功能与设计理念，创造出一个理想的光的世界和娱乐的天堂，以达到艺术性、娱乐性的设计效果。装饰和灯光是本娱乐会所最引人入胜的部分，各种灯光遥相呼应，有主有辅，形成一个靓丽的灯光对比效果。色彩搭配落落大方，令人犹如身处富丽堂皇的金殿之中，又如置身于光影交织的梦幻迷宫之内，豪华的环境氛围取得了完美的空间效果，达到了放松休闲的娱乐目的。

3
4
5
6
7
8
11
15
20
21
75946
J
H
2/G
G
1/F
F
E
D
2/C
C
B
A
A'
B'
C'
D'
L'
K'
J'
H'
G'
F'
57720
1/3
9
4/9
10
12
13
14
16
19

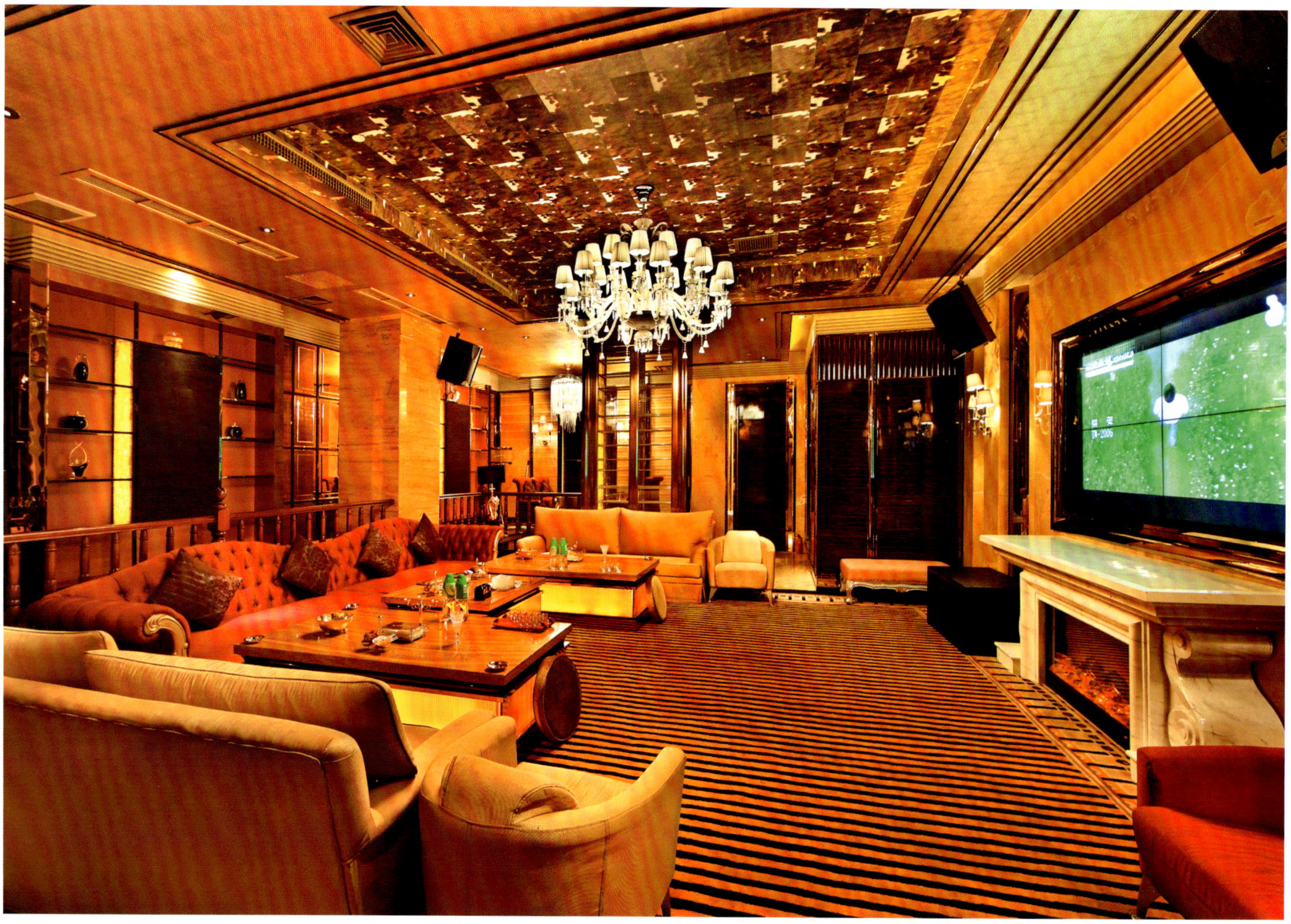

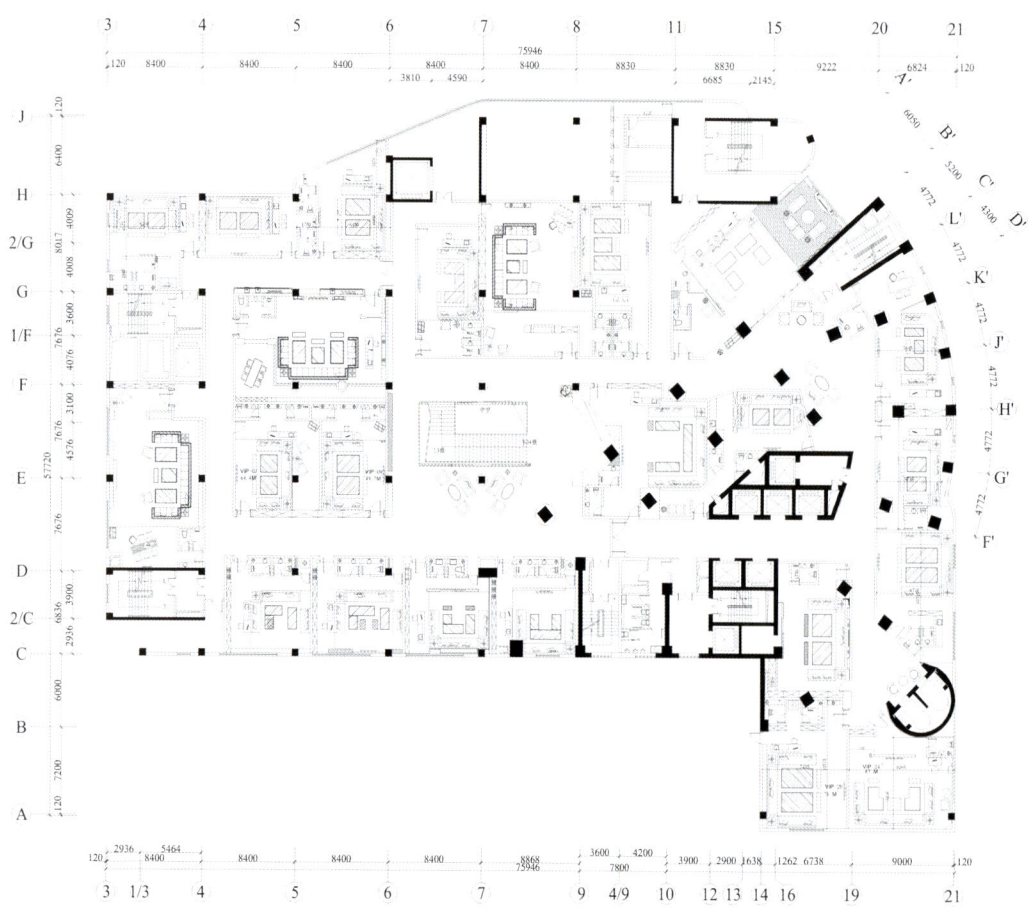

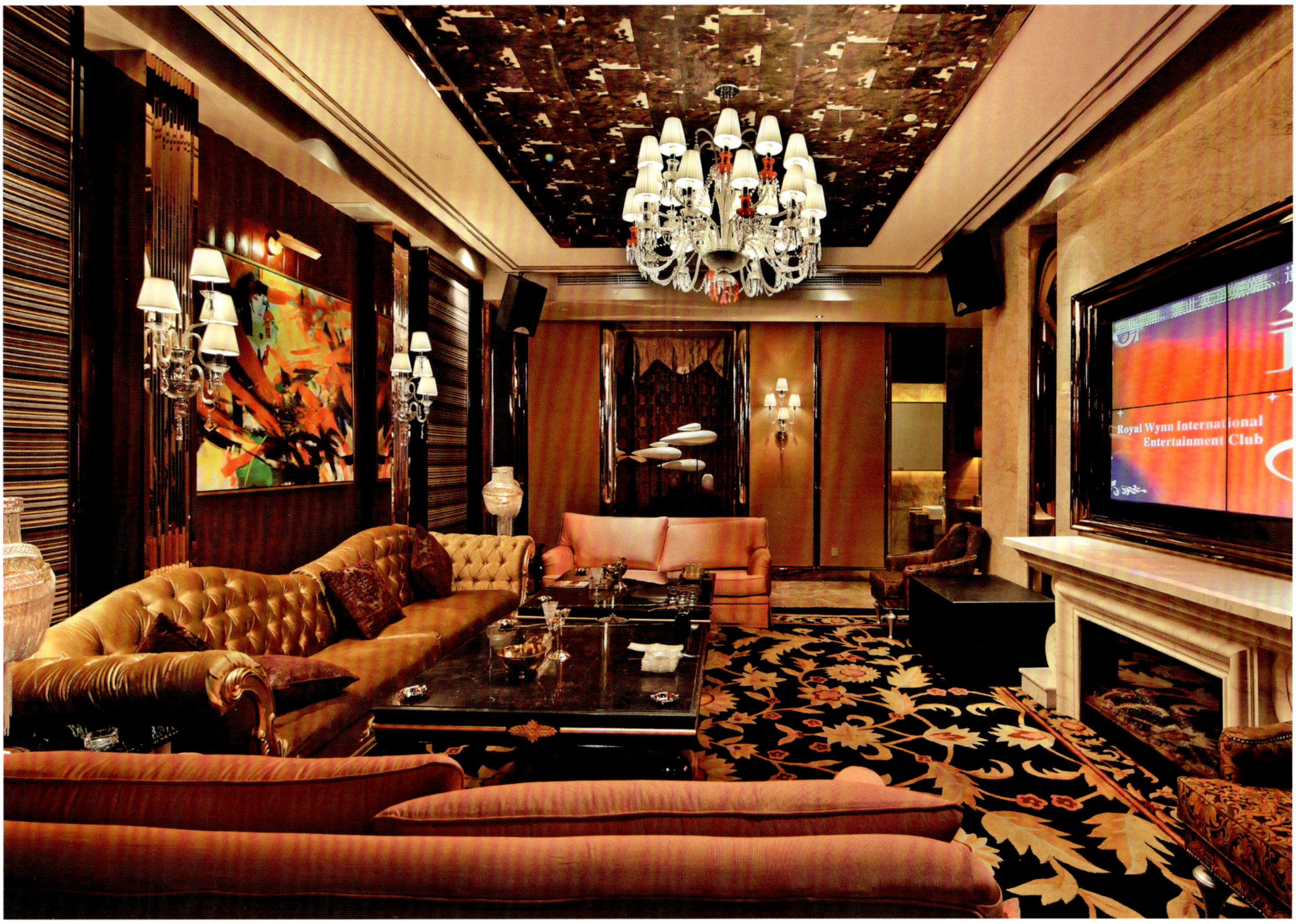
Royal Wynn International
Entertainment Club

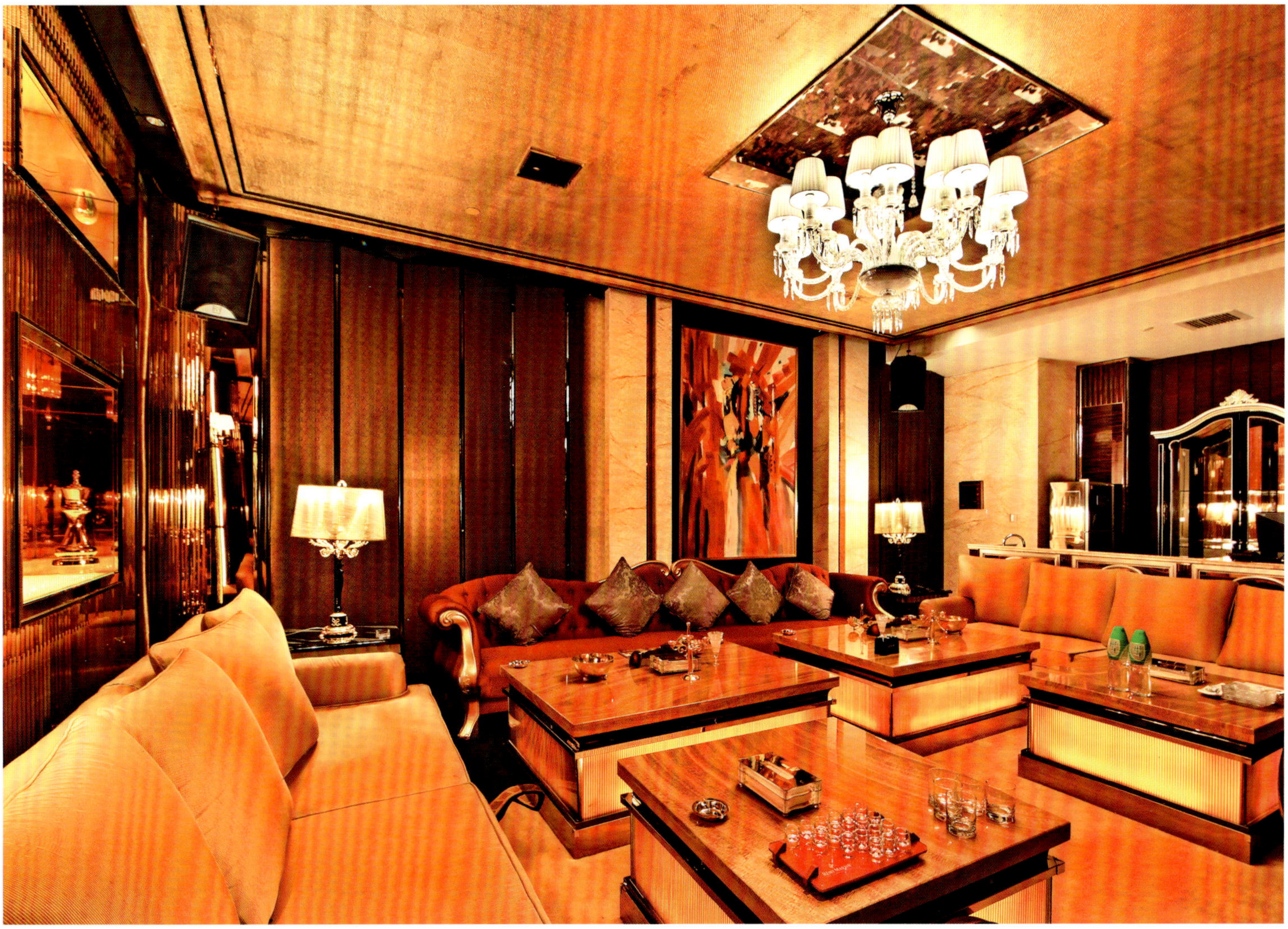

地点	面积	设计师	设计公司	主要材料	摄影
/ 西安	/ 3000m²	/ 王永	/ 北京建极峰上大宅装饰西安分公司	/ 雅士白石材、 黑木纹石、灰木纹石、布纹石、防火板、玻璃砖、皮革硬包、灰镜、茶镜、钢化玻璃、亚克力、钨钢、不锈钢、乳胶漆、LED 灯等	/ 张小明

Sugar Cube Wholesale KTV

—— 方糖量贩 KTV

The project features the designing core of "health, fashion, joy, and elegance," with efforts to create a pleasant place for releasing emotion and dispersing the pressure of urban life. The biggest designing character is to abandon the enchanting, mysterious and dark space atmosphere in previous nightclubs, but focusing on the creation of dignified and noble temperament. The public area introduces an extensive use of Ariston white stone material and gray mirror, with the match of white and gray to show a fashionable and healthy feeling.

While meeting the needs of multiple spaces, the designer creates an environment with elegant mood through the processing of space rhythm, sequence and level, integrating each space together and set the wholesale area and business area to meet the needs of different groups of people. In the design of private rooms, the changes and combination of gray mirror, leather hard decoration and fire-proof plates are introduced to reflect the design concept of "health, fashion, joy and elegantce ".

本案以"健康、时尚、欢乐、高雅"为设计核心，努力打造出一个释放情感、驱散都市生活压力的惬意之所。其最大设计特点在于摒弃以往夜场妖娆、神秘、昏暗的空间氛围，转而注重端庄、高贵气质的营造。公共区域大量运用雅士白石材及灰镜，以白与灰的搭配来表现一种时尚、健康的感受。

设计师在满足多重空间需求的同时，通过对空间节奏、序列、层次的处理，塑造出意境高雅的环境，把各个空间融入其中，并设置量贩区和商务区，满足不同人群的需求。在包间的设计中，通过灰镜、皮革硬包、防火板等材质的变化与结合，体现出"健康、时尚、欢乐、高雅"的设计理念。

地点	面积	设计公司	照明设计	主要材料	摄影
/ 北京	/ 1500m²	/ 睿智汇设计	/ 睿智汇设计	/ 玫瑰金不锈钢、粉镜、黑镜、水晶、绒布、石材、冰砖、LED 灯	/ 孙翔宇

New Yuesheng KTV Club

—— 新乐圣 KTV 会所

Entertainment Trend of 3D Pixel

The designers draw inspiration from the "pixel" form in digital image and present them at the entrance, lobby and waiting areas. In order to break the old and boring traditional mode, the design team defines the wall facing the entrance as the first visual focal point, and one side is designed with hundreds of stainless steel rose gold coupled with LED light source, coupled with the dynamic three-dimensional structures to enliven the space atmosphere and touch the curiosity of consumers. Our eyes reach the peak surface along with the dynamic pixel, and a power filling the whole space breaks out and extends in wavy shape, like the entertainment influence of the space, featuring the visual space with the most popular elements to bring consumers an extraordinary first-hand experience. New Yuesheng KTV space design presents the sense of technology, storm and pixel, and consumers not only watch, but also feel and participate in the interaction, to experience the joy and shock from the space.

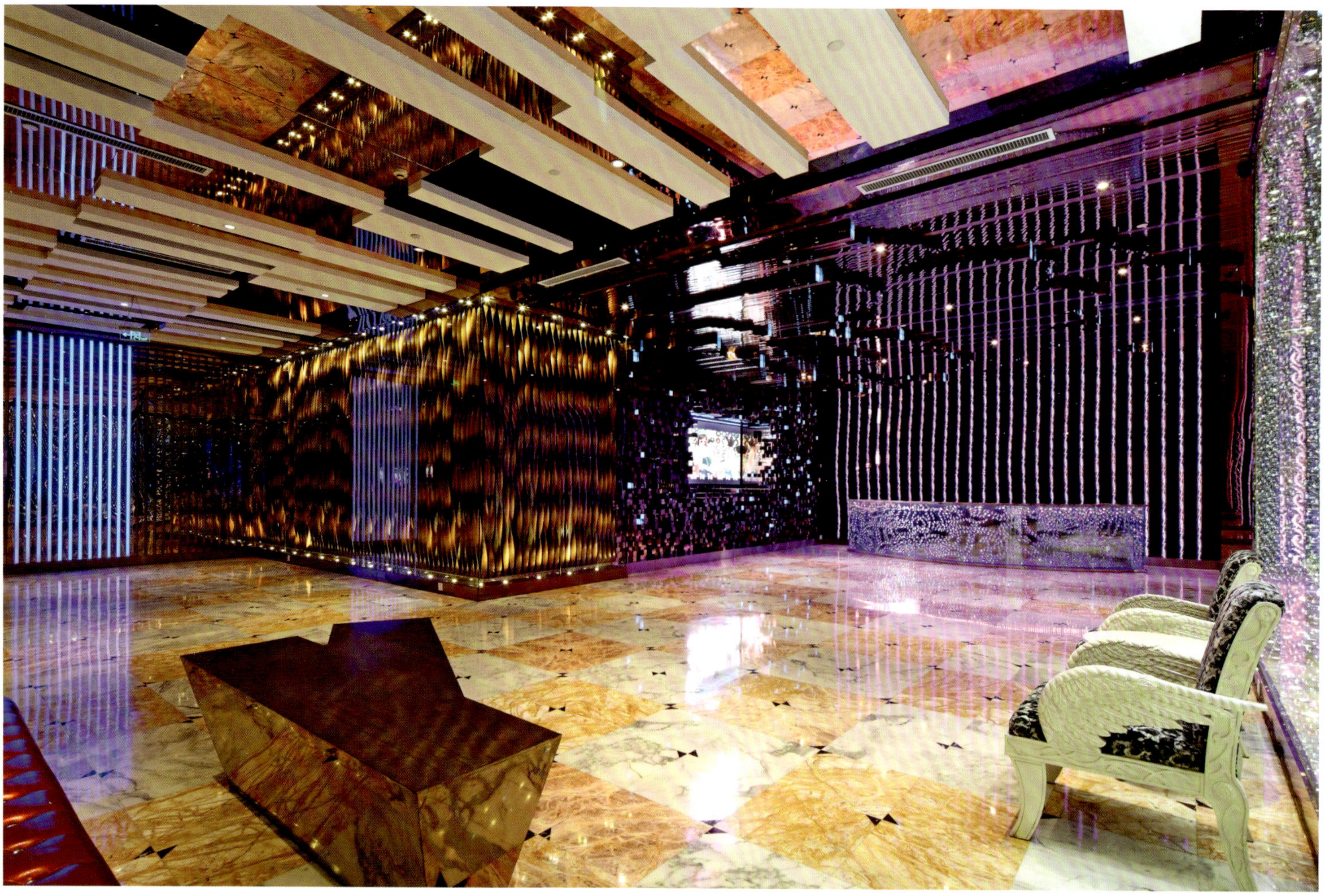

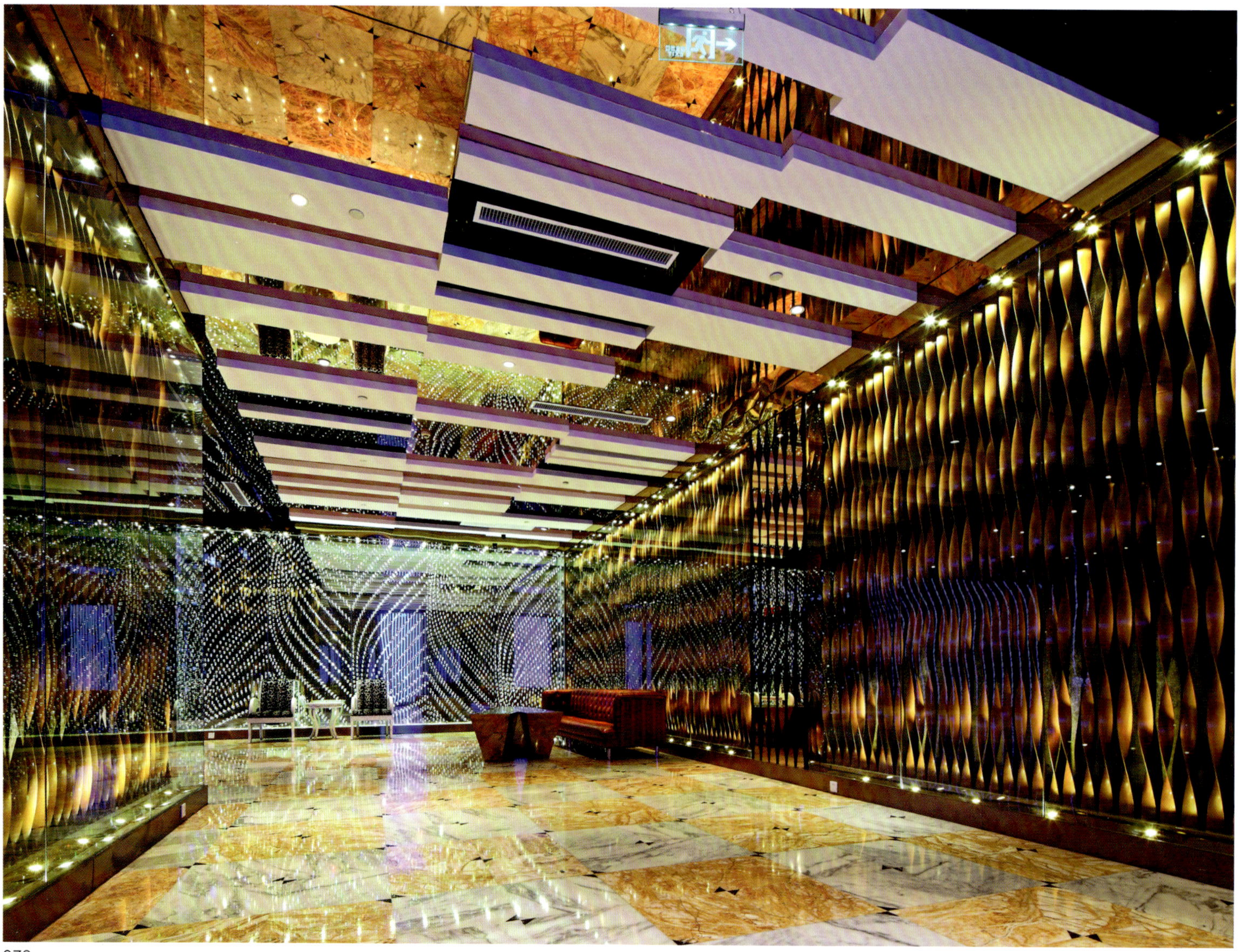

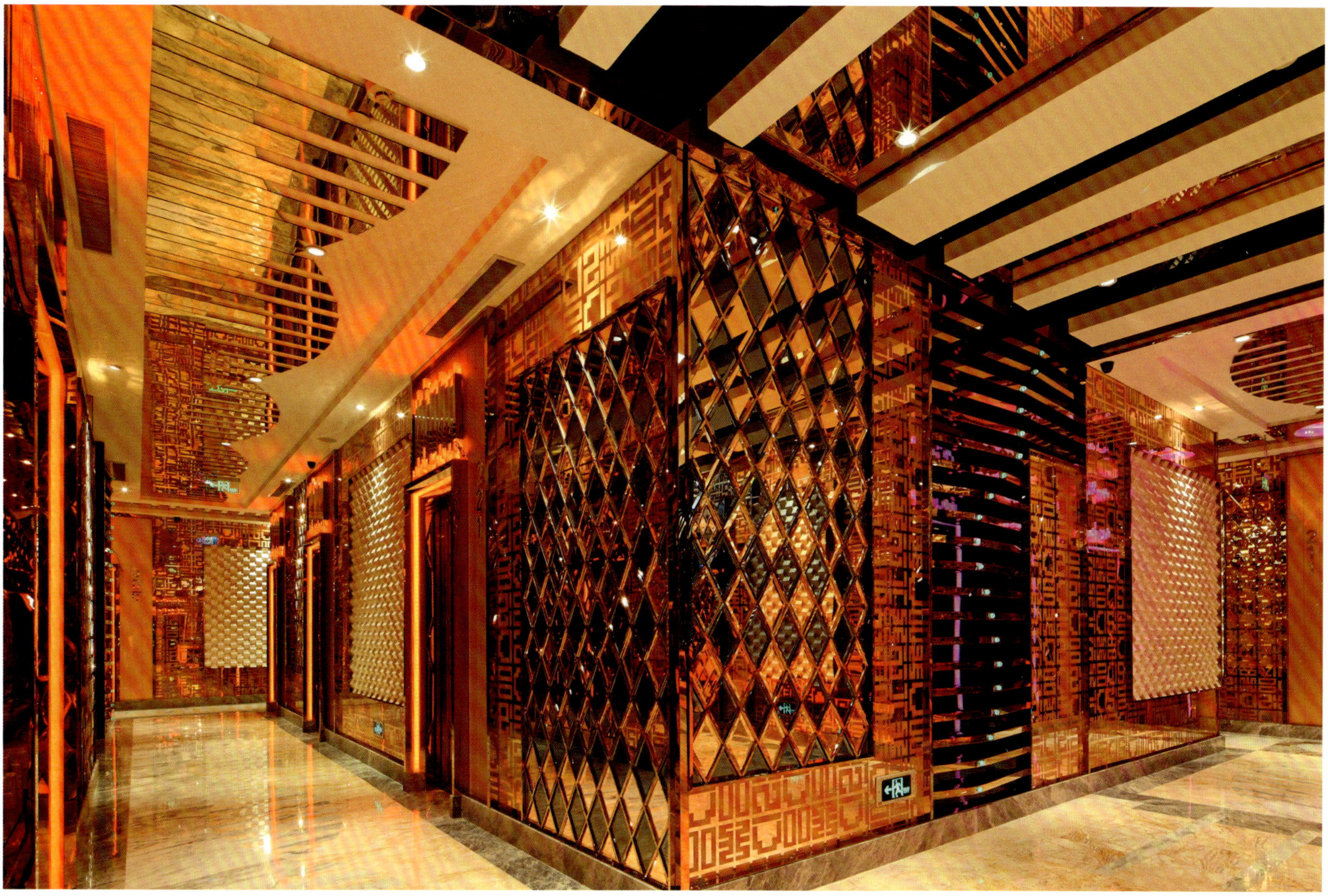

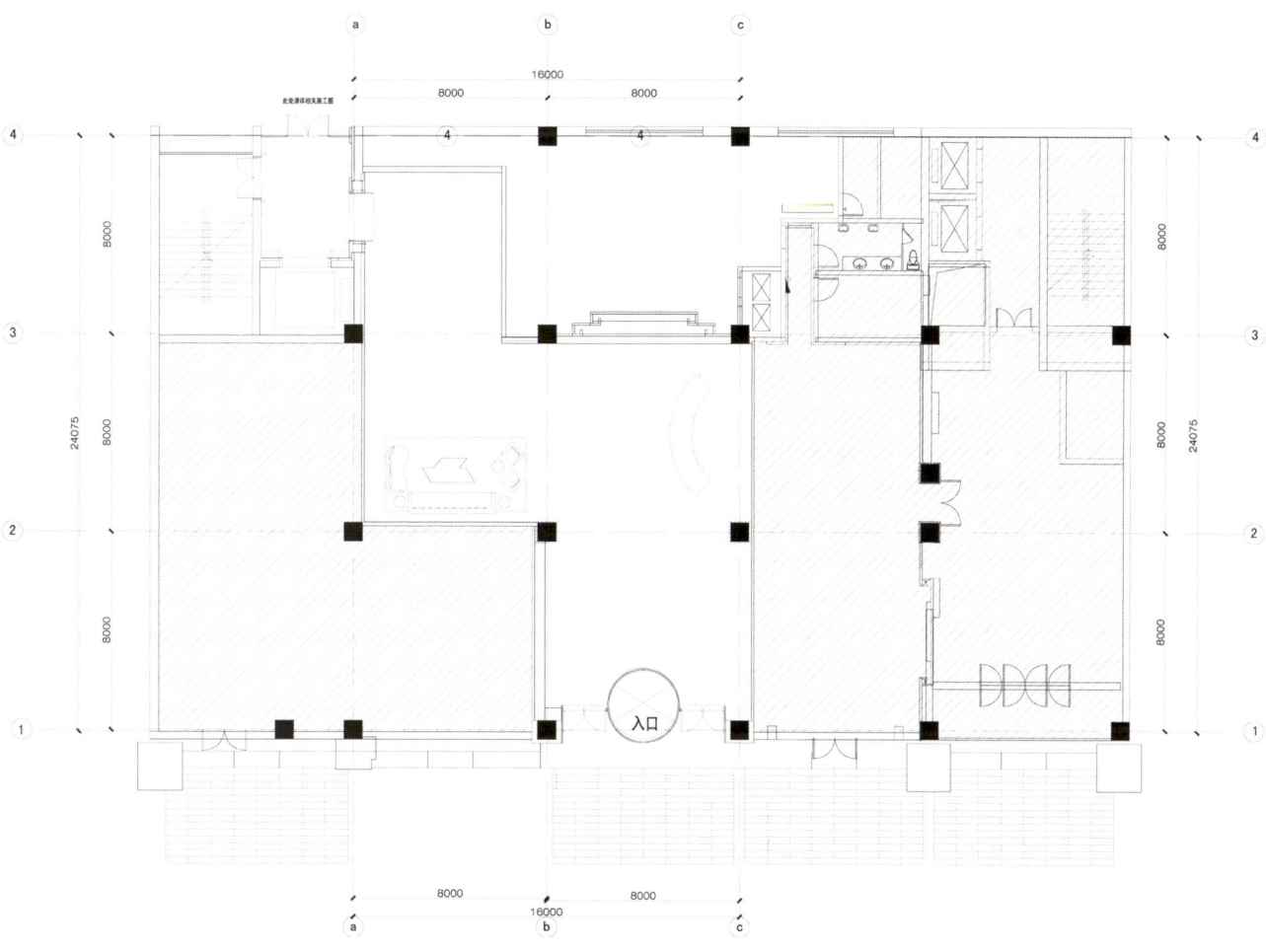

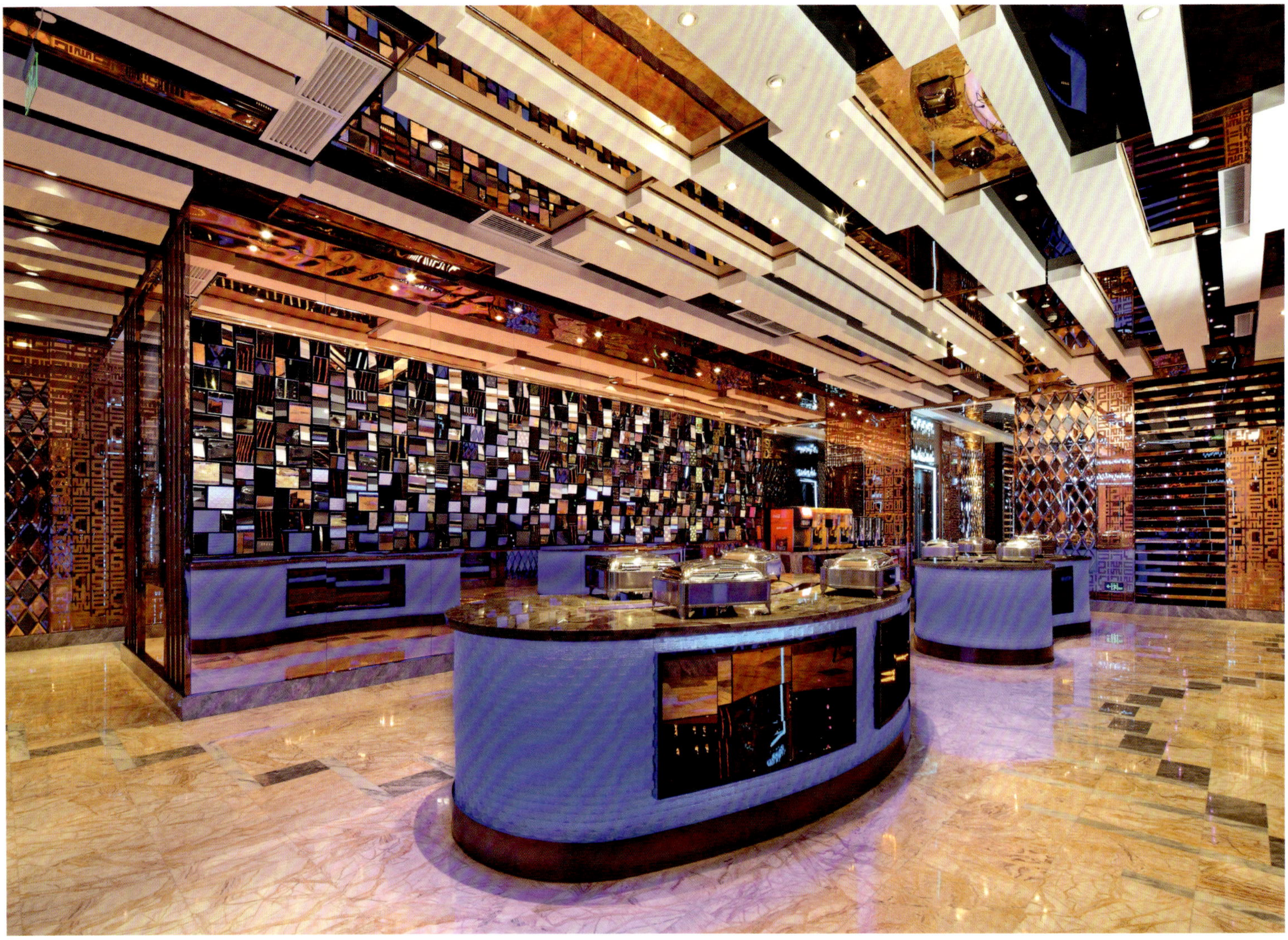

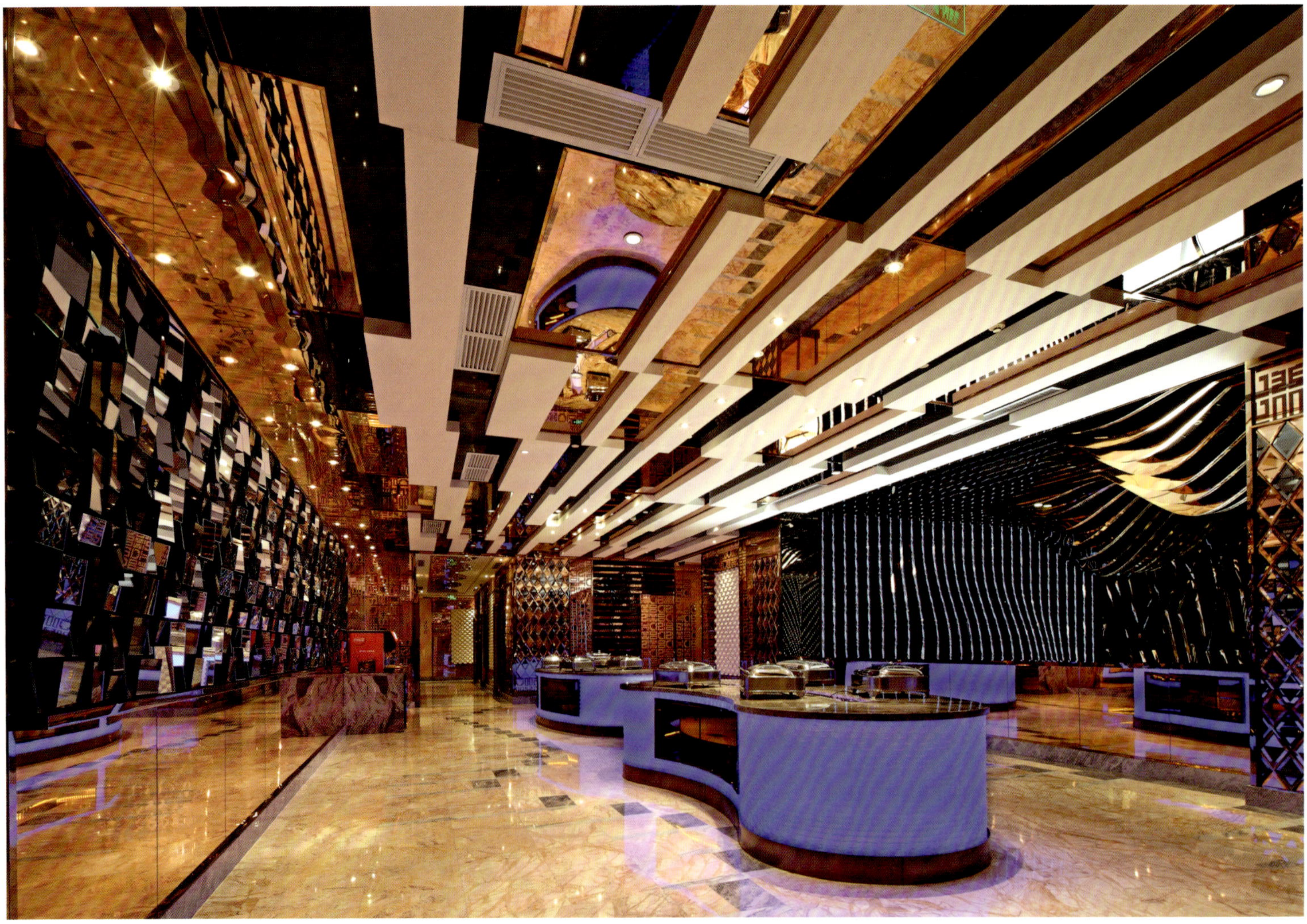

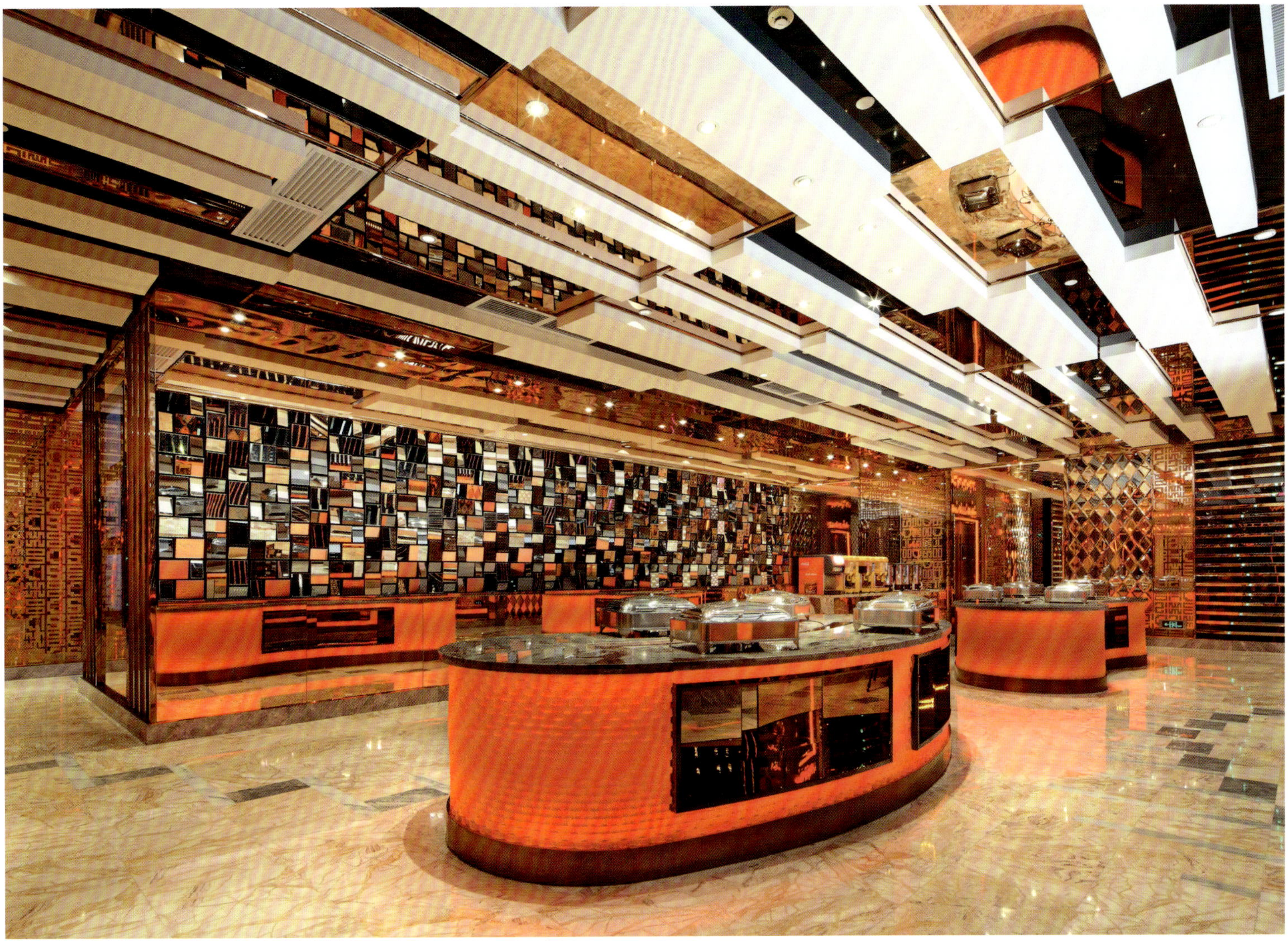

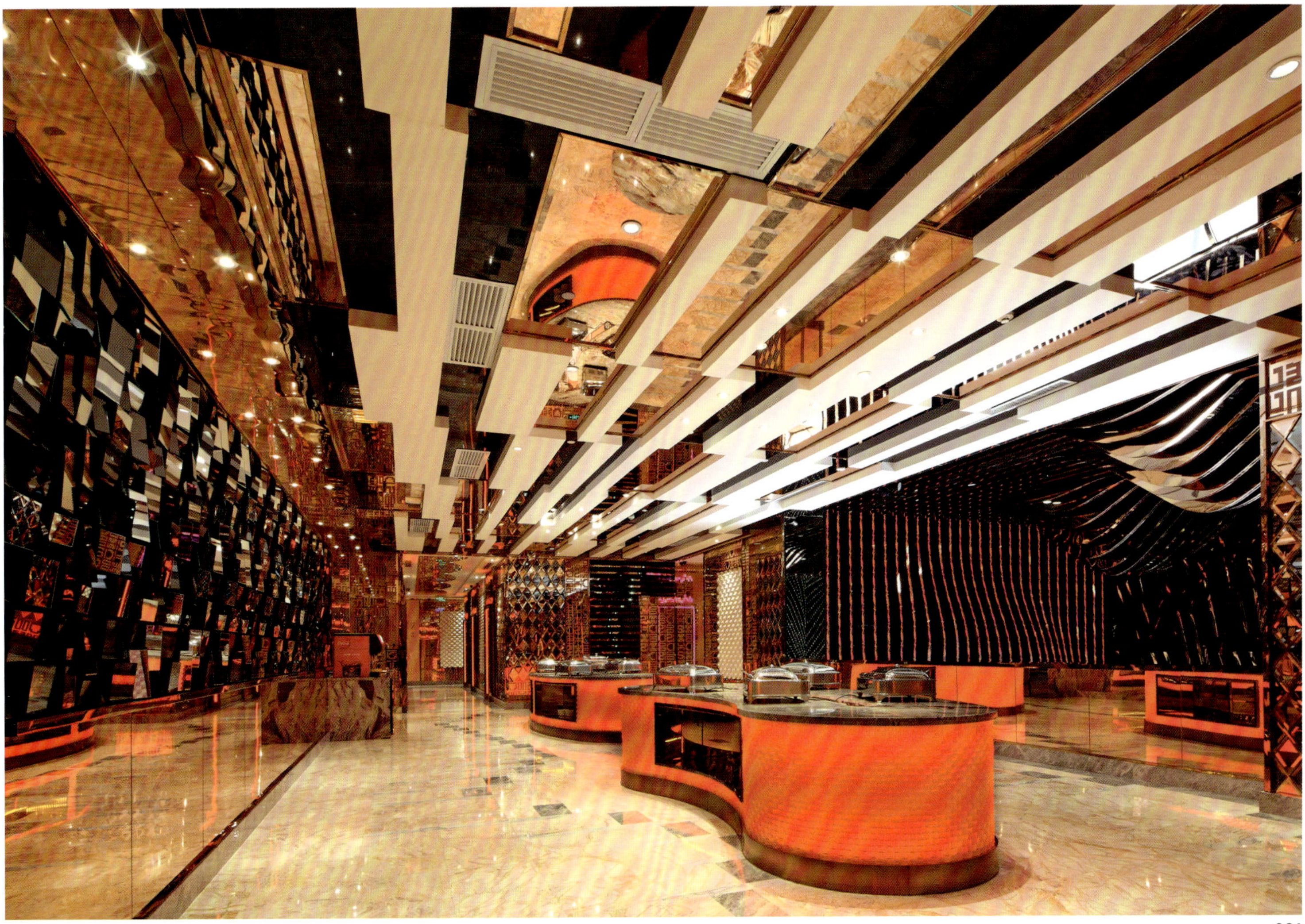

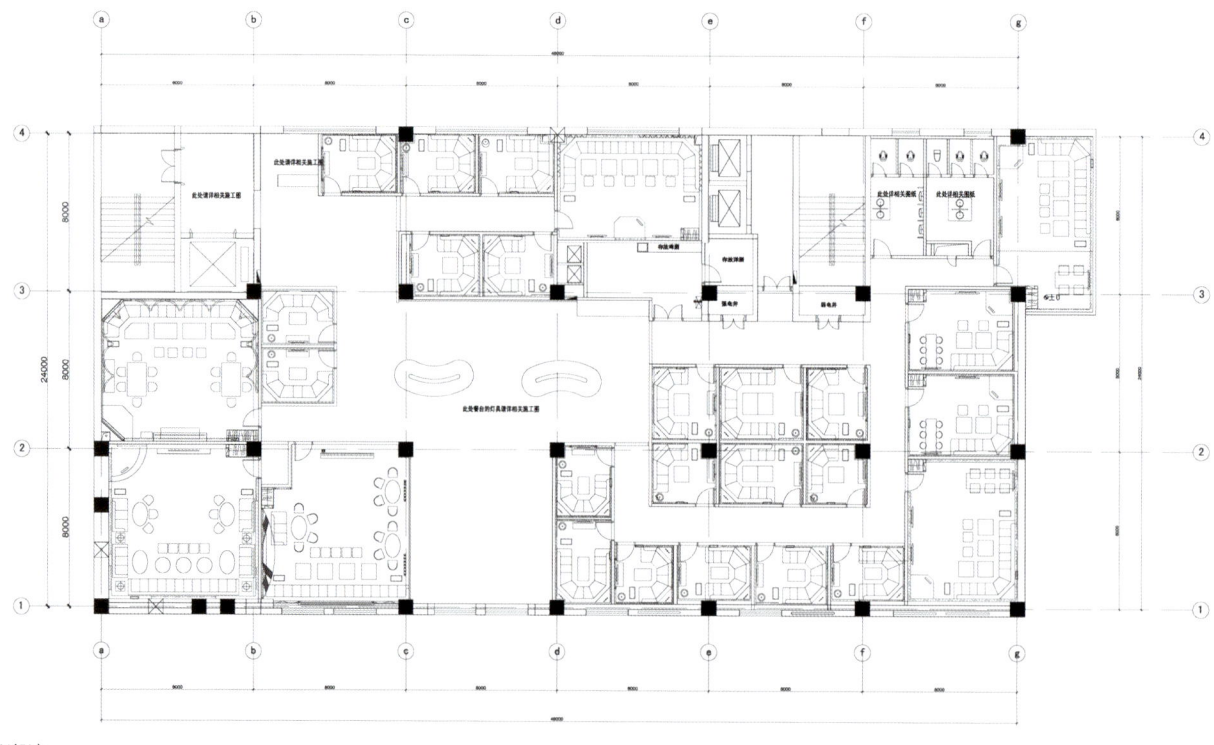

3D 像素的娱乐潮流

设计师从数码影像的"像素"形态中汲取灵感，并呈现于入口、大厅和等候区域。为了打破陈旧与乏味的传统模式，设计团队将面对入口的墙面设定为第一视觉聚焦点，一个由数百个不锈钢材质的玫瑰金配合LED光源而成的组合物，加上动感十足的立体构成，活跃了空间气氛，触动了消费者的好奇心。目光跟随着动感像素延伸到顶面，一股波及整个空间的力量呈波浪形态爆发、延展出去，如这个空间的娱乐影响力，拥有最流行元素的视觉空间带给消费者超乎寻常的亲身体验。

新乐圣KTV空间设计将科技感、风暴感、像素感等概念呈现出来，让消费者不仅是观看，更多的是感受并参与到互动中，体会空间带来的愉悦和震撼。

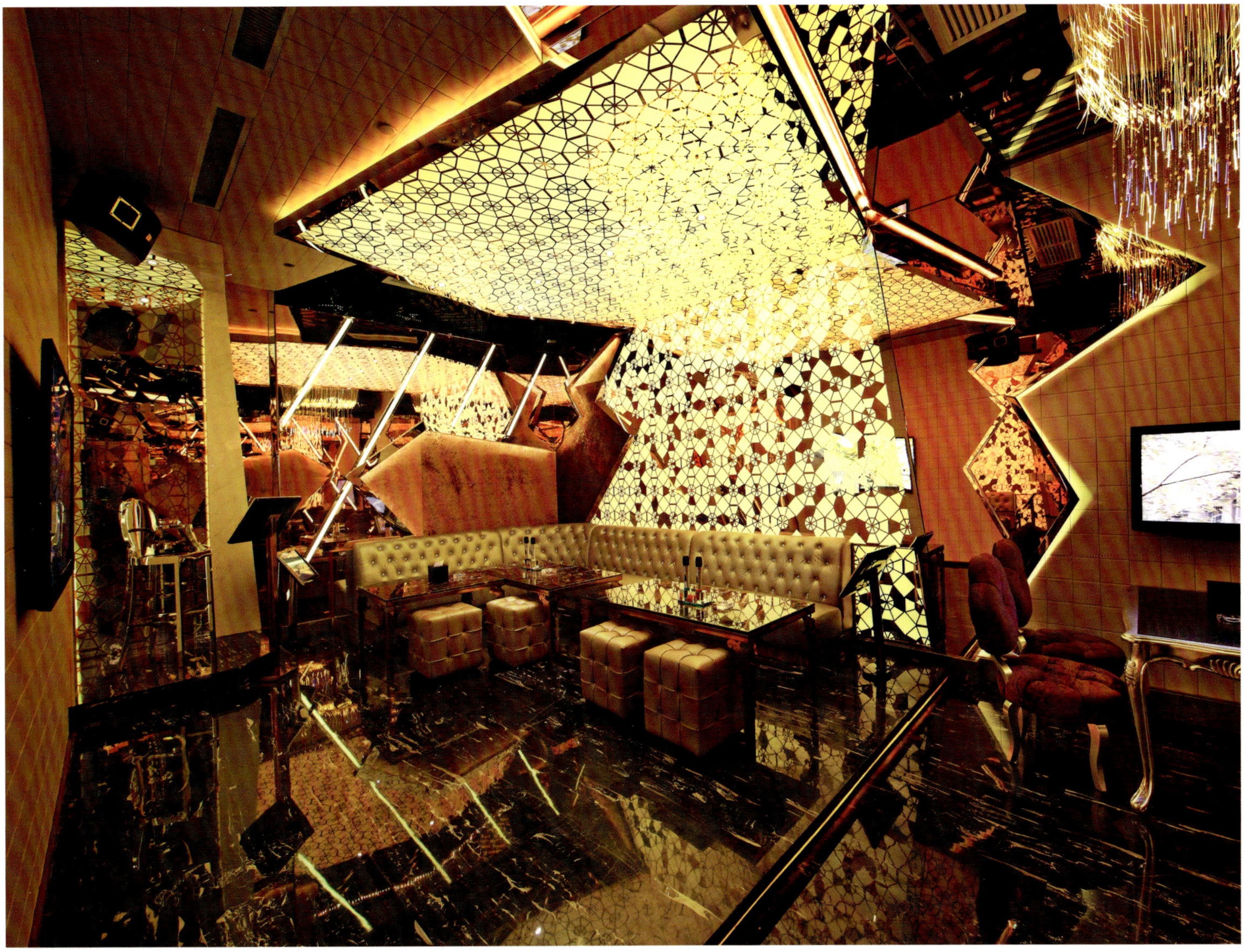

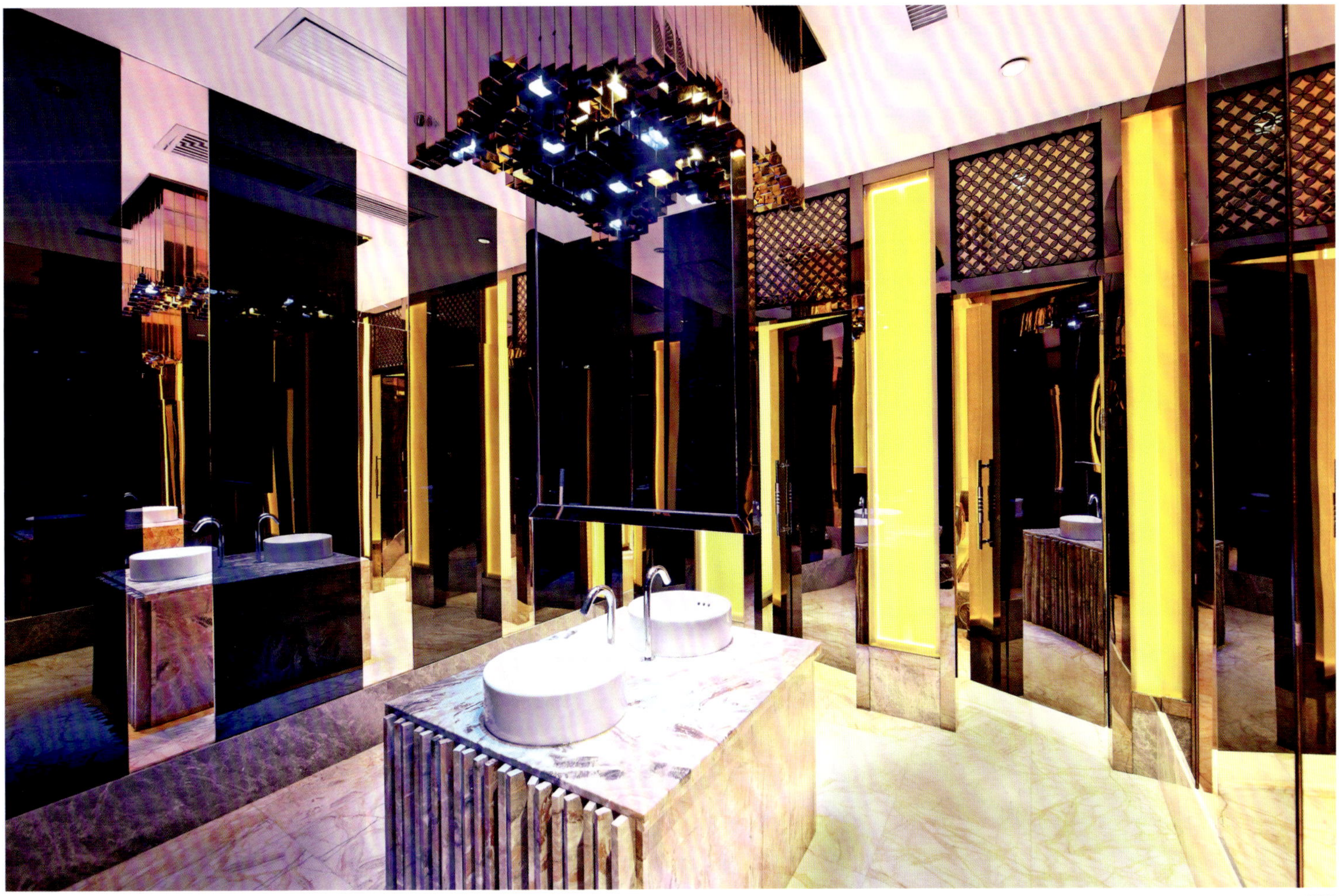

地点	面积	设计师	设计公司
/ 郑州	/ 1500m^2	/ 罗国春	/ 罗一博装饰设计有限公司

Night of Zhengzhou · Kigo Reception

—— 郑州之夜·凯歌酒会

The project is located in a prime location at the intersection of Jinshui Road and Jingsan Road. It is one of the high-end entertainment clubs in Zhengzhou. The club is positioned as a five-star entertainment place with a small number of private rooms featuring different styles. The designer defines the space as a luxurious European style, integrating fashionable modern elements to provide an innovative design to create a generous, luxurious and elegant atmosphere, which let people enjoy the illusion through time and space at the moment of entering the space, as if into a medieval European palace, and the luxurious and elegant decoration is amazing.

The space is not large, so the designer makes full use of every inch of it to fully guarantee the privacy of each room, truly allow customers to relax and feel at ease. The space design is bold and exaggerating, a luxury shown out, and the colors match is generous and gorgeous, harmoniously integrated with decorative elements and space atmosphere, creating a palace-like luxurious experience and allowing customers to enjoy the entertainment and imperial sense of respect.

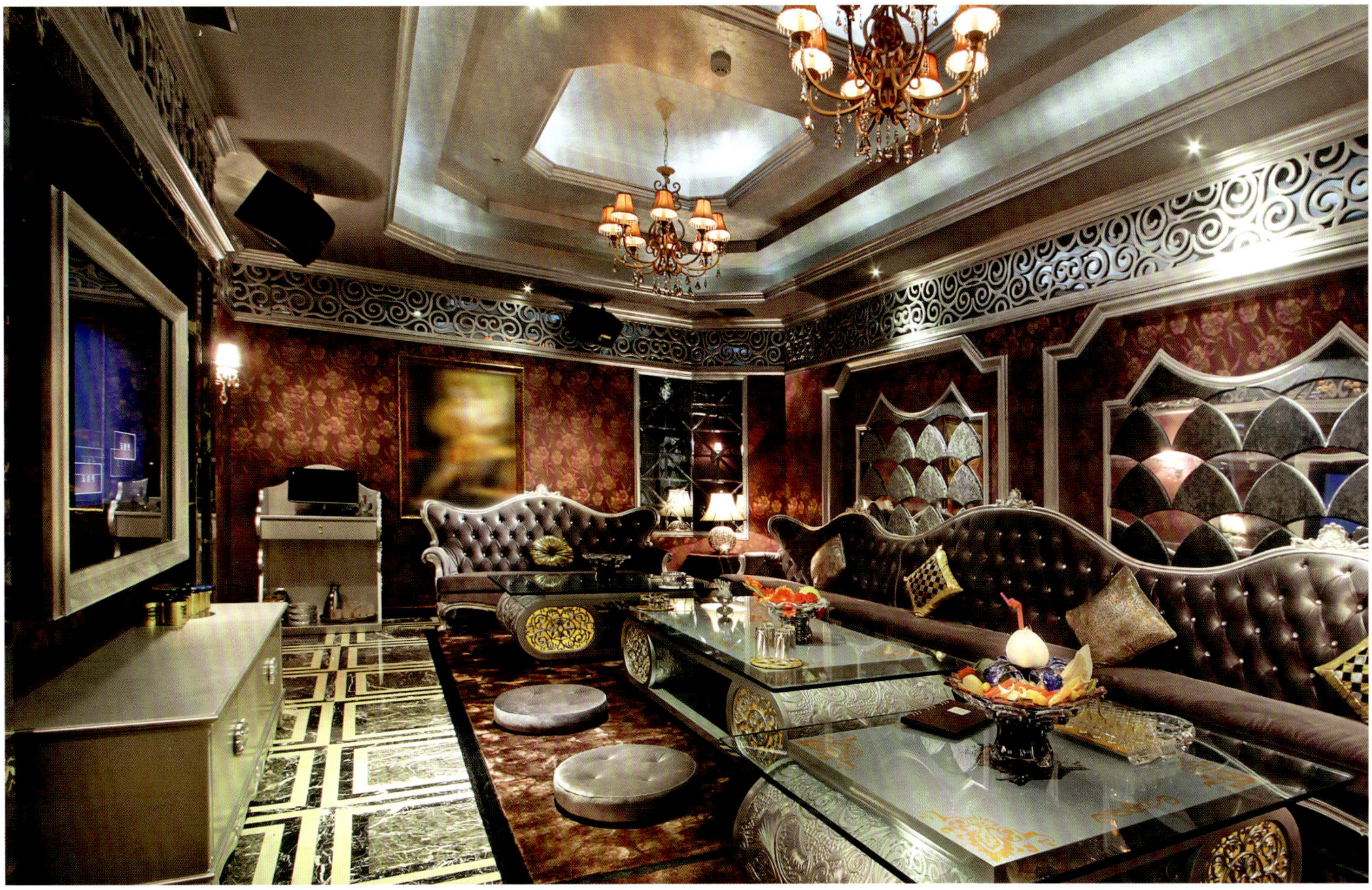

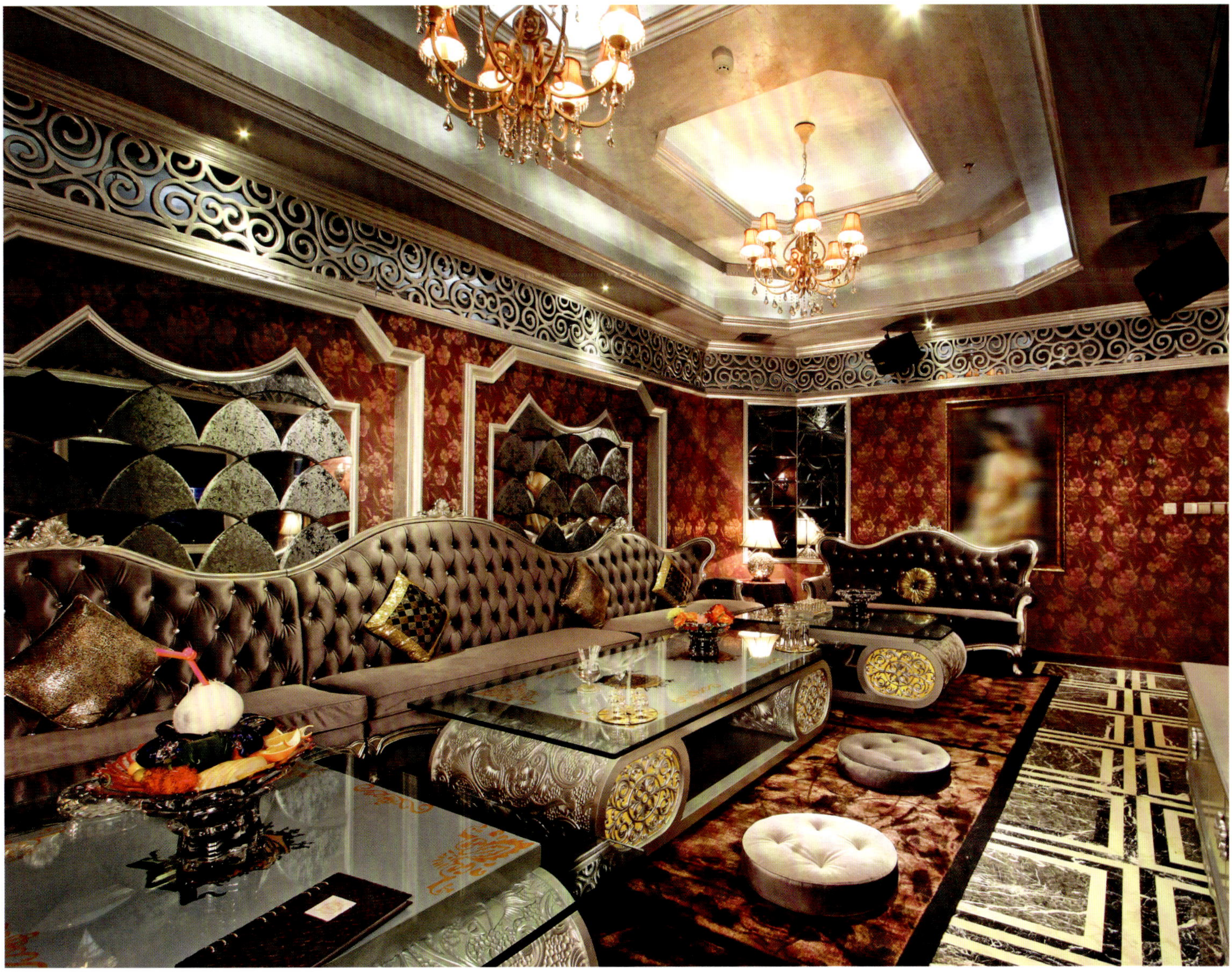

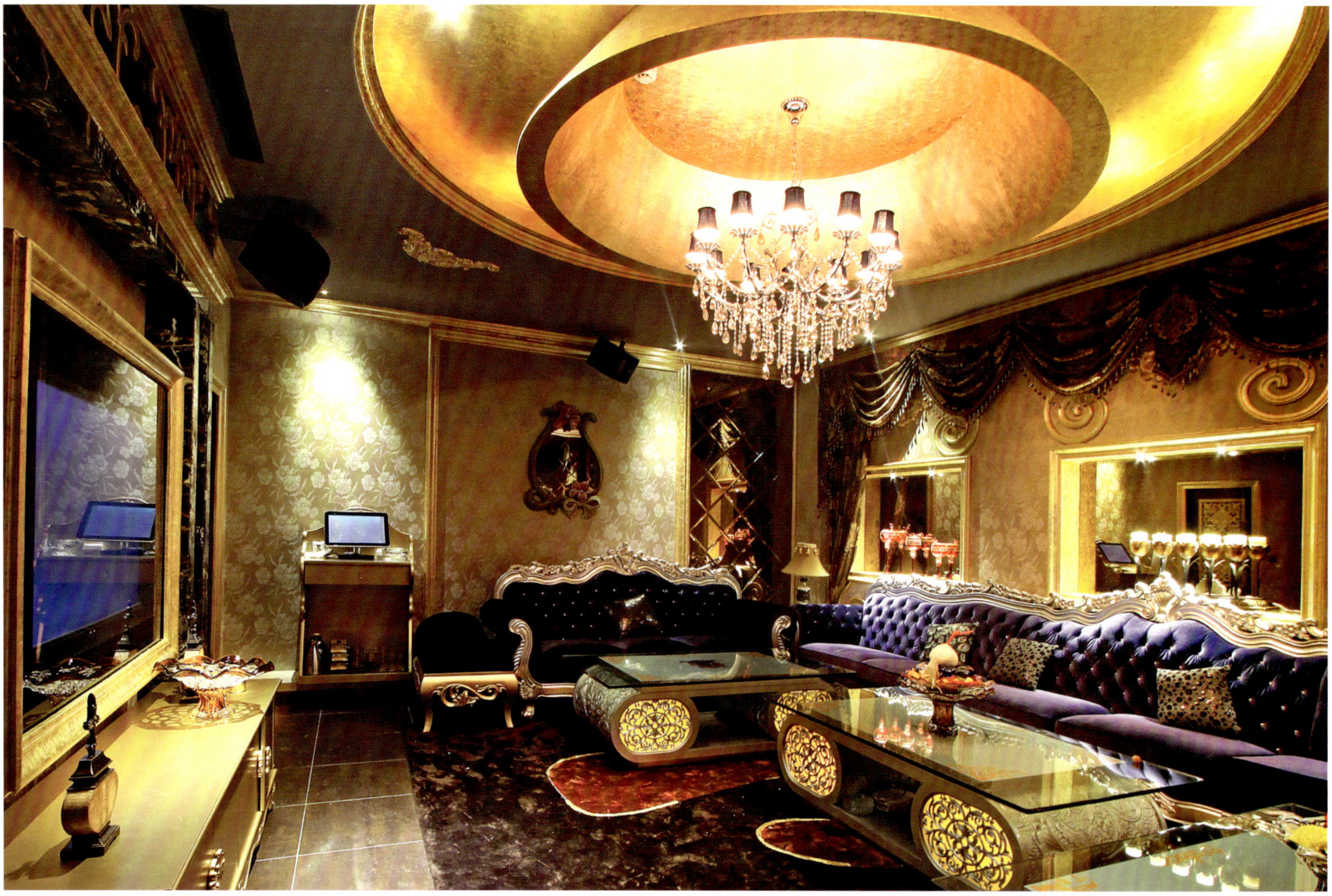

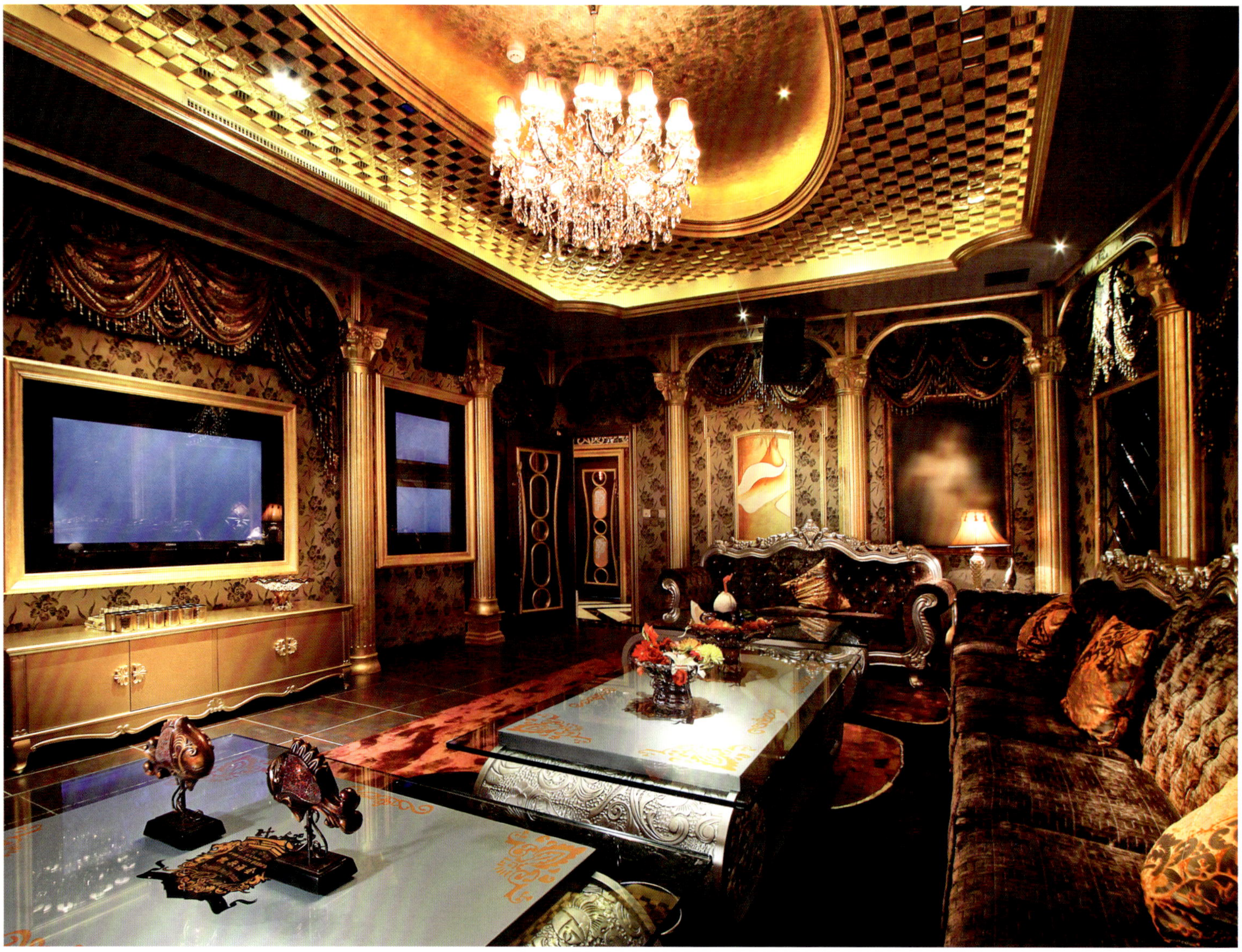

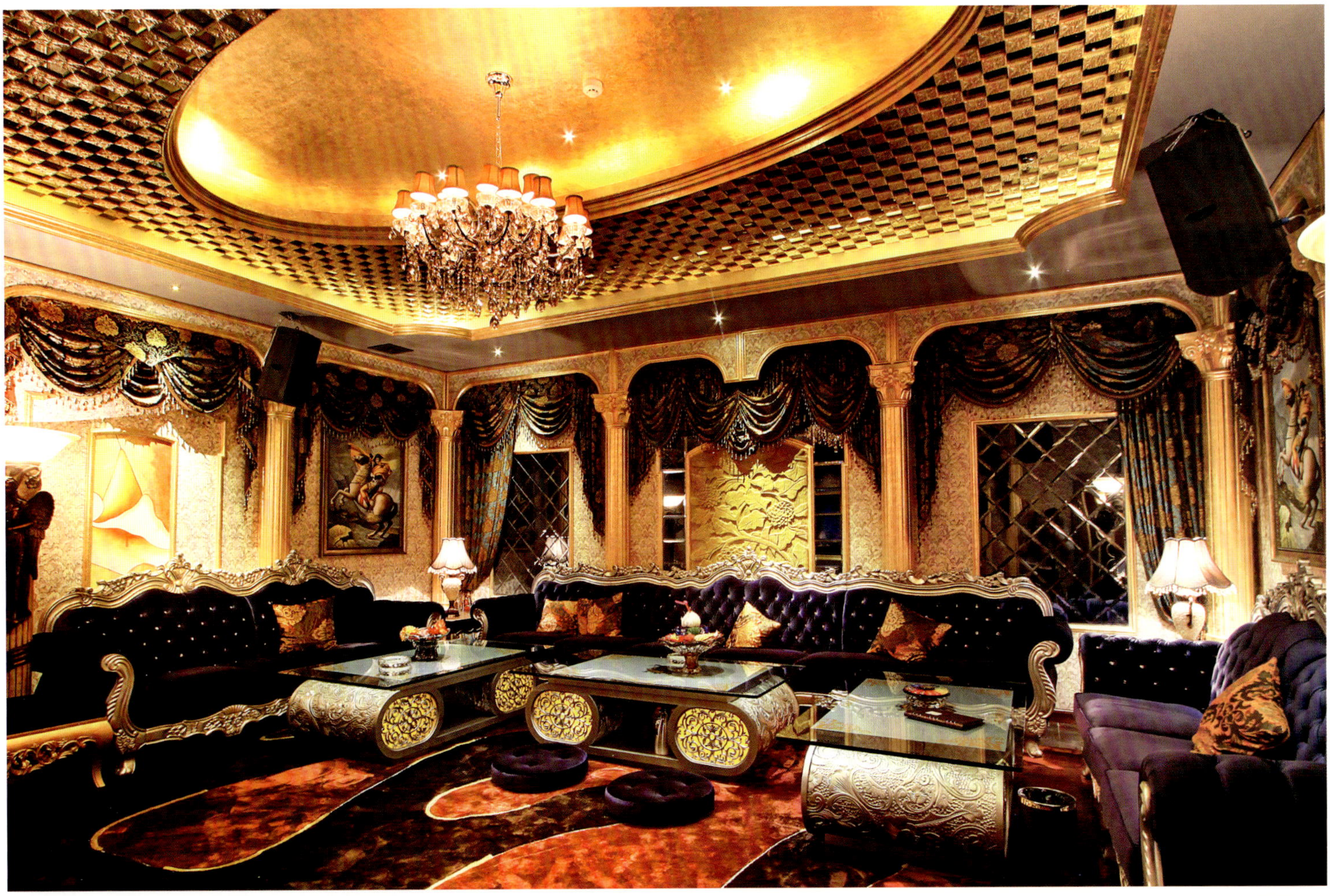

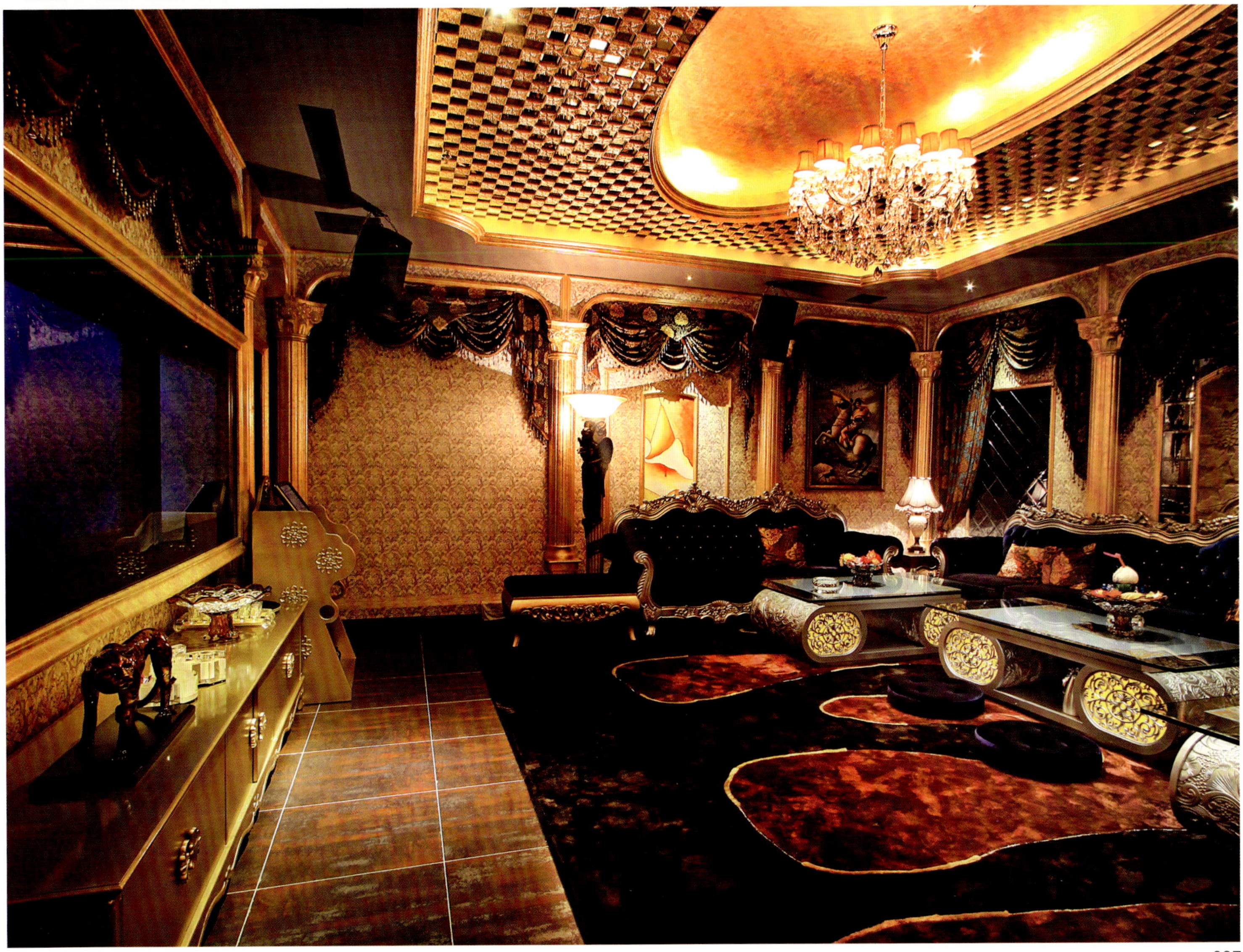

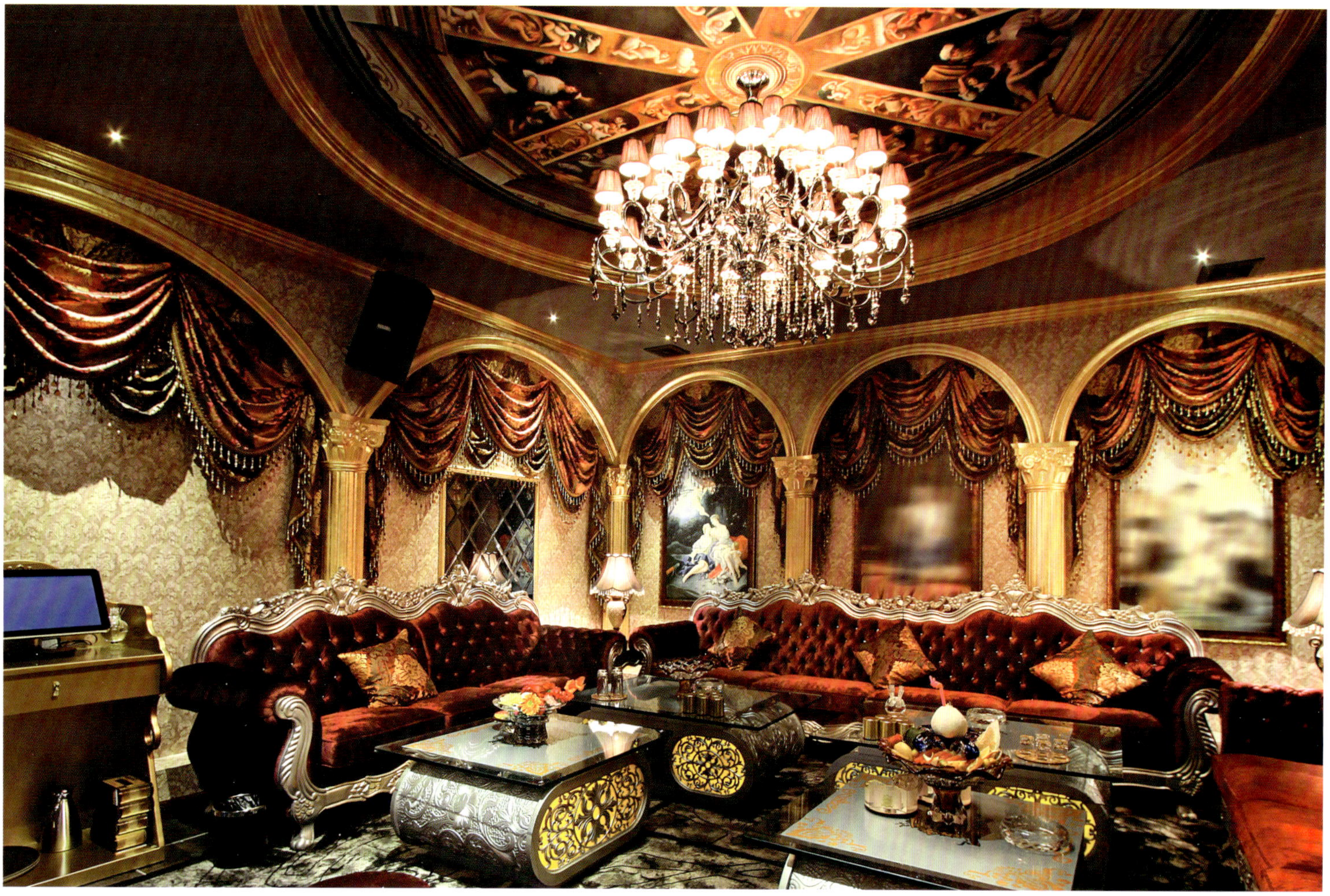

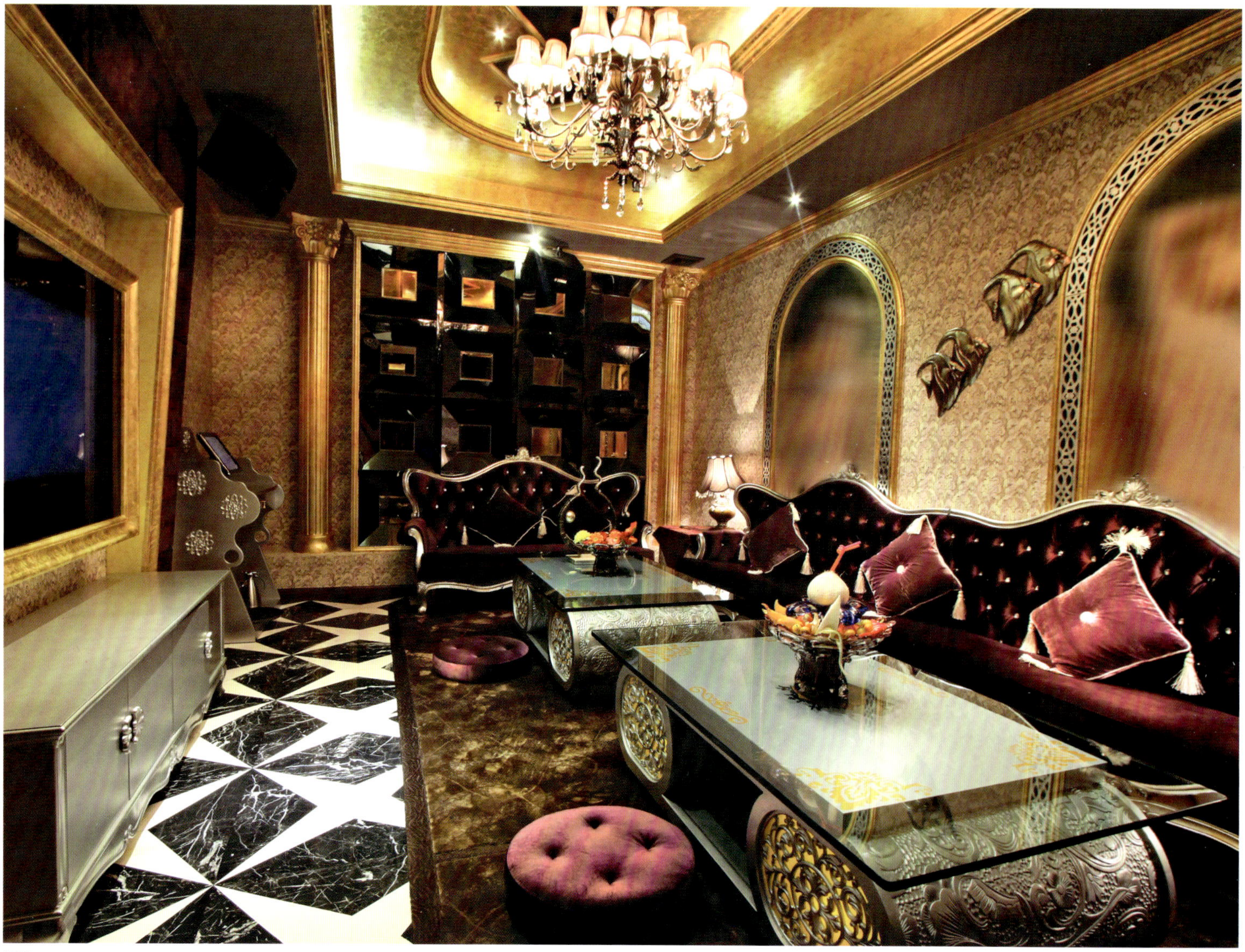

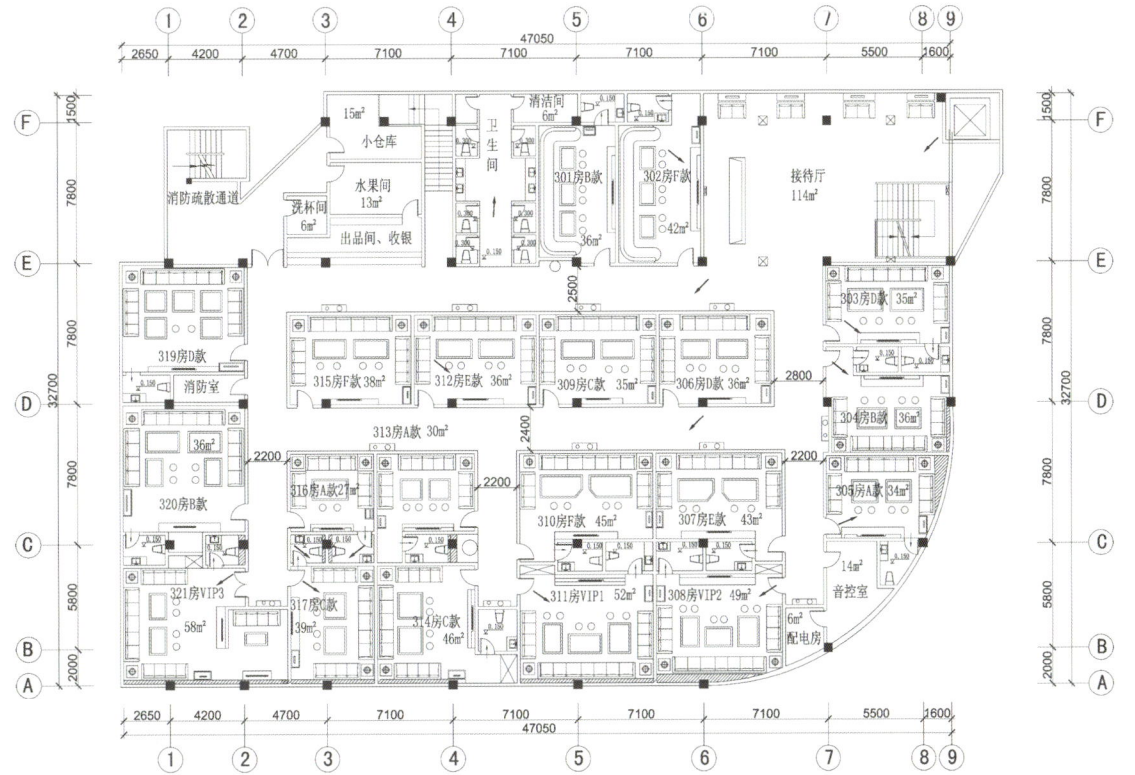

本案位于金水路和经三路交叉口的黄金地段，是郑州市高端娱乐会所之一。会所定位为五星级消费场所，包房数量不多，但却各具特色。设计师以欧式奢华风格定义空间，并融入了现代时尚元素，进行创新设计，从而营造出大气、华贵、优雅的环境氛围，让人在进入空间的瞬间就有了穿越时空的错觉，仿佛进入到中世纪的欧式宫廷，奢华、典雅的装饰让人惊叹不已。

空间面积不大，因此，设计师充分利用好每一寸空间，充分保证各房间的私密性，真正做到让顾客安心、放松。空间的设计大胆而夸张，展现出张扬在外的奢华，色彩搭配大方而华贵，与装饰元素和空间氛围协调融合，创造出宫殿般的奢华体验，让顾客在尽情娱乐的同时，享受到帝王般的尊崇感。

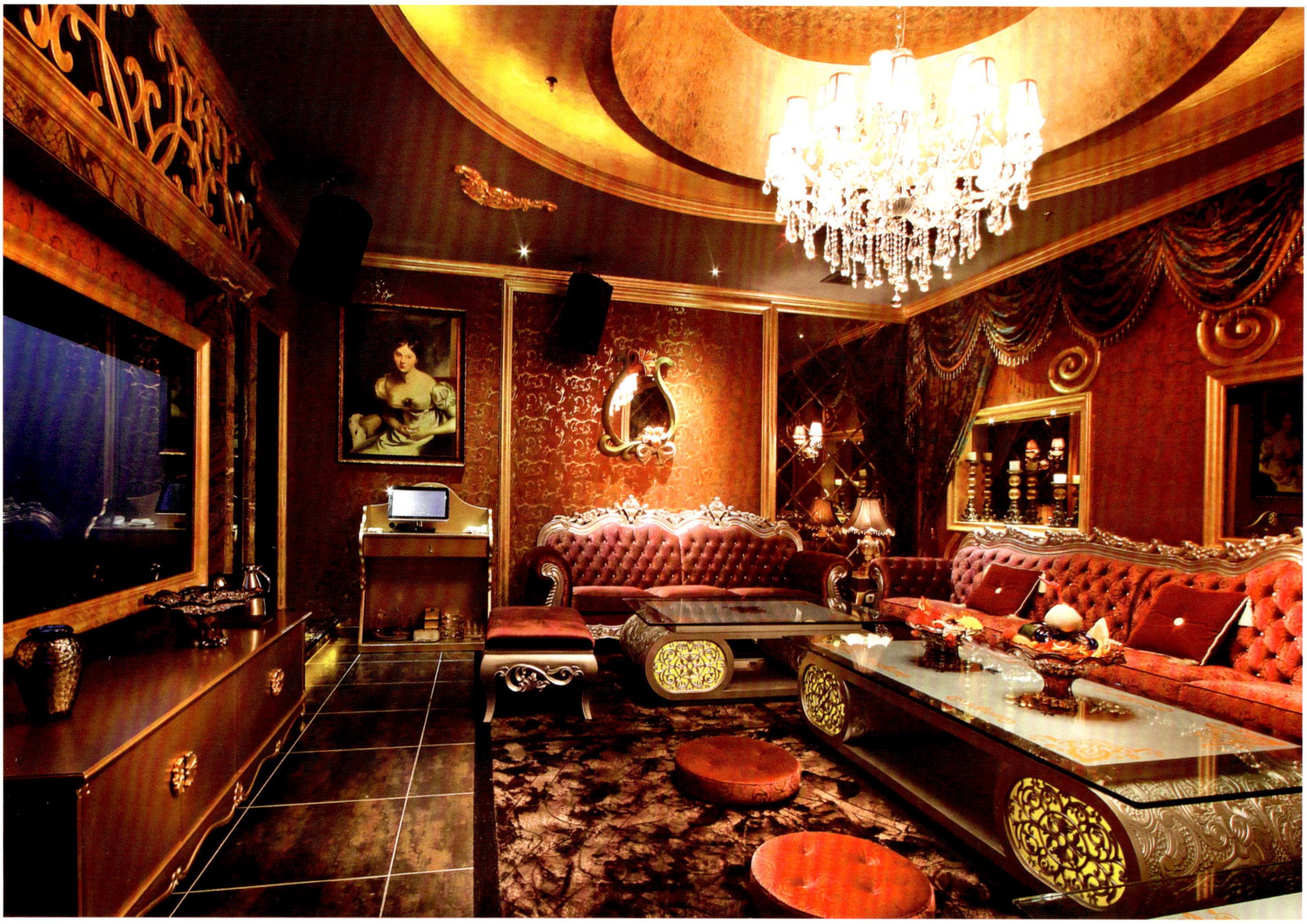
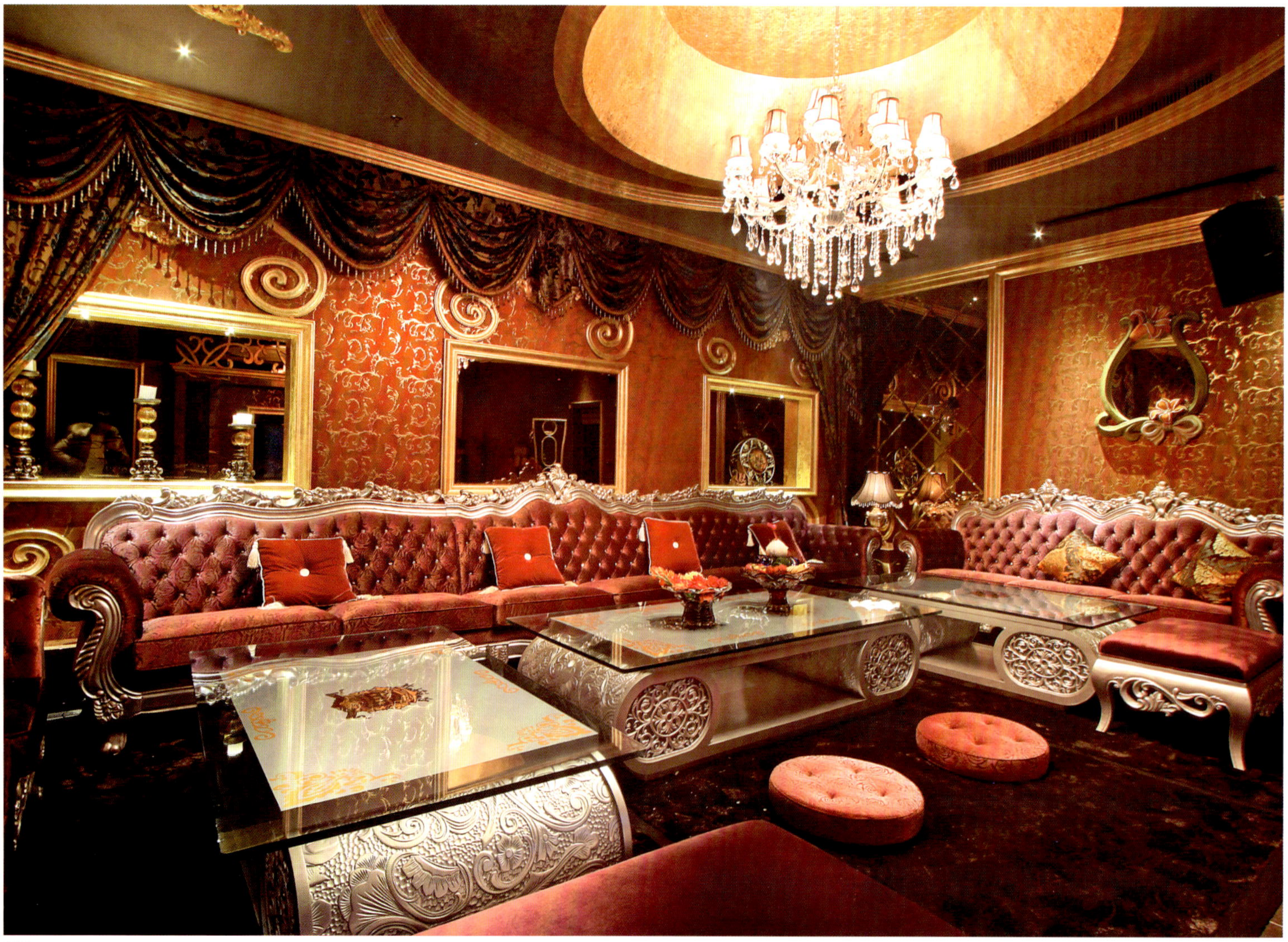

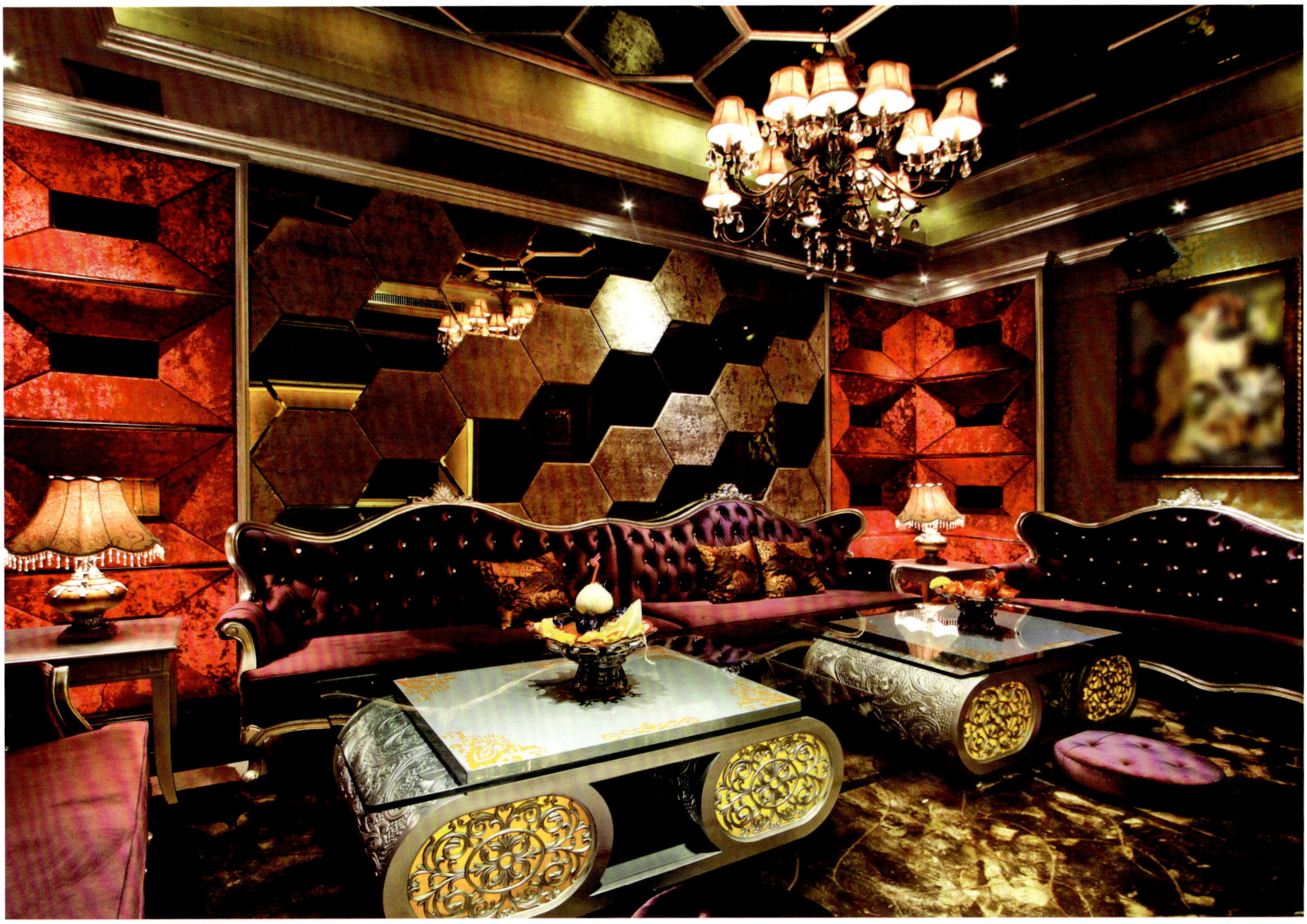

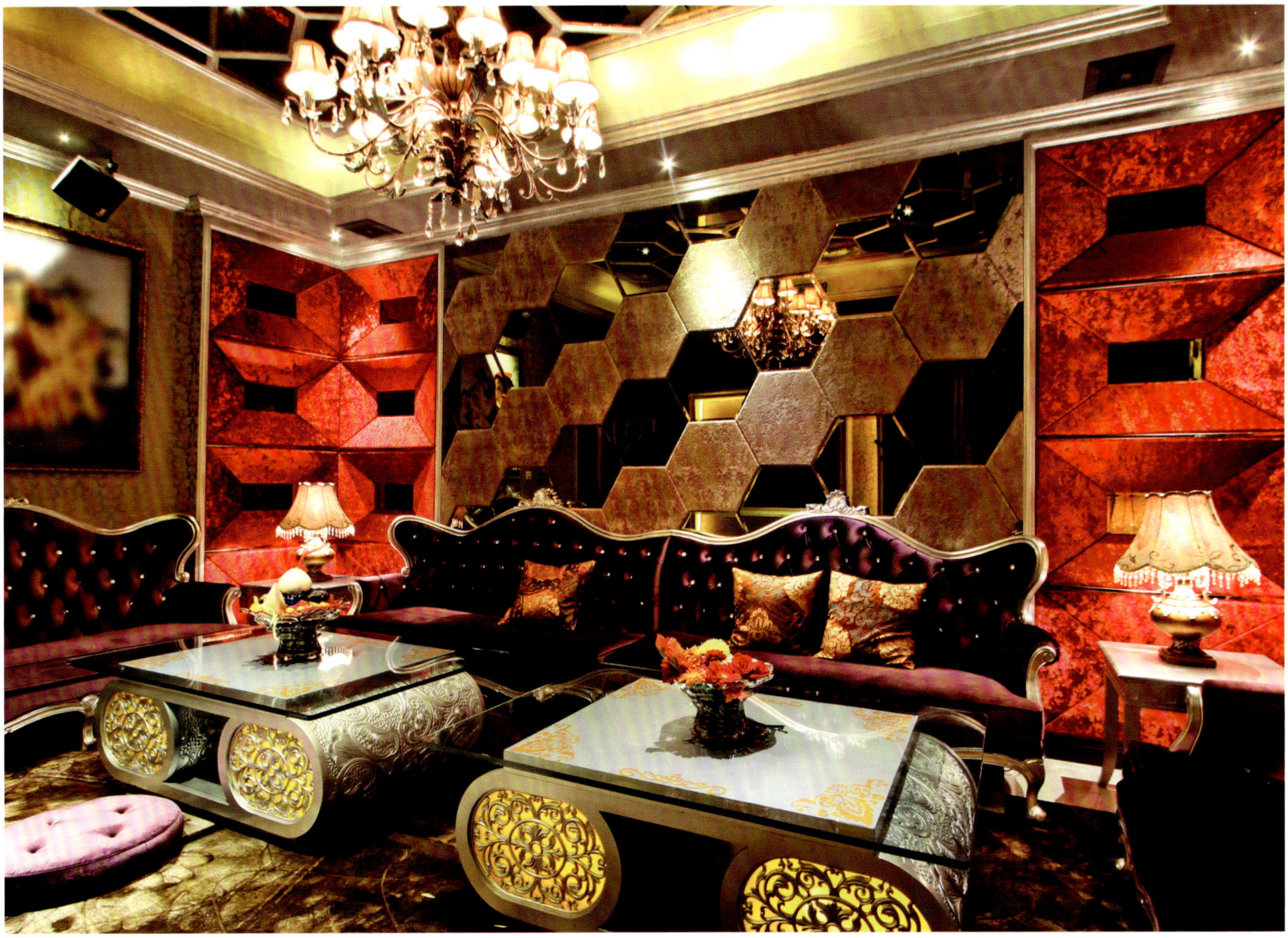

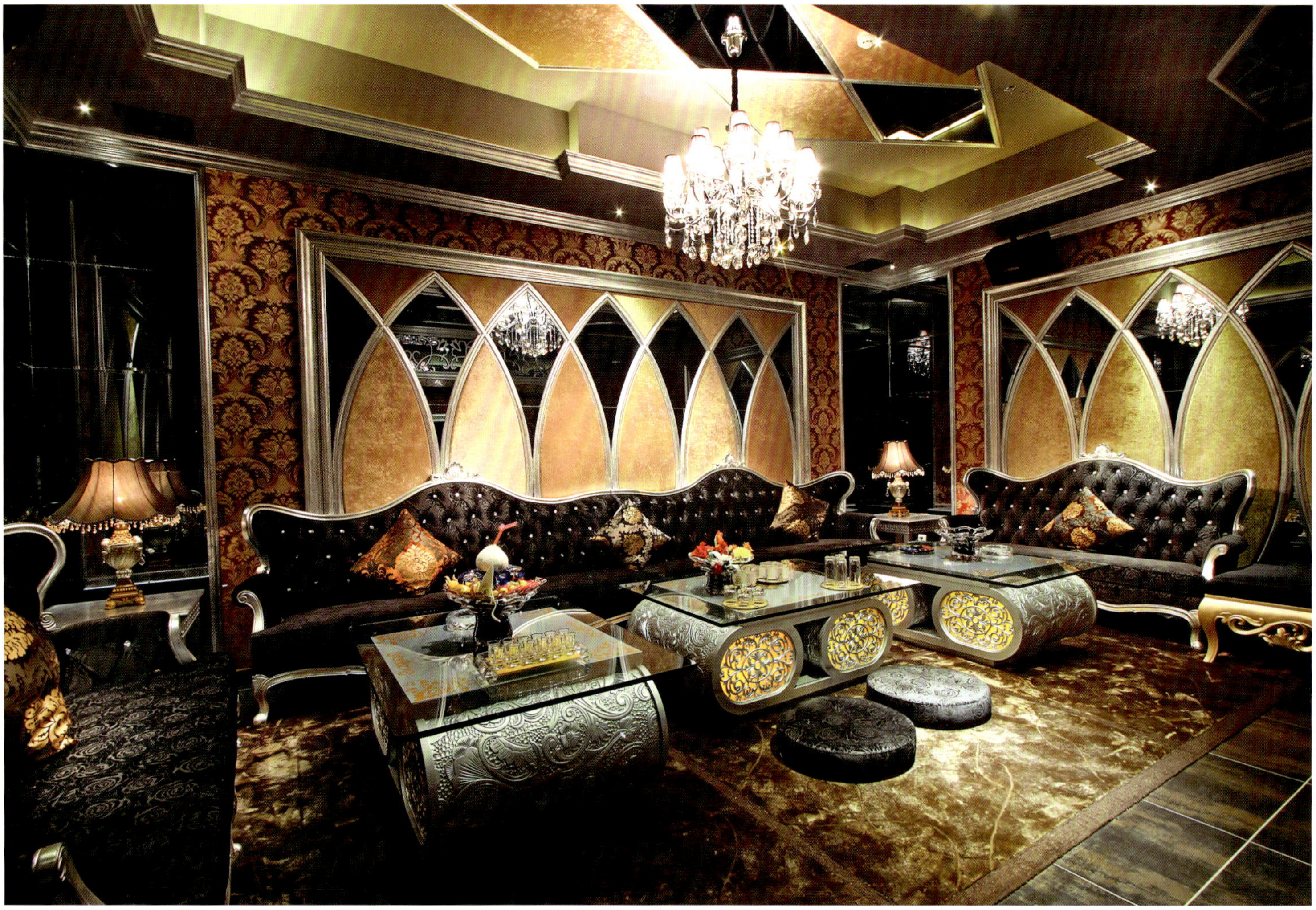

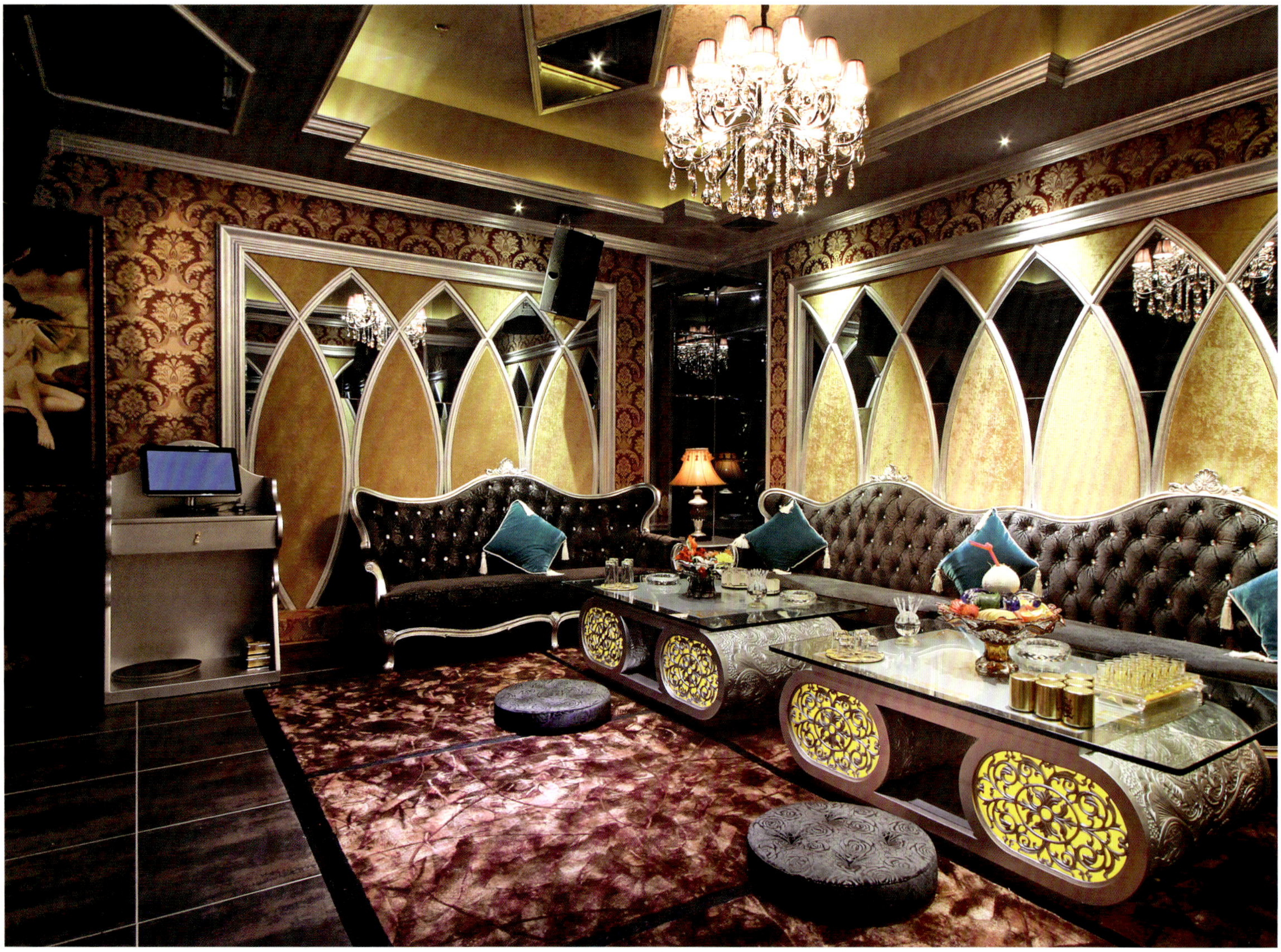

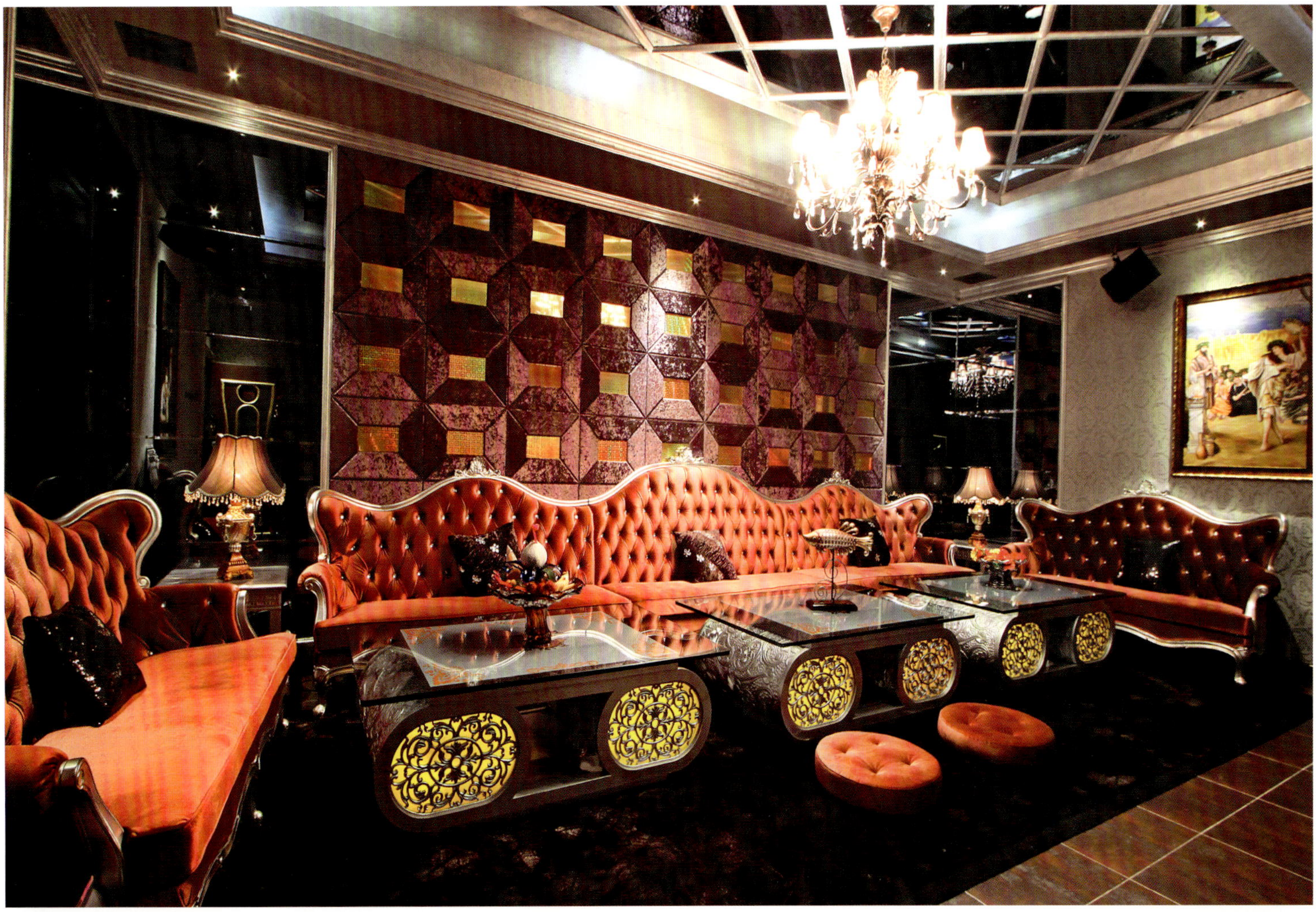

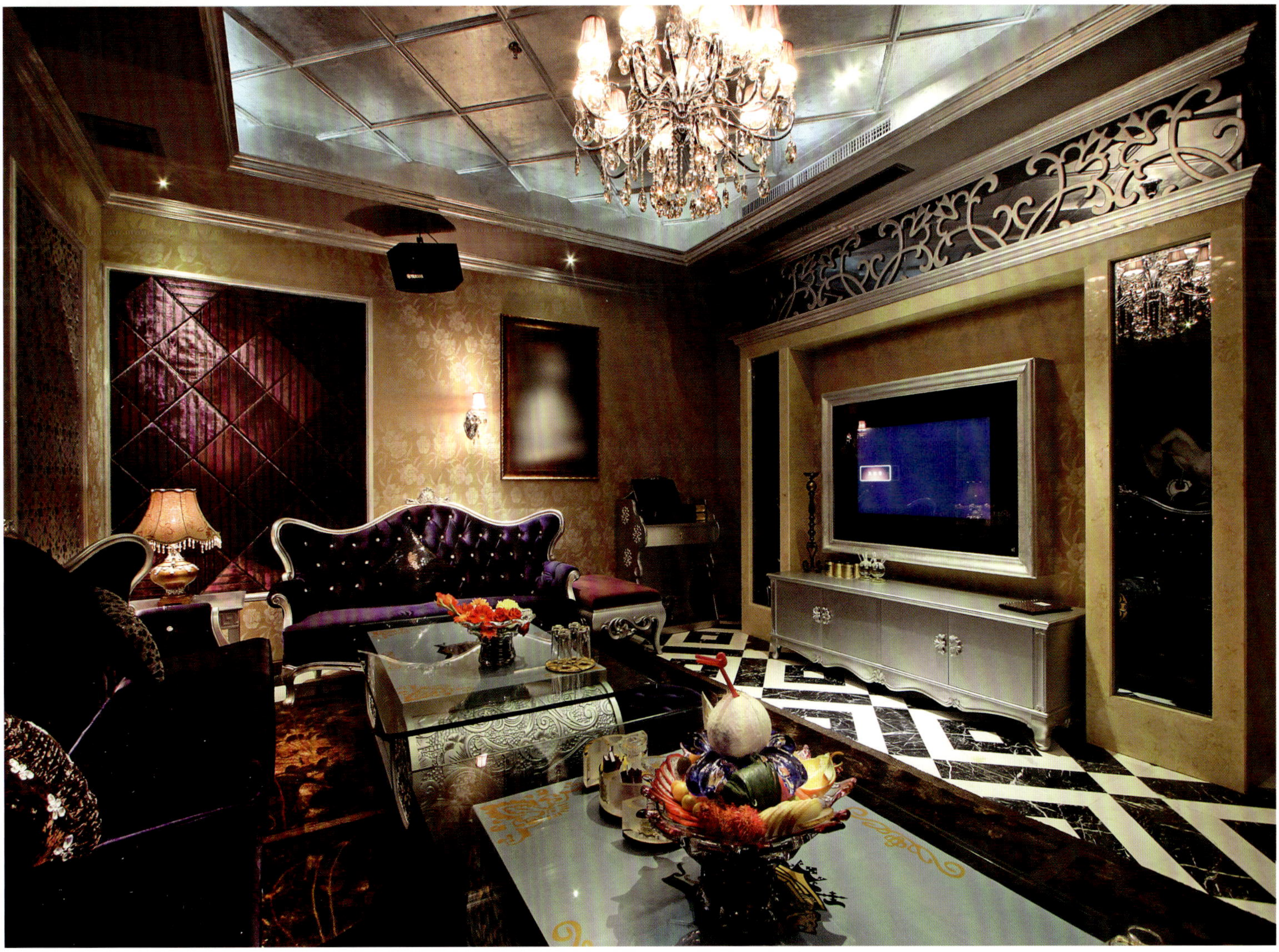

地 点	面 积	设计师	设计公司	主要材料
/ 深圳福田	/ 1400m²	/ 袁静、钟建福	/ 朗昇国际商业设计	/ 大理石、地砖、马赛克、防火涂料、玻璃等

Happy Zone KTV

—— 欢乐盛开 KTV

"Happy Zone" is the first world style themed KTV in Shenzhen. It is not large, and the owner wants to create a high-end KTV in Shenzhen. Through repeated study and scrutiny between the design team and the owner, they finally decide to design this KTV with the theme of geographical characteristic.

The 15 themed rooms are based on the unique history, culture, folk custom, architecture, customs and other elements of 15 countries and regions, integrated, or reconstructed, or decomposed, finally creating a series of different shapes, images and effects of decoration match. It can be an Egypt style with ancient pyramids and the Pharaohs, or primitive tropical forest Kenya in Africa; it can be a beautiful and pure ocean style of Greek Aegean Sea, or Chinese courtyard full of strong Beijing flavor; it can be a Denmark fairy tale style with childhood fun, or a heroic and luxurious Las Vegas worth of billions of dollars...a unique room provides a unique entertainment experience!

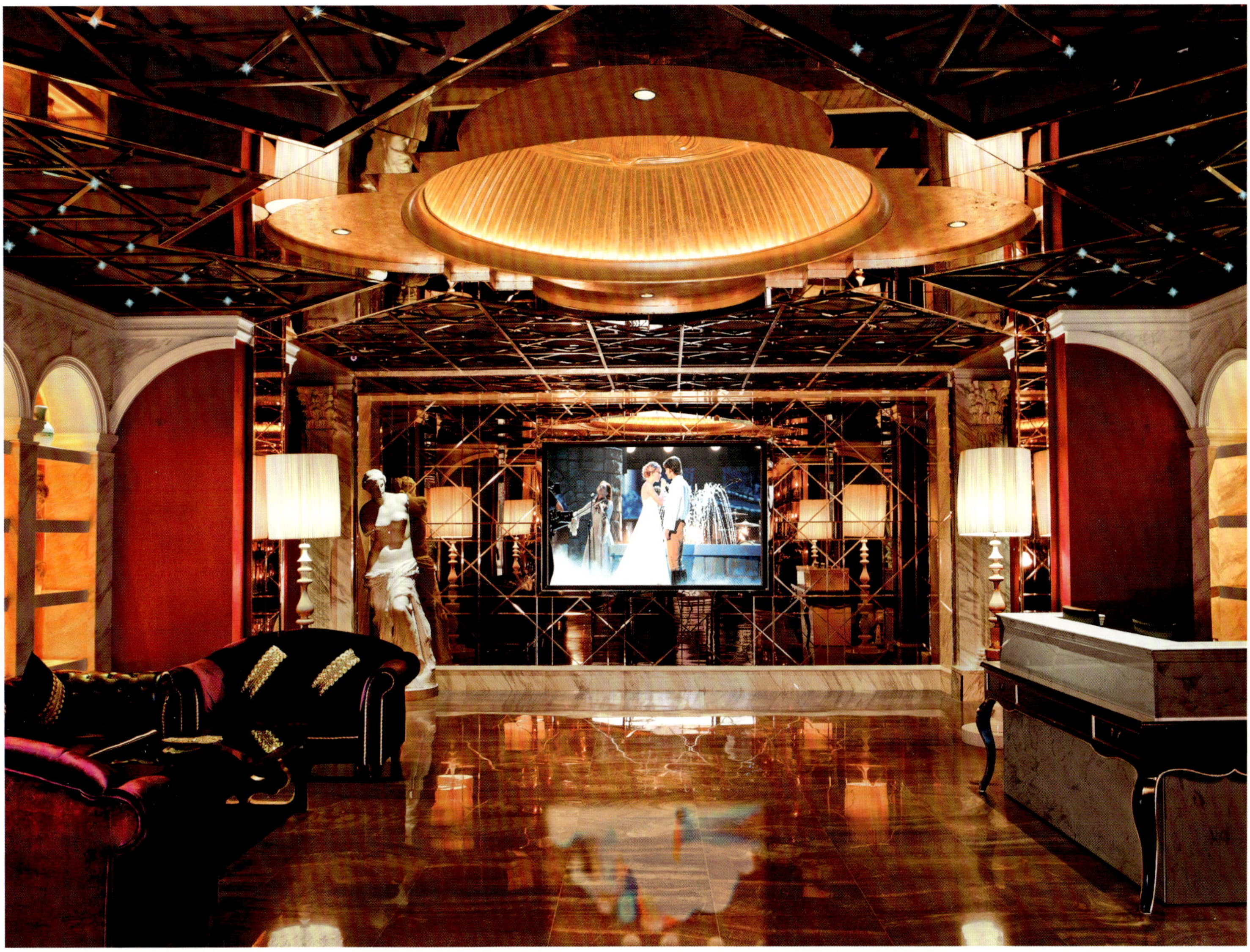

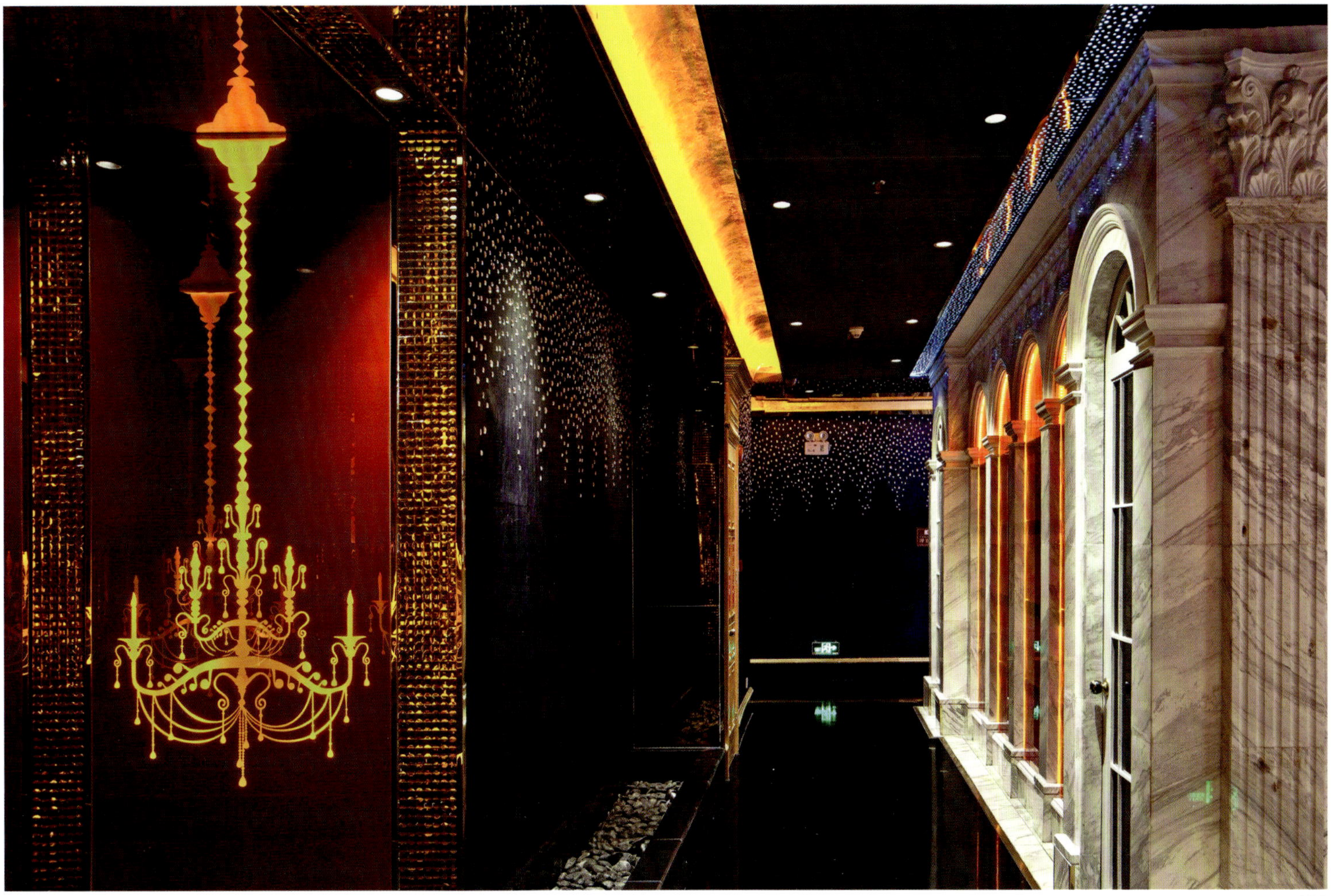

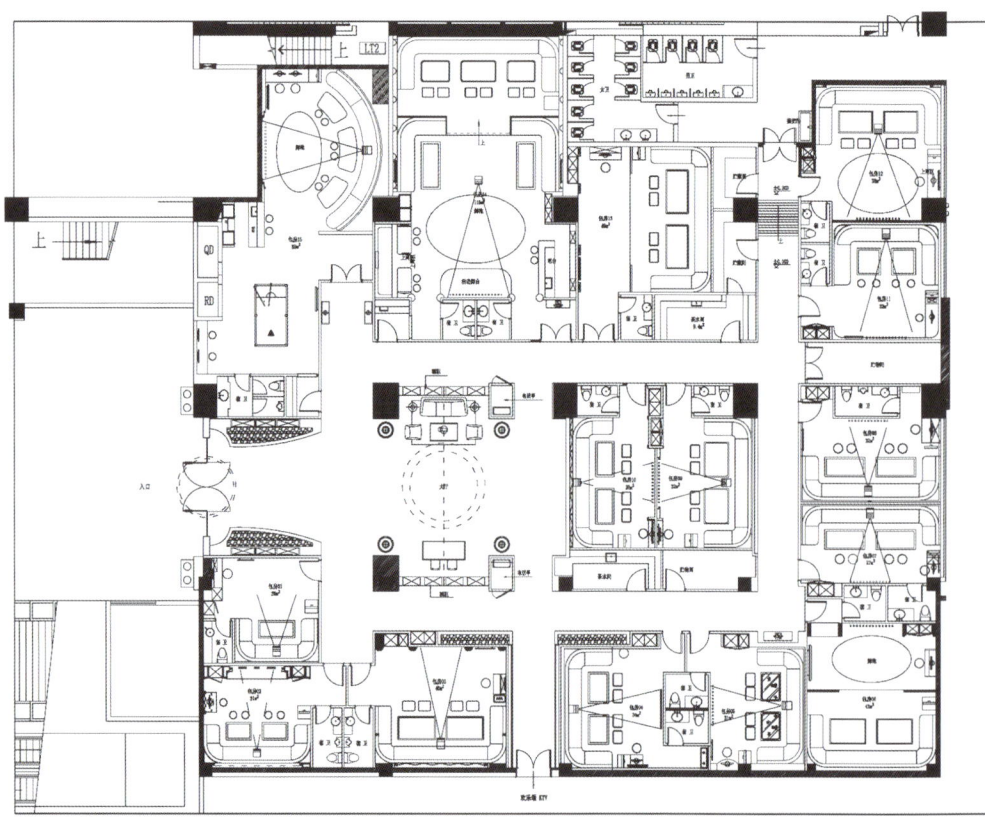

　　"欢乐盛开"是深圳首家世界风情主题式 KTV。建筑面积不大，业主主要将其打造成深圳高端的 KTV。经过设计团队与业主的多番研究与推敲，最终决定以地域特色主题来打造这一 KTV。

　　15 间主题房以 15 个国家和地区特有的历史、文化、民俗、建筑、风情等元素为基础，或融合，或重构，或分解，最终创造出一系列不同的造型、意象以及配饰搭配效果。可以是古老金字塔和法老王的埃及风情，也可以是非洲肯尼亚原始热带森林；可以是美丽纯洁的希腊爱琴海海洋风情，也可以是老北京味儿十足的中国四合院；可以是充满童真趣味的丹麦童话风情，也可以是一掷千金、豪迈奢华的赌城拉斯维加斯……独一无二的房间，带给人独一无二的娱乐体验！

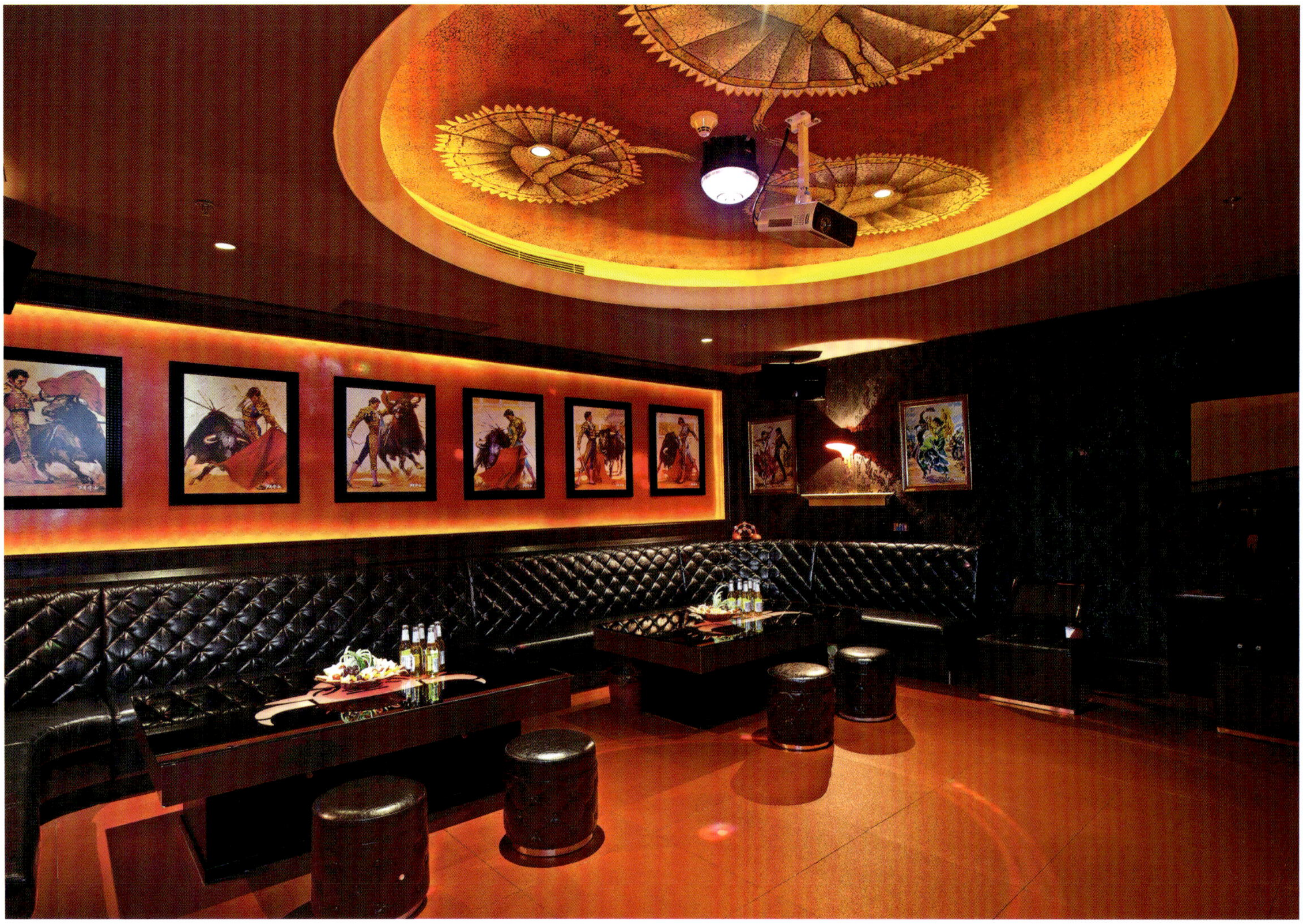

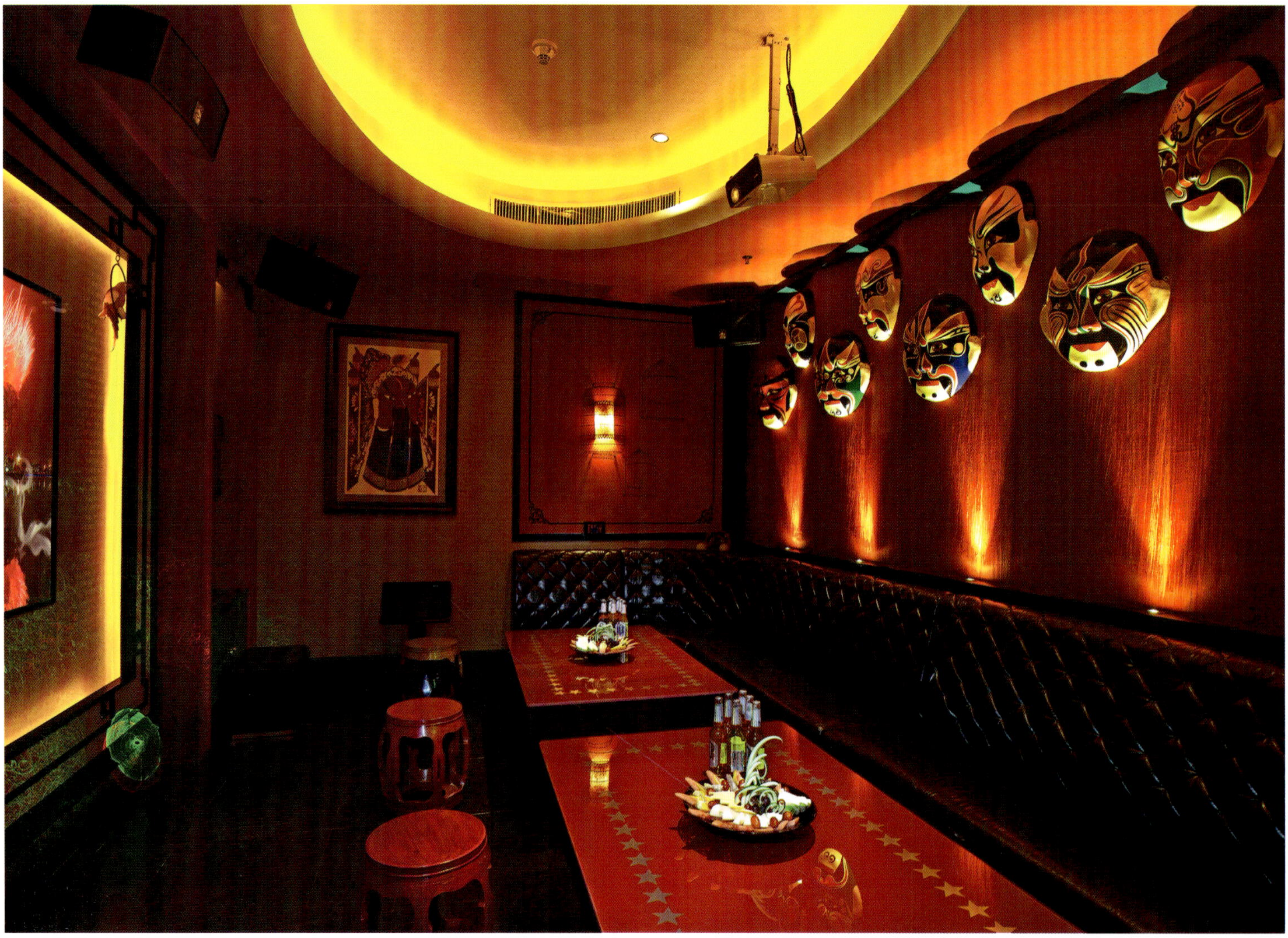

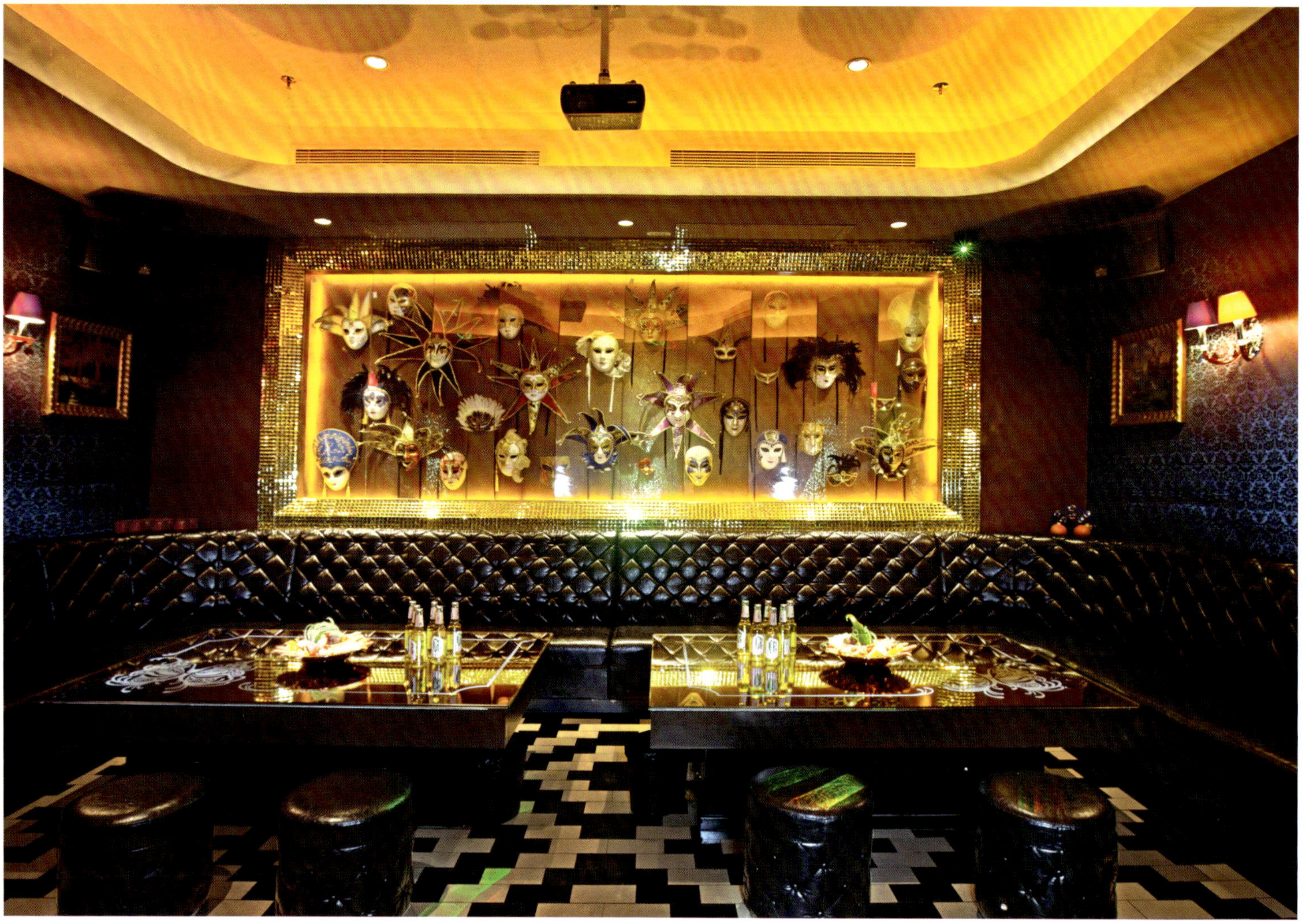

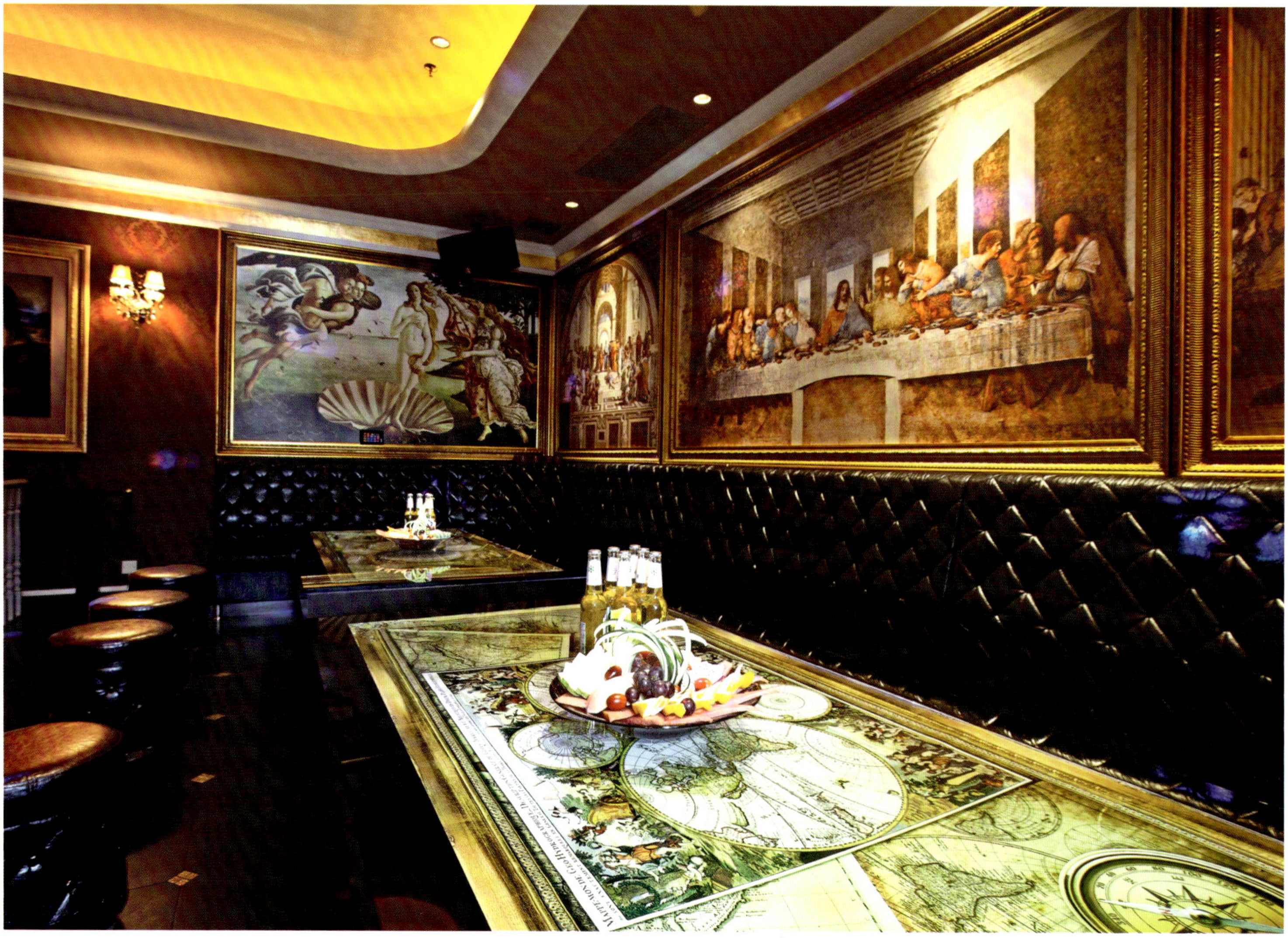

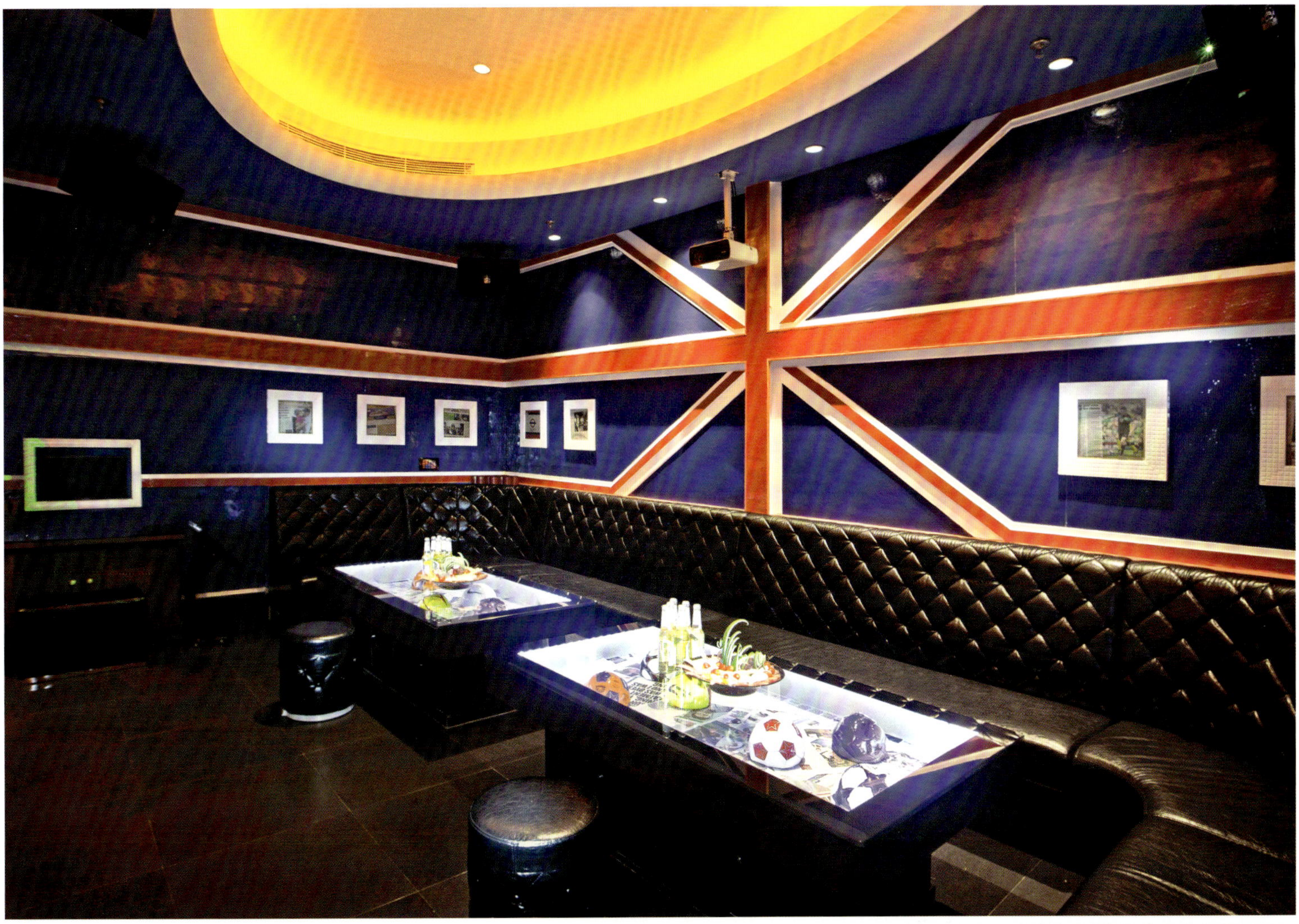

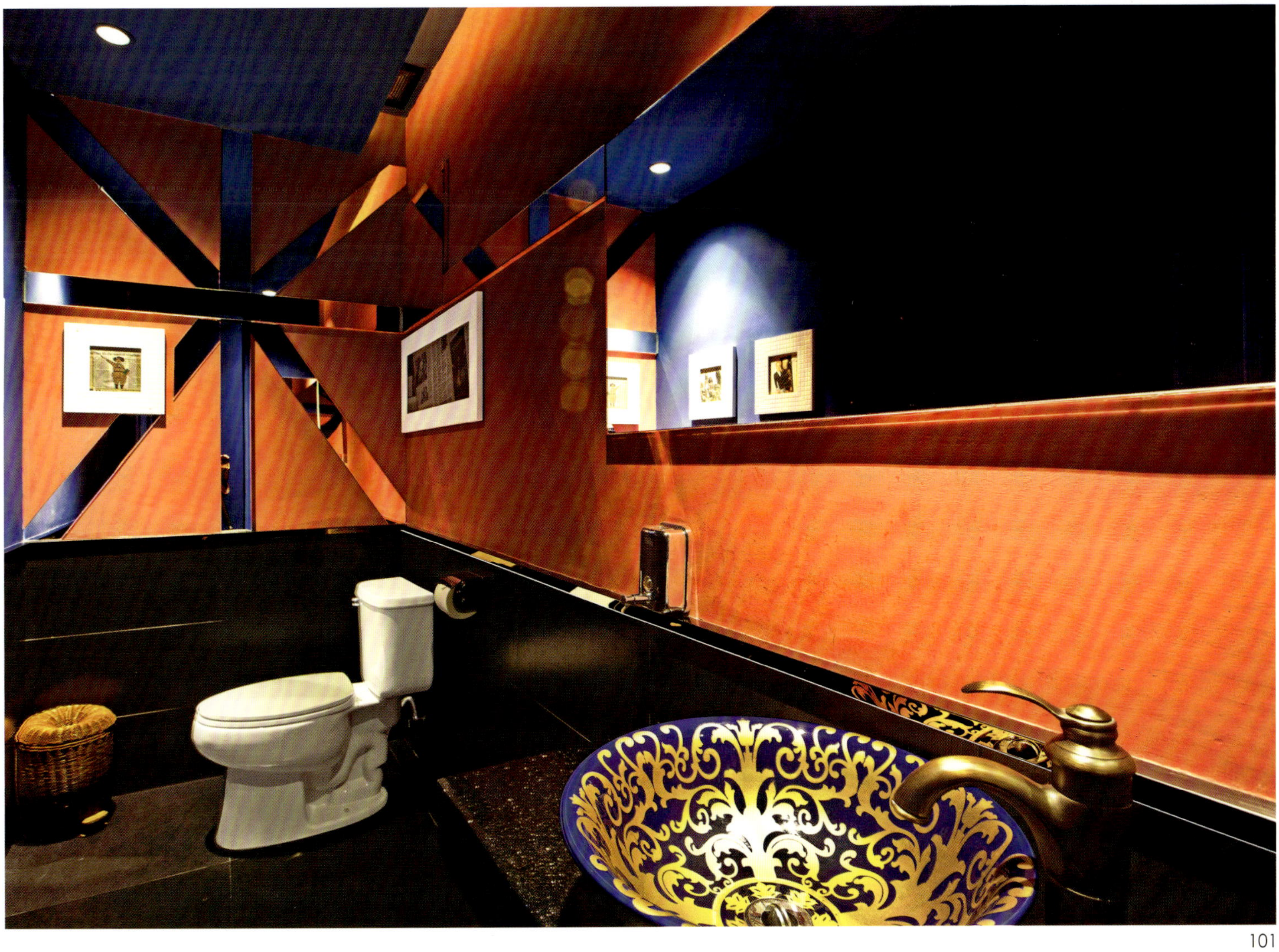

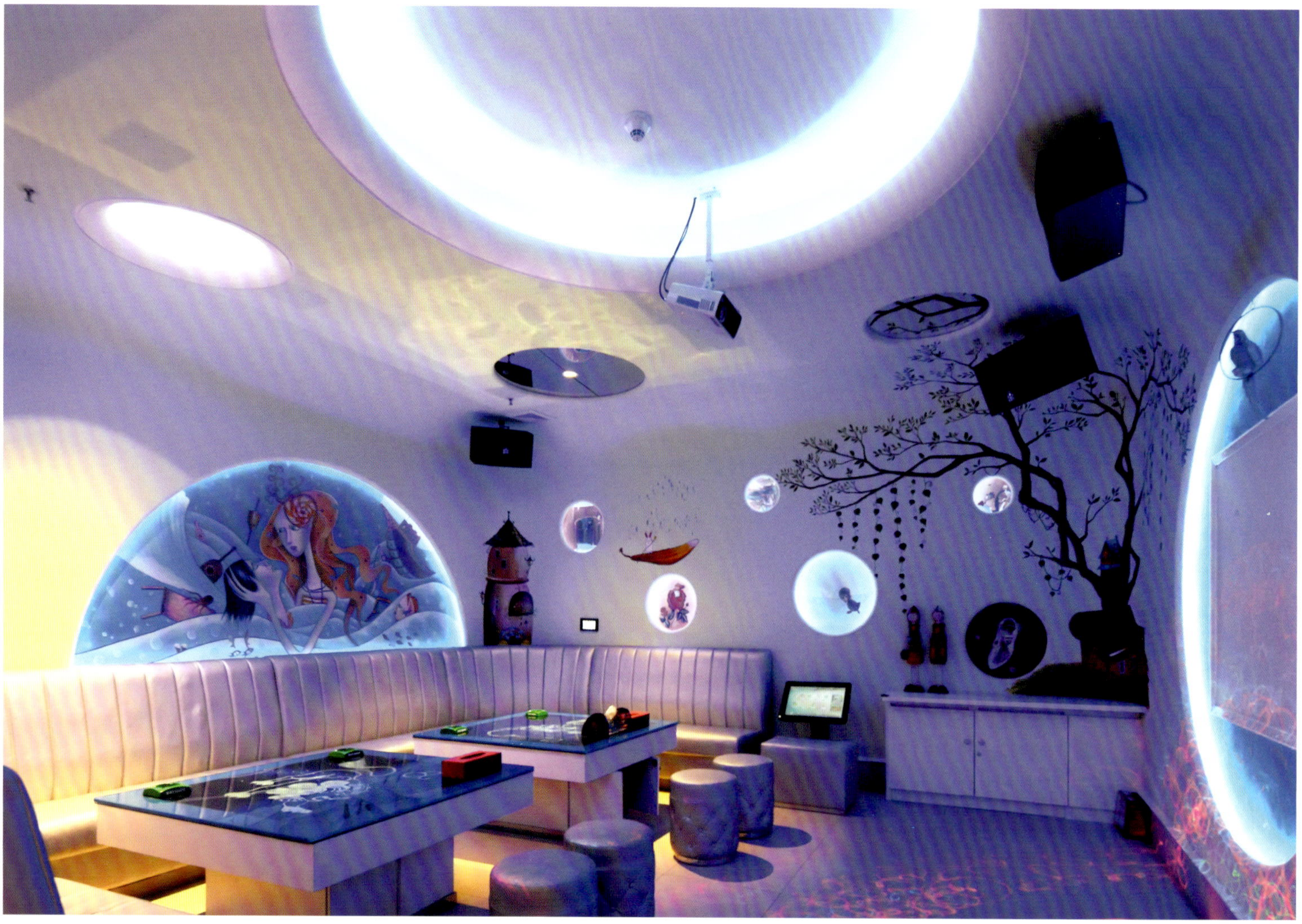

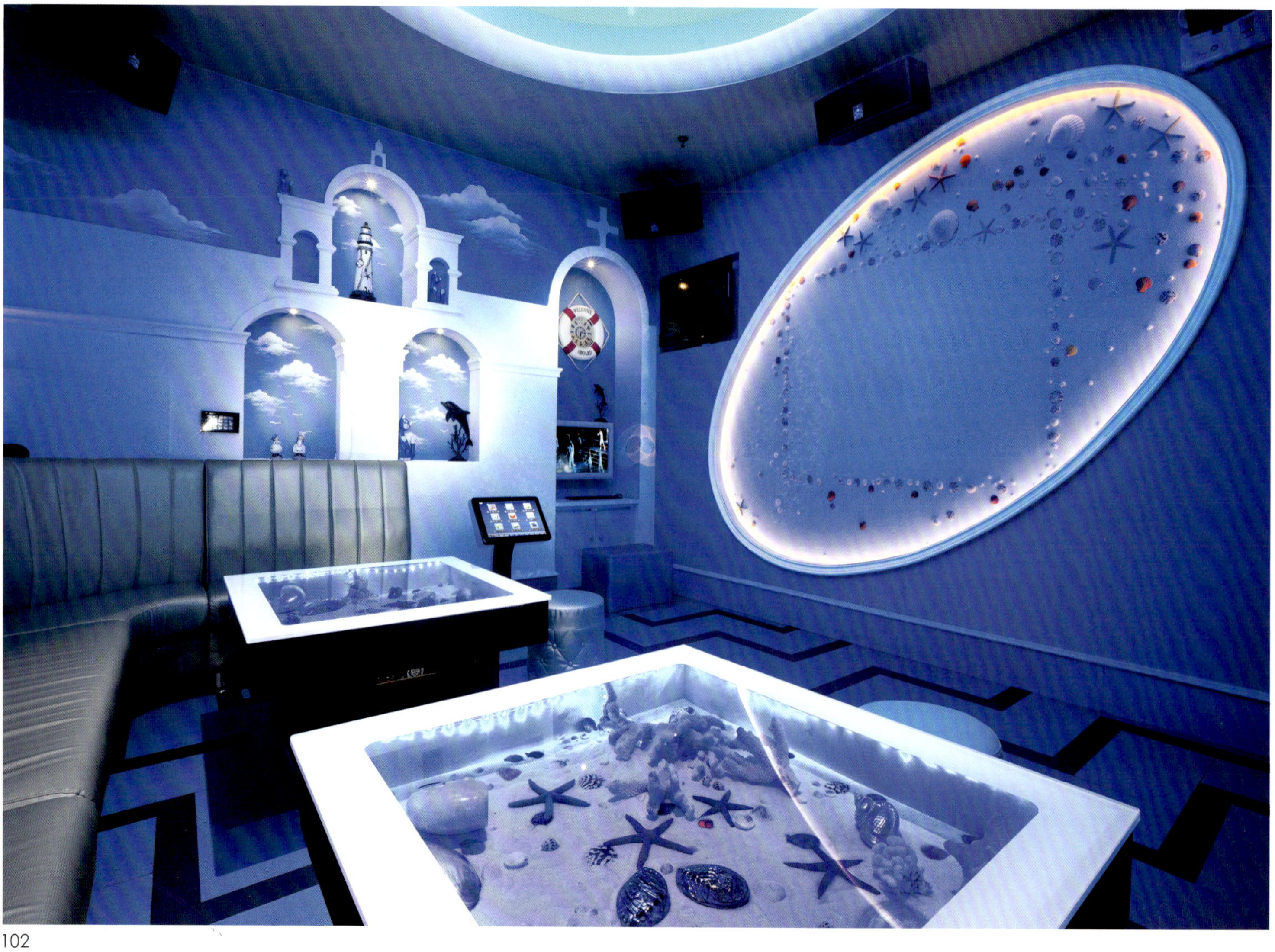

地点	面积	设计师	设计公司
/ 深圳	/ 2800m²	/ 贺玮、肖枫	/ 深圳风荷印象室内设计有限公司

Focus KTV (Headquarters in Longhua)

焦点KTV（龙华总店）

The project strives to create a fashionable varied space with bright colors, and the whole space display is well-proportioned, showing out an artistic illusion of superposition, dislocation and deviation. The designers make use of separating, linking, and cutting design techniques to lead a variety of design forms, linking or separating each space, so the existing design and the original space achieve the effect of mutual penetrating and excellent matching.

The designers introduce a large number of slowly changing LED lights, magically creating a cool and varied temperament for the space; through the reflecting materials of mirror stainless steel and glass to do a verified radiation treatment, the whole space seems like waving in a dreamland, attracting us.

A fantasy and a passion, the designers interpret the entertaining passion of Focus KTV through a new momentum. Here, the space is divorced from materials to a certain extent, entering into a spiritual level, leading and stimulating the release of emotions, which are natural and sincere.

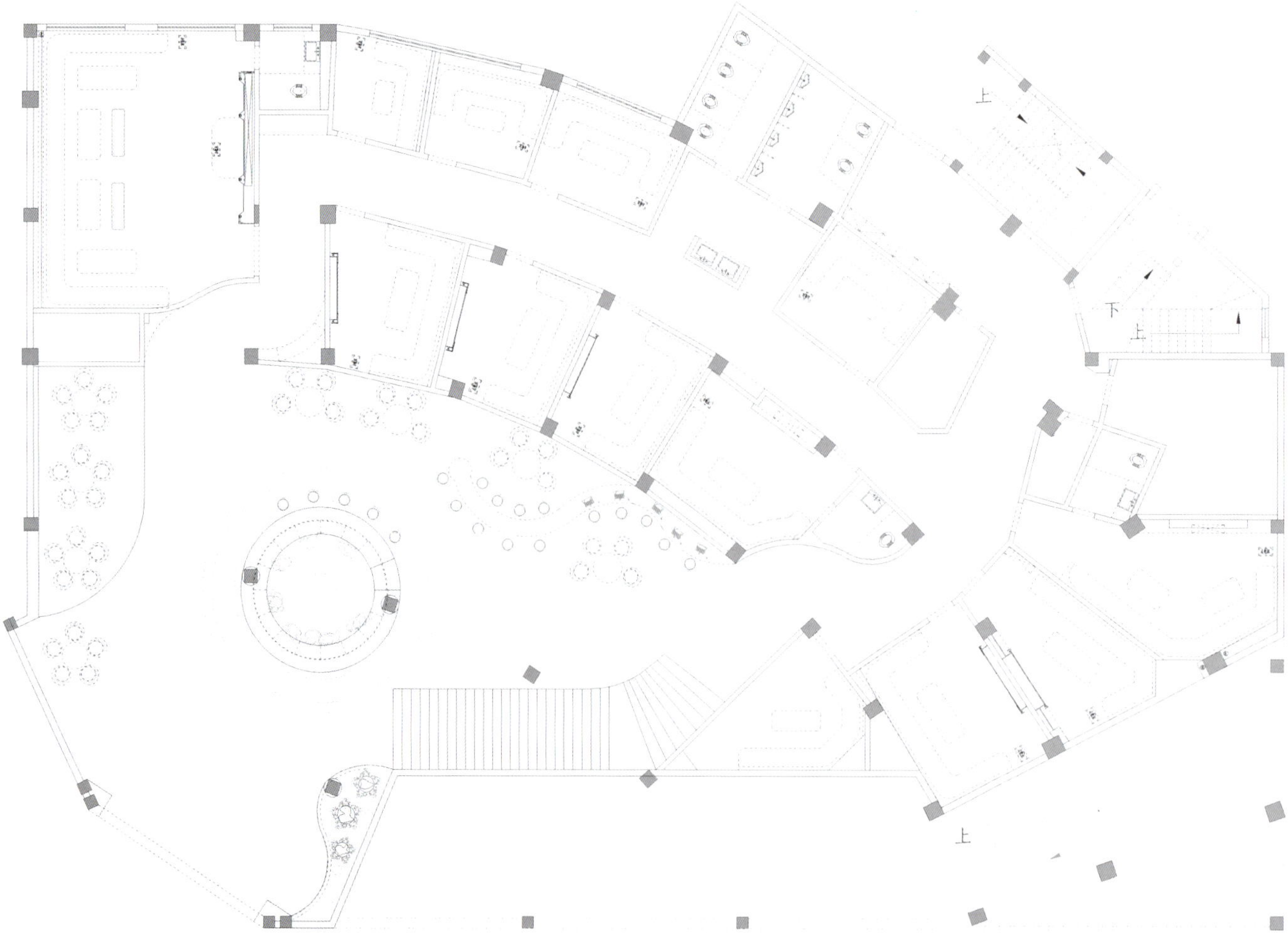

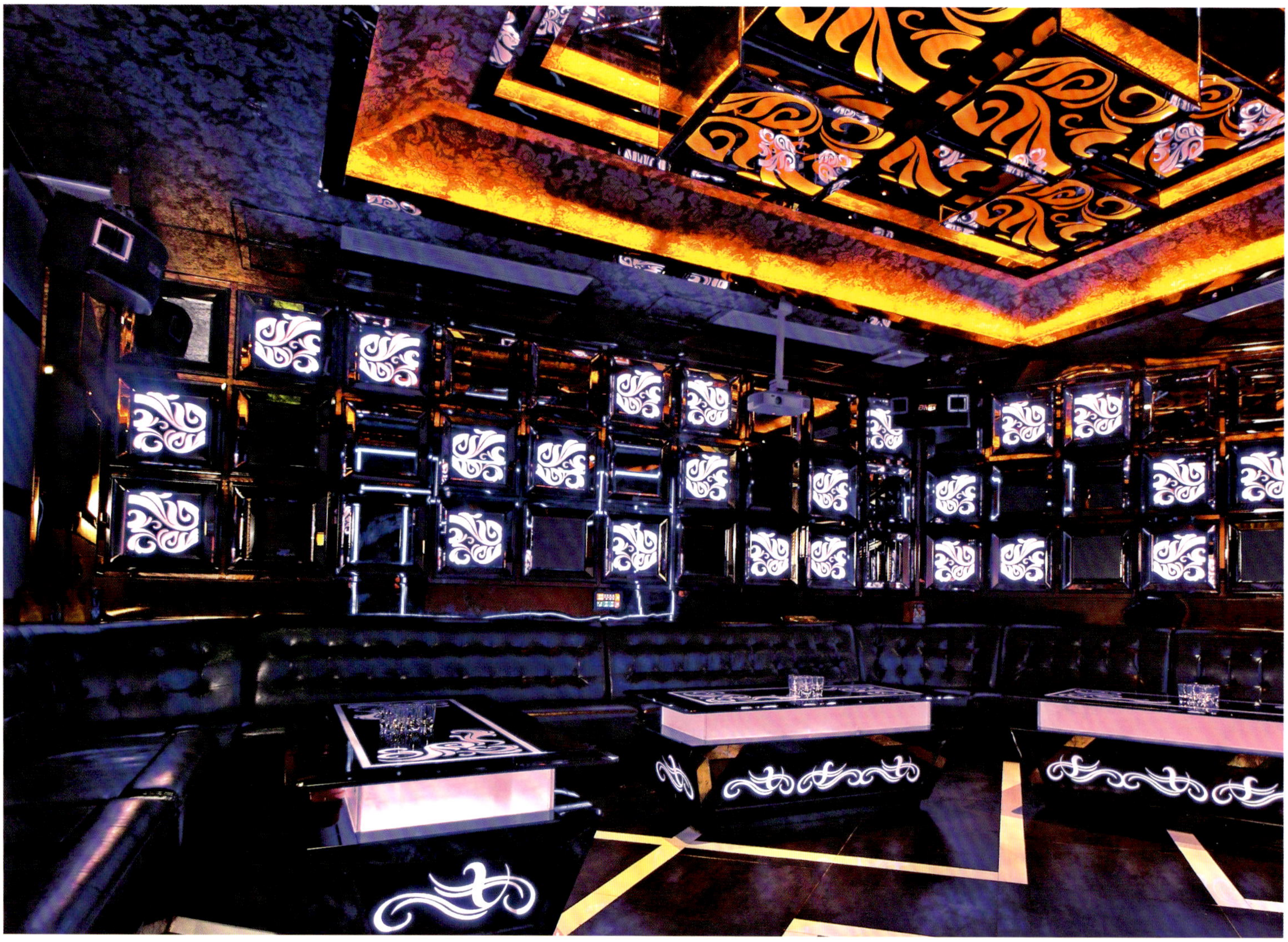

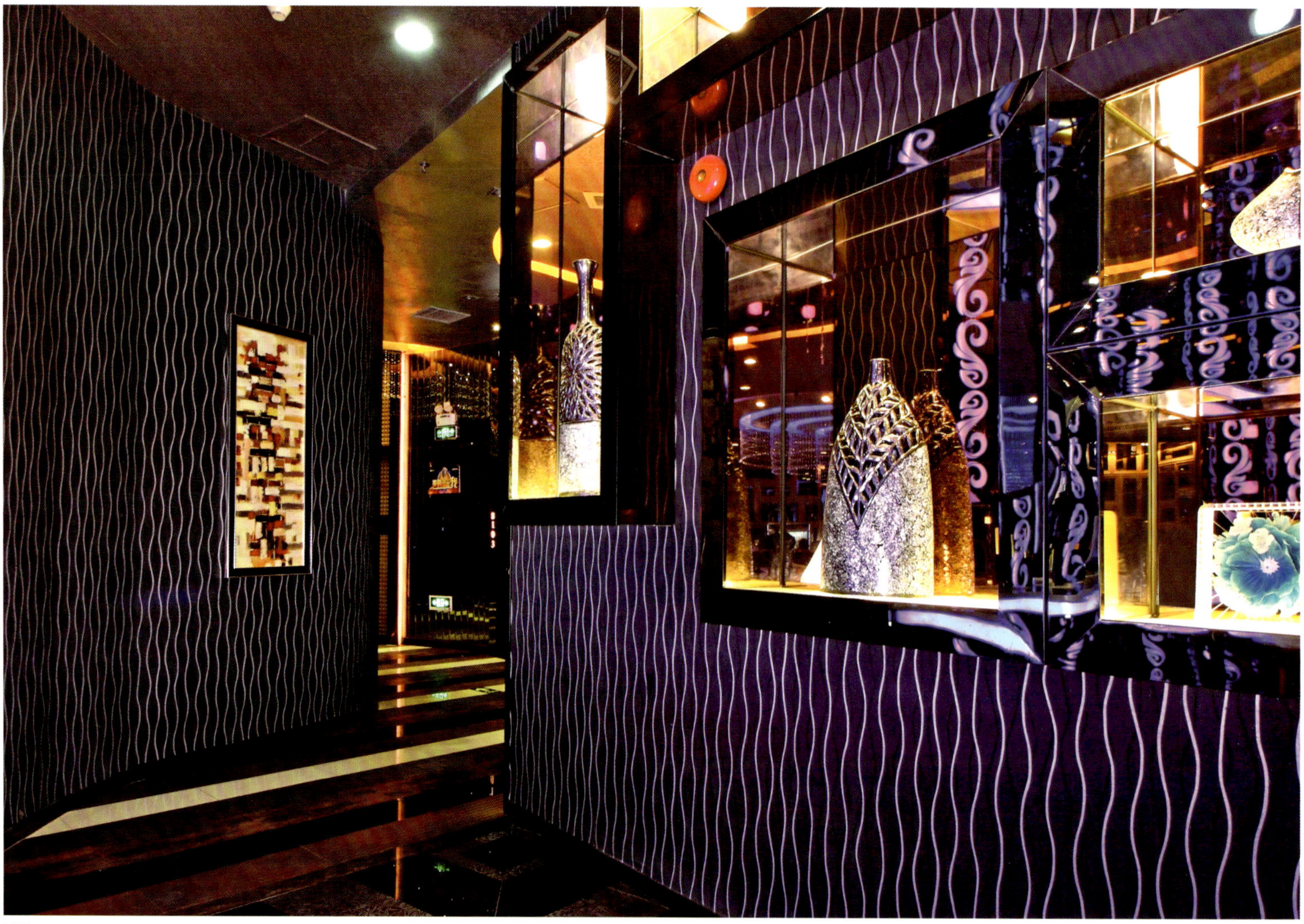

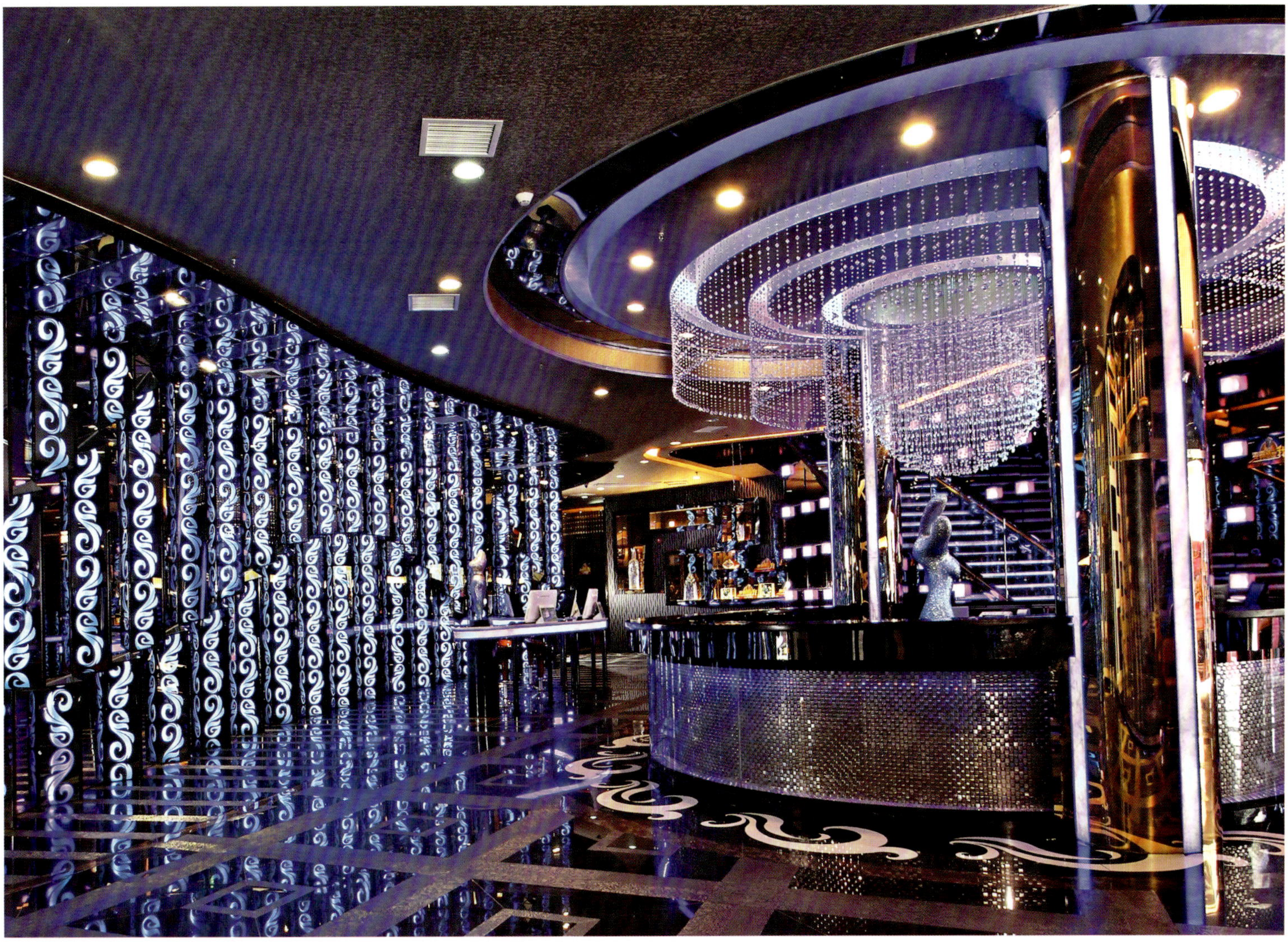

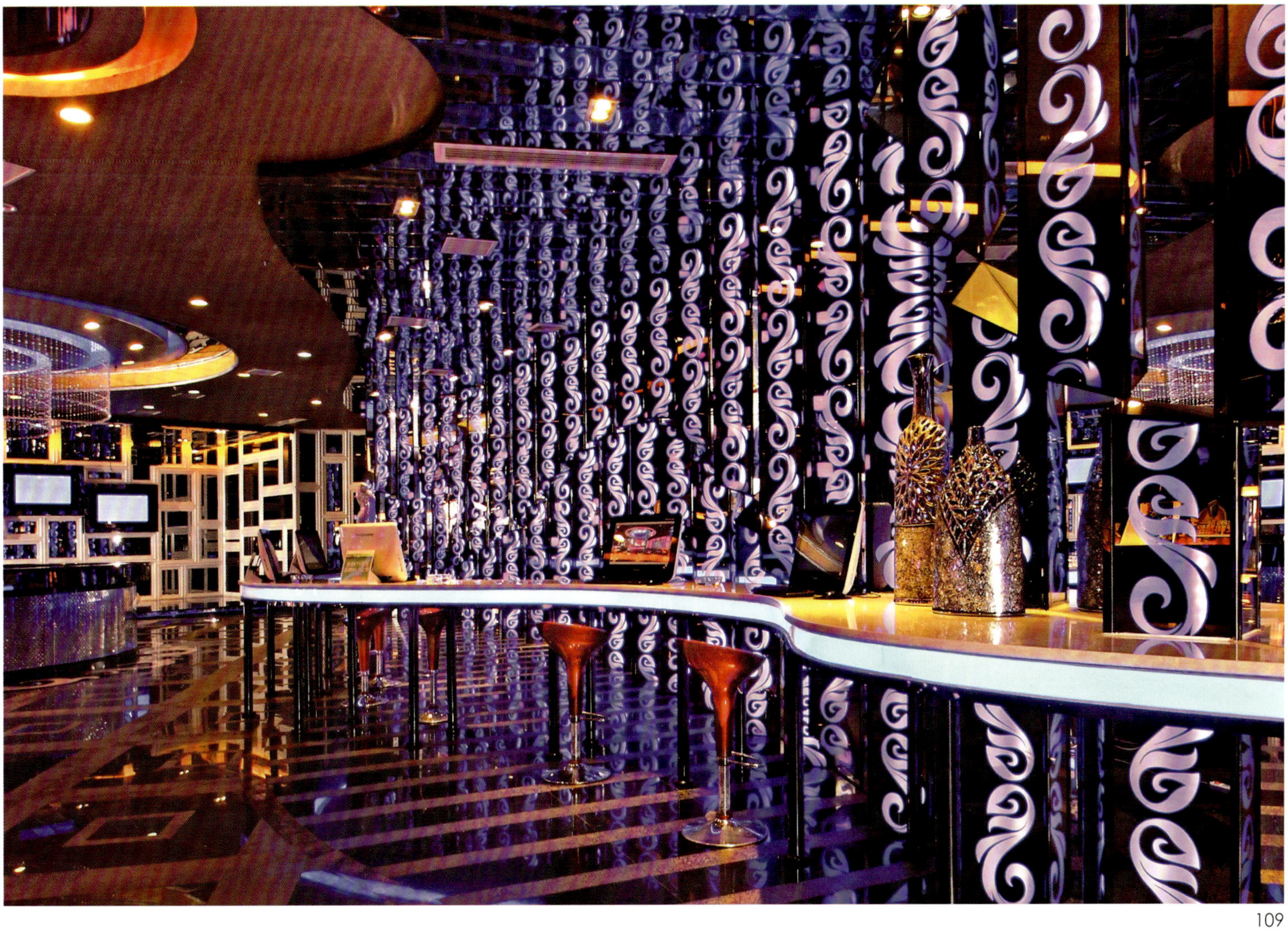

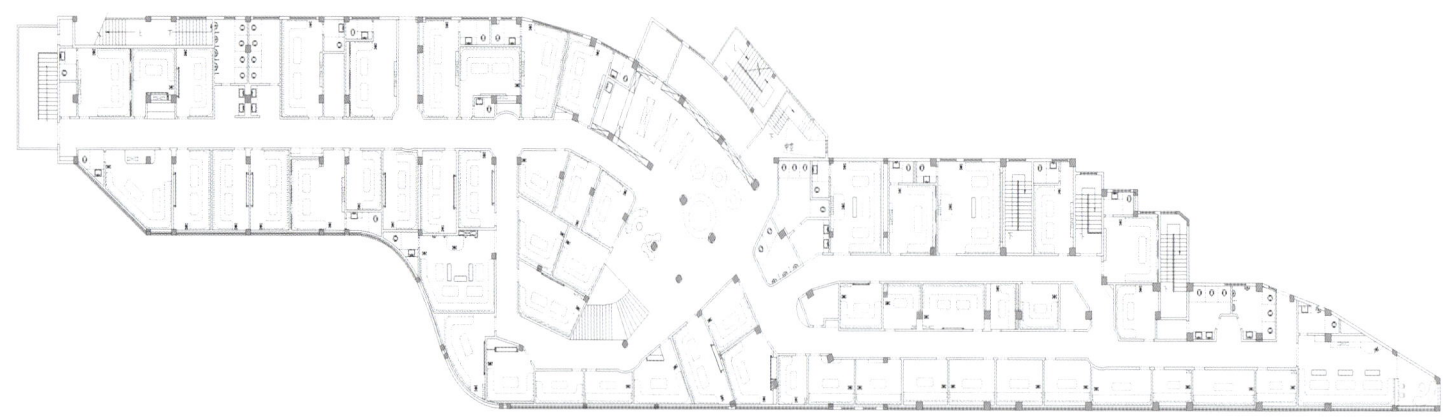

　　本案力求打造一个时尚多变的绚彩空间，整个空间陈列错落有致，呈现出重叠、错位、偏离的艺术错觉。设计师运用独立、串联、切割等设计手法牵引各种设计形态，连接或区分各个空间，使现有设计与原有空间达到相互渗透、珠联璧合的效果。

　　设计师采用大量缓变 LED 灯，魔术般地为空间打造出酷而多变的气质；再通过镜面不锈钢、玻璃等反光材料将灯光做多元化的放射处理，使整个空间犹如荡漾在梦境之中，使人沉醉。

　　一丝梦幻，一份激情，设计师以全新的气势诠释出焦点 KTV 的娱乐激情。在这里，空间在一定程度上脱离了物质性，进入了精神层次，带动并激发人的情绪释放，一切进行得自然而真诚。

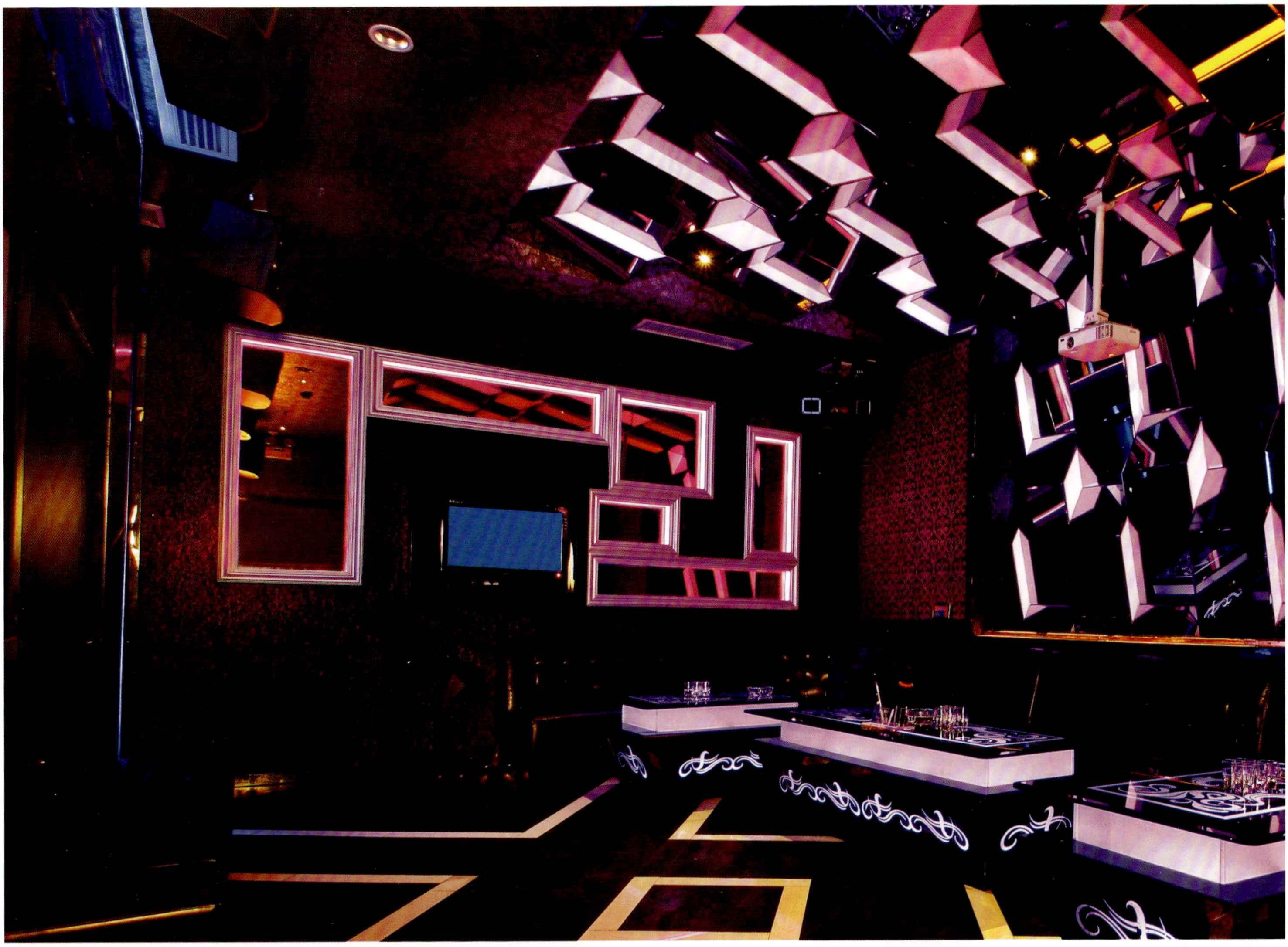

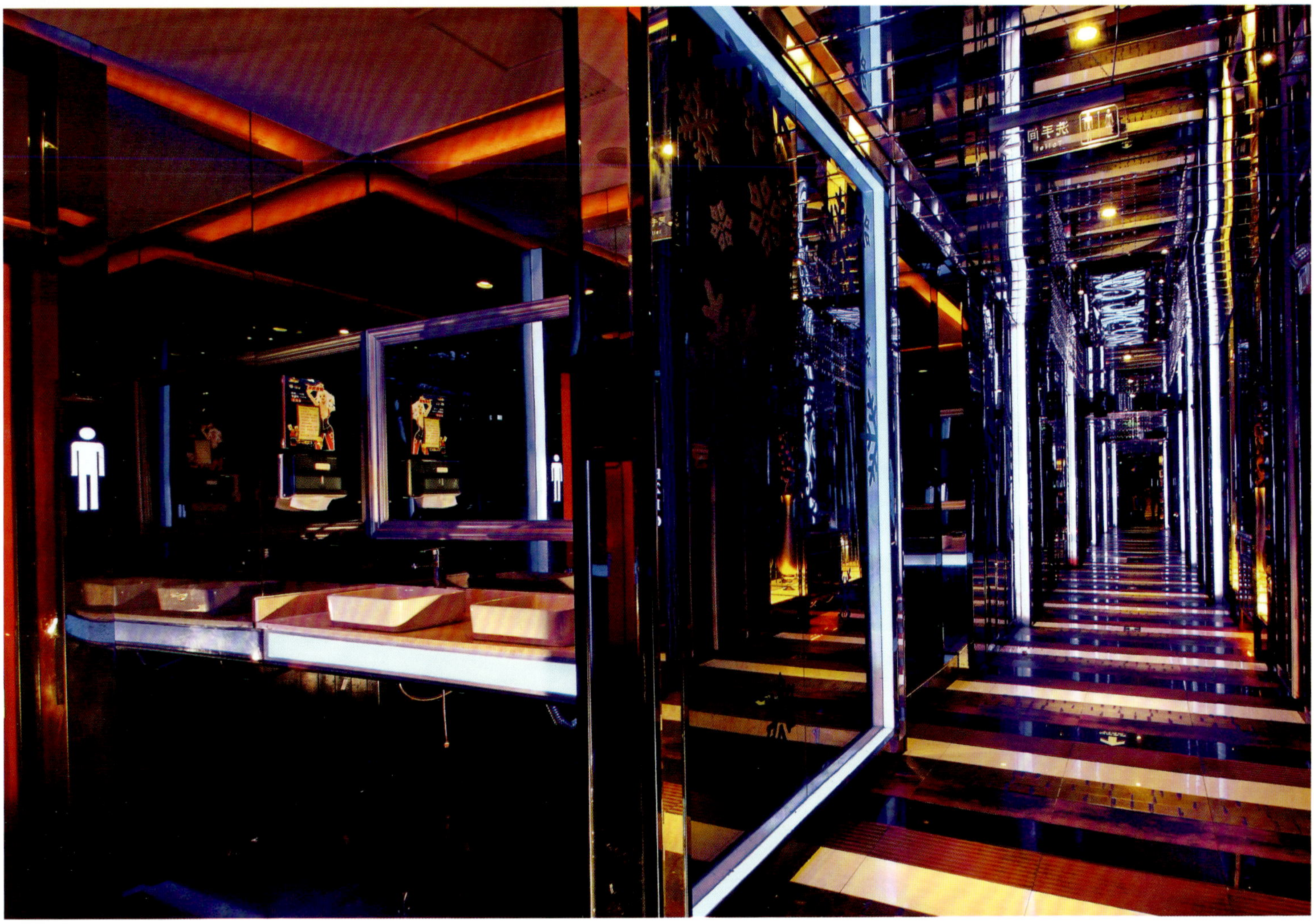

地点	面积	设计师	设计公司	主要材料
/ 福建	/ 3000m²	/ 吴华林、林彬	/ 香港丰华设计策划有限公司	/ 合成大理石、黑色不锈钢、布艺软包、雕花钢板、珠光马赛克

Cross-Straits Beauty Entertainment Club

—— 两岸佳人娱乐会所

The project tends to create a high-grade club for the group of high-end consumers in city, so the designers abandon the magnificent decorating style and gorgeous nightclub atmosphere, but to create a bright, clean and elegant leisure and entertainment place.

The entertainment place is designed with neo-classical European style and modern fashion style, without contrast of brilliant light and bright colors, just featuring the open and clean space layout and elegant and generous furniture combination to show out the temperament and style of the space. In the space modeling, the simplified Roman columns make the space more open and generous, coupled with concise lines and grand and gorgeous chandeliers to define a cool, pretty and elegant space. In the aspect of furniture, the comfortable and gorgeous sofa is decorated with edges of gold and silver carvings, and in the mutual reflect and shining of marble floor and velvet wallpaper, showing an elegant temperament. The introduction of European murals and patterns allow the artistic temperament hidden in the space, featuring an elegant club culture.

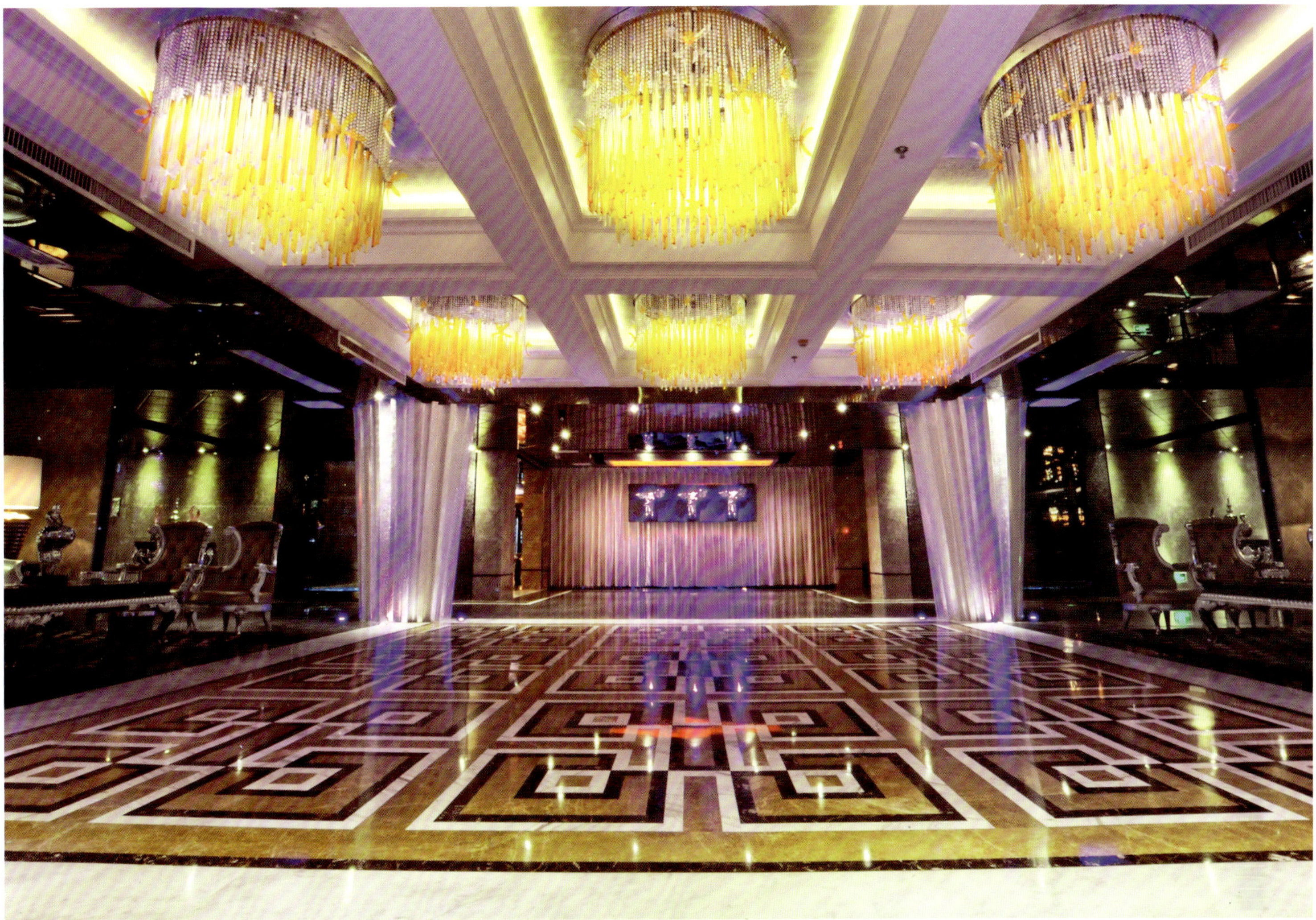

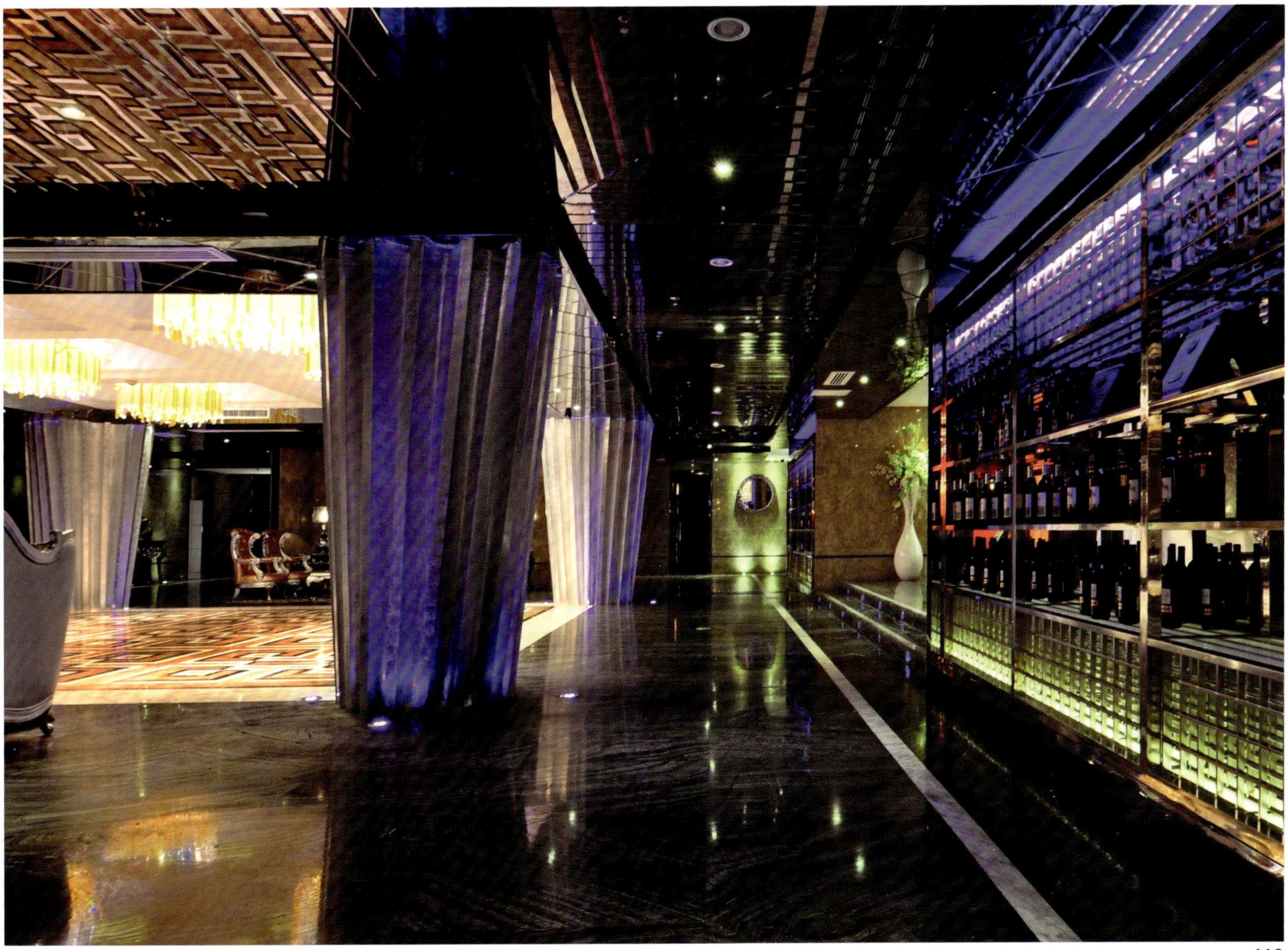

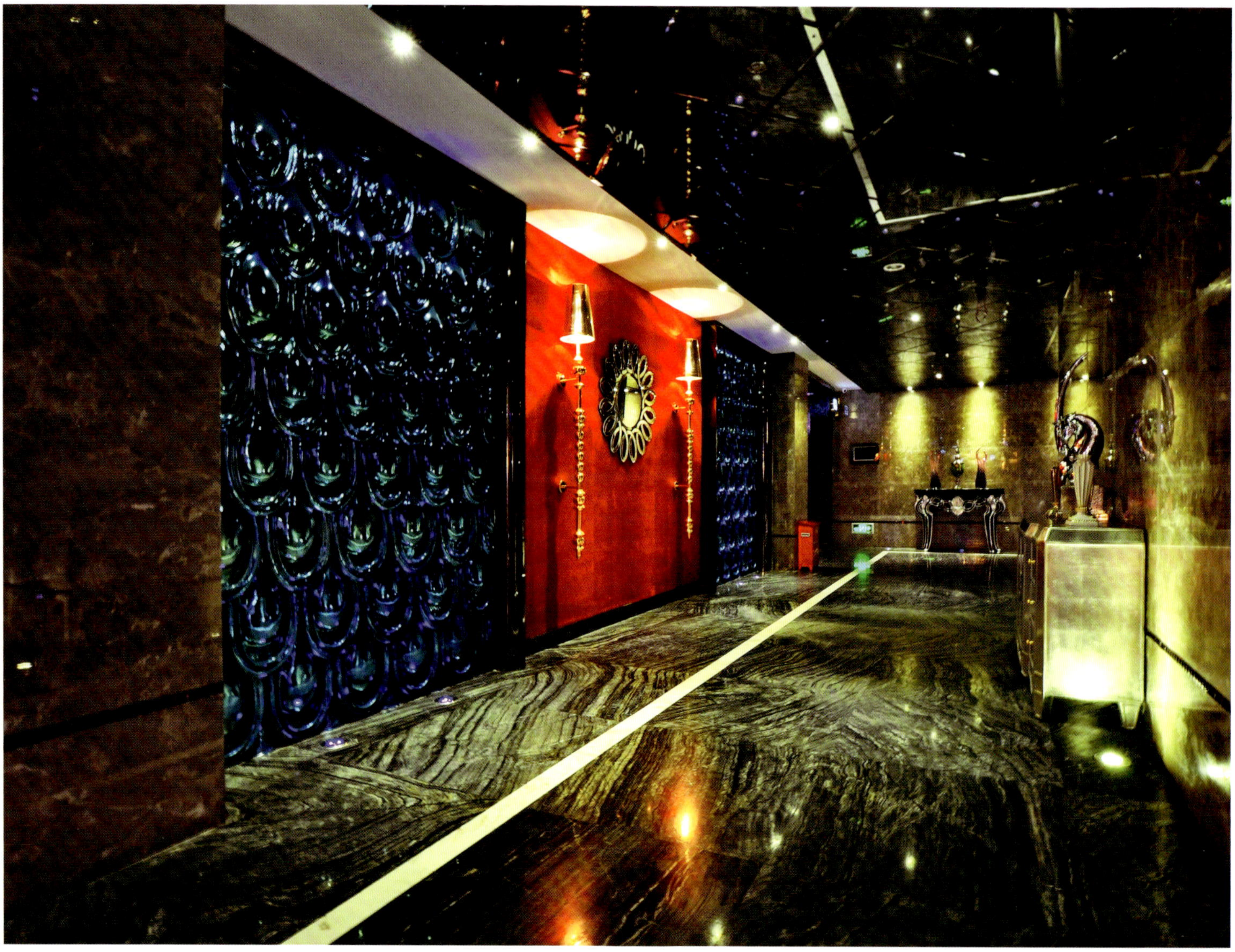

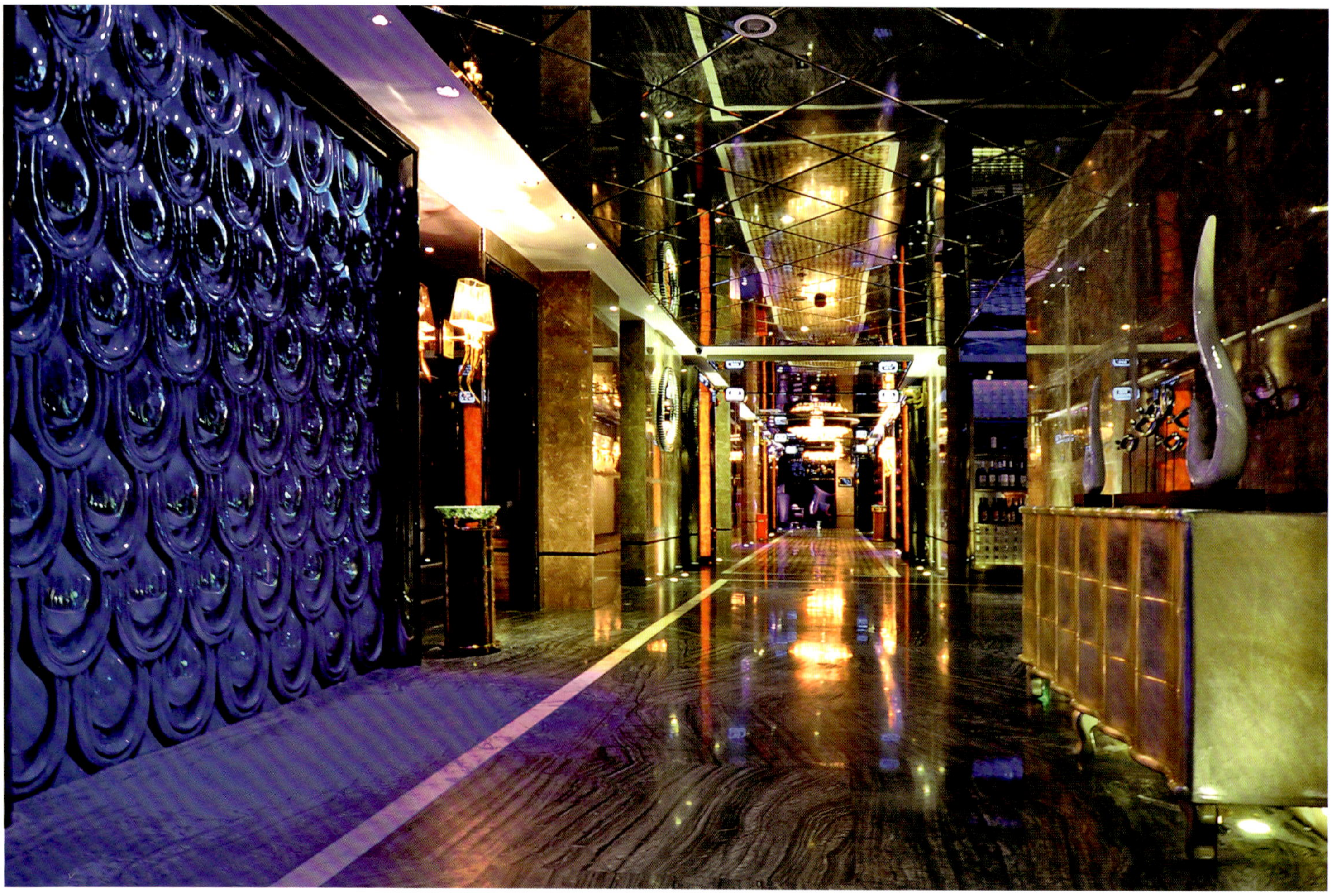

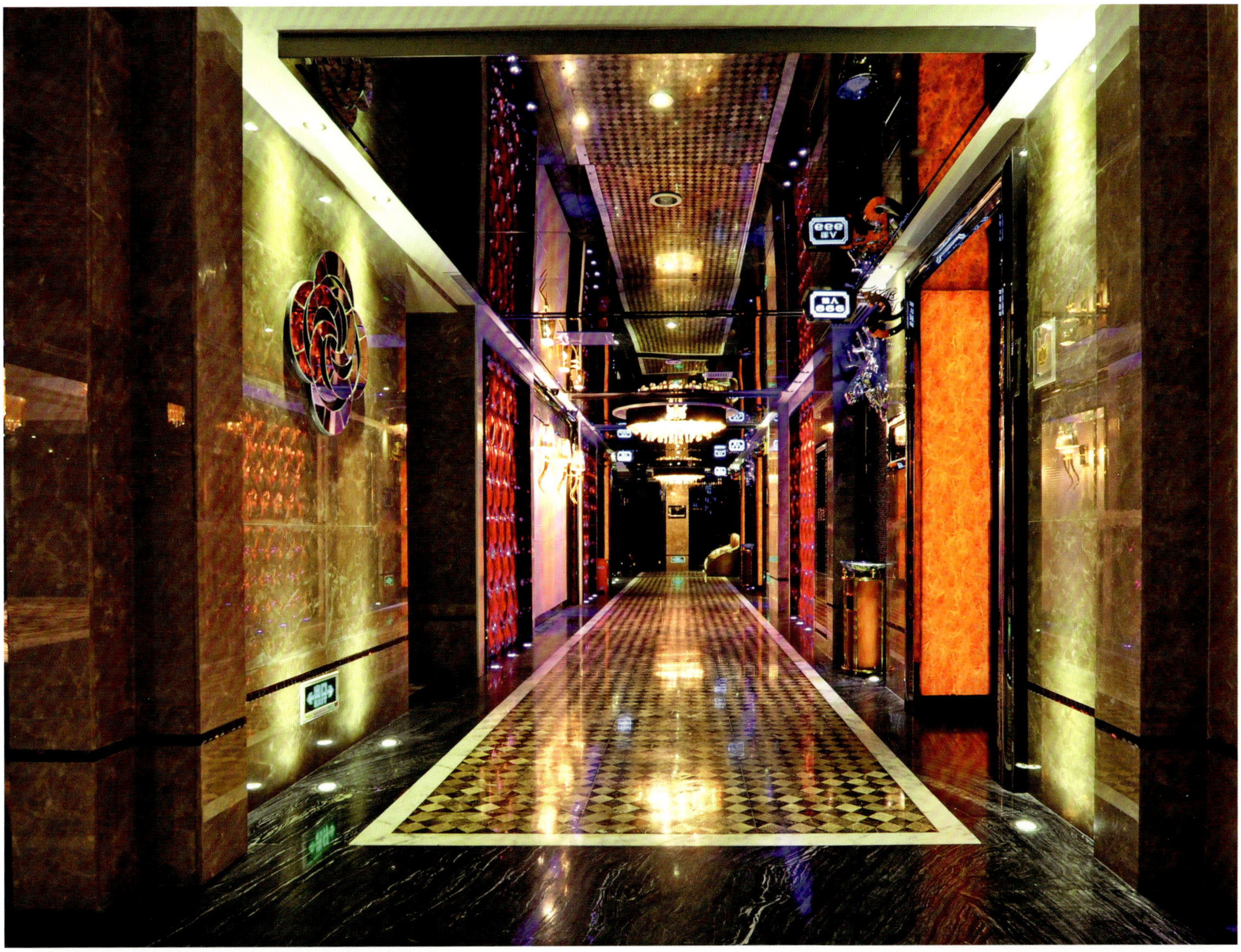

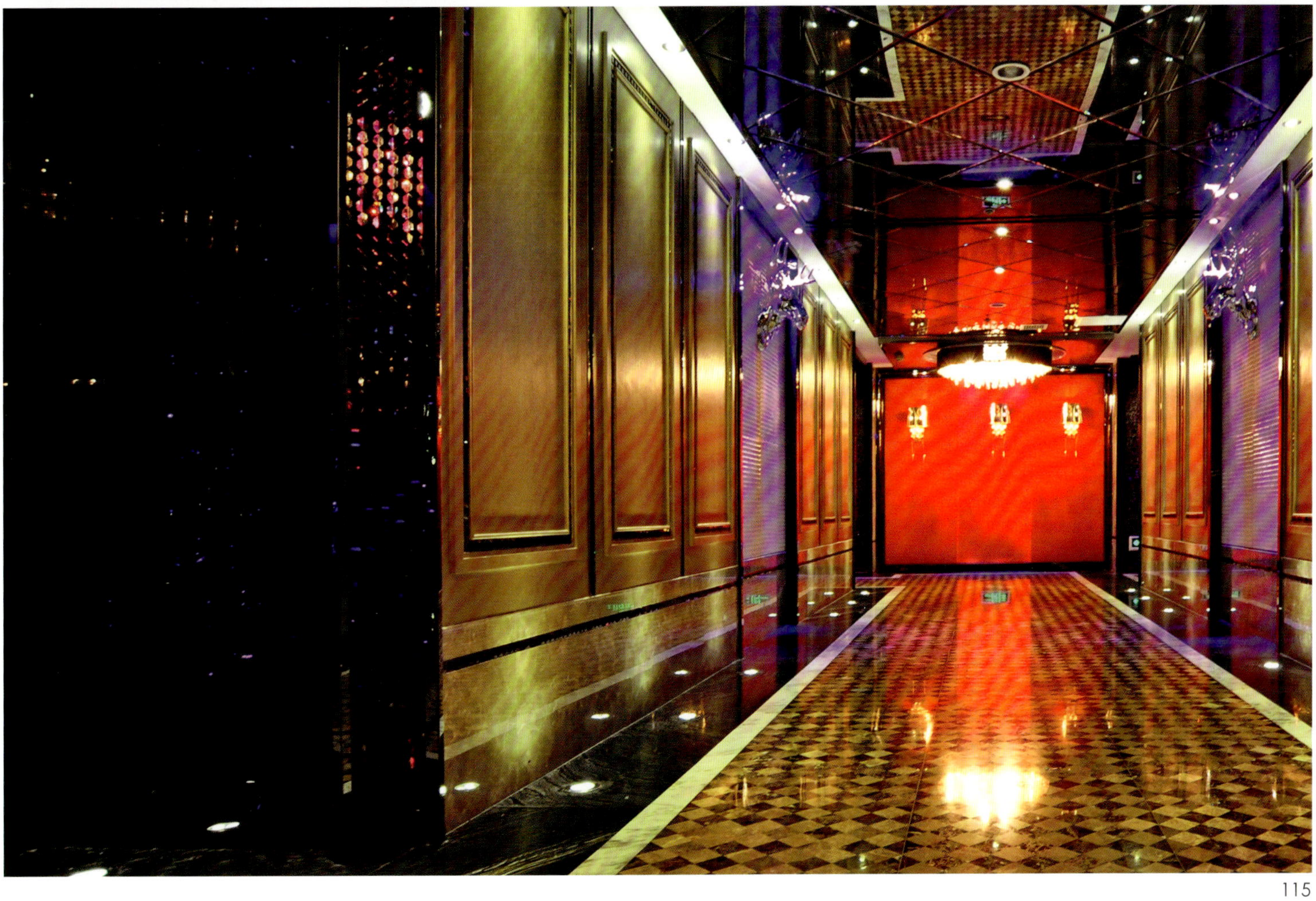

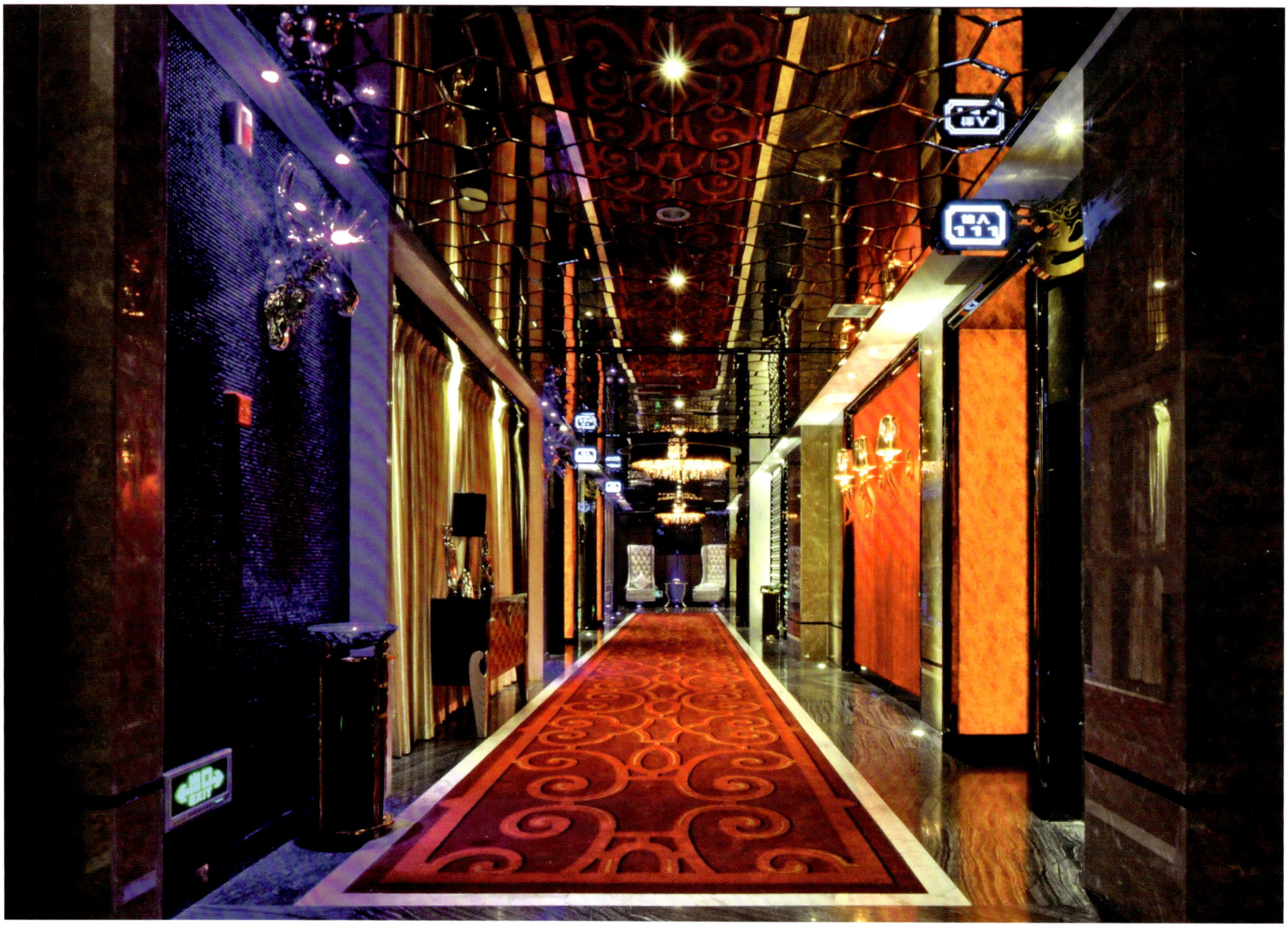

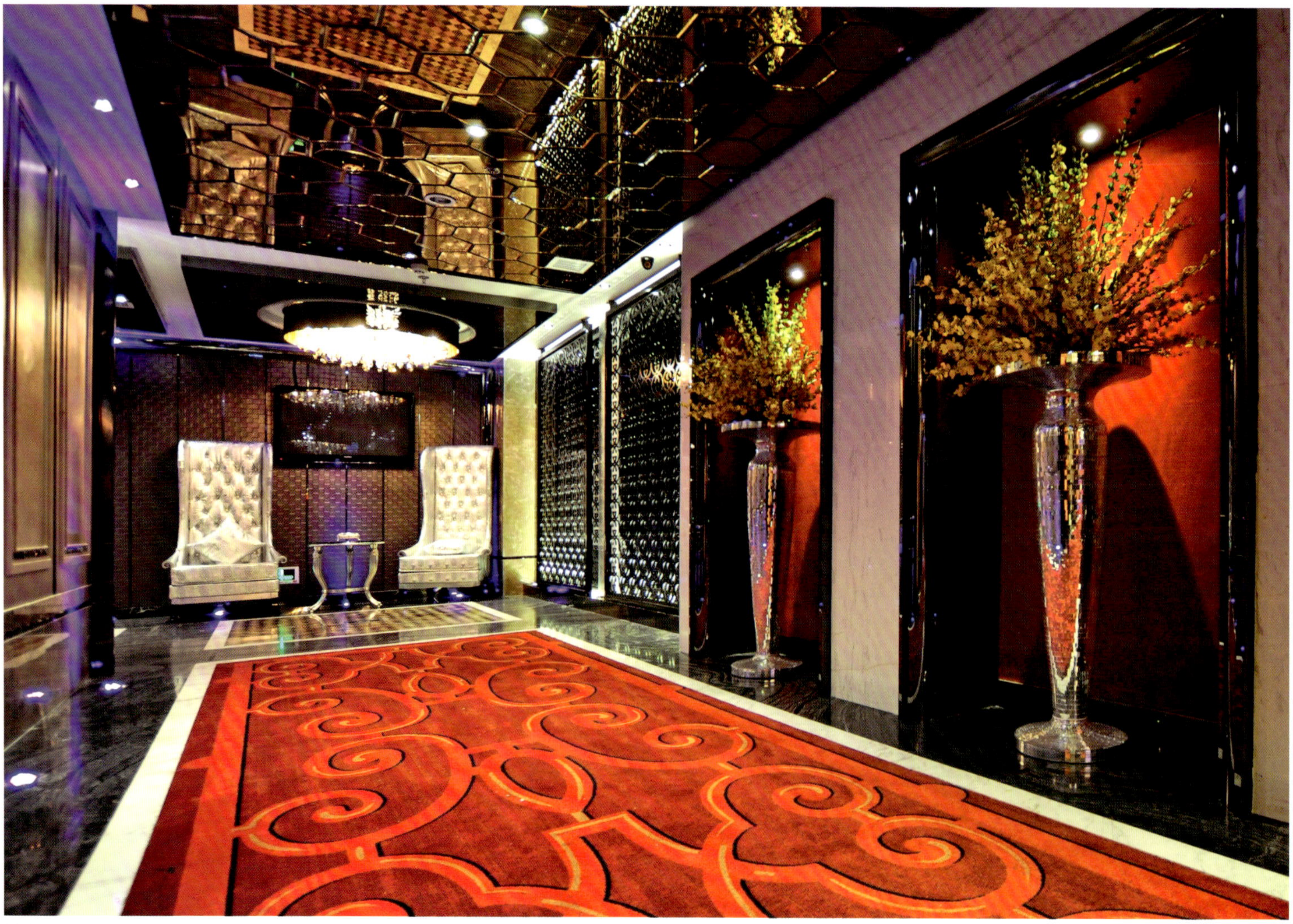

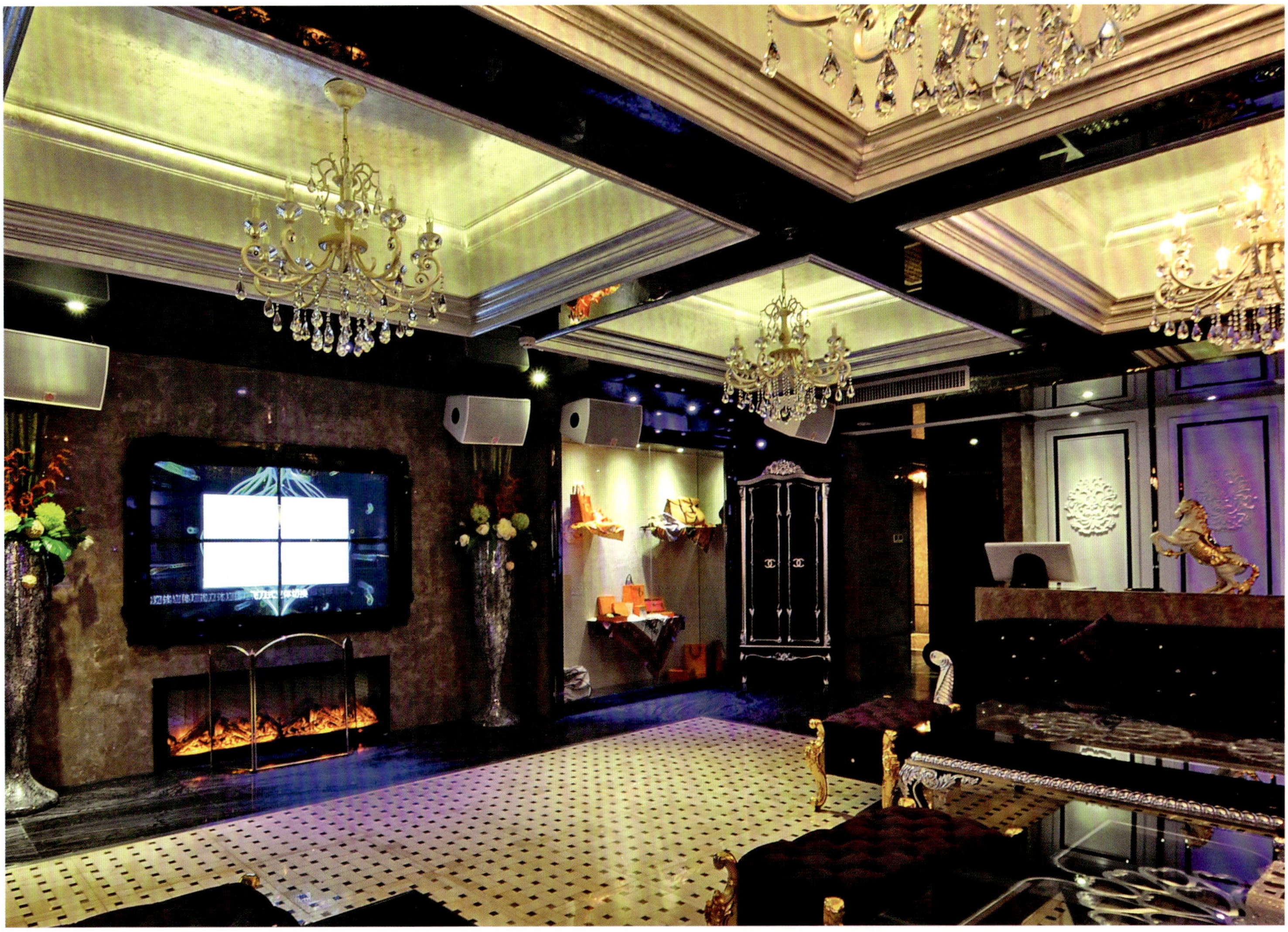

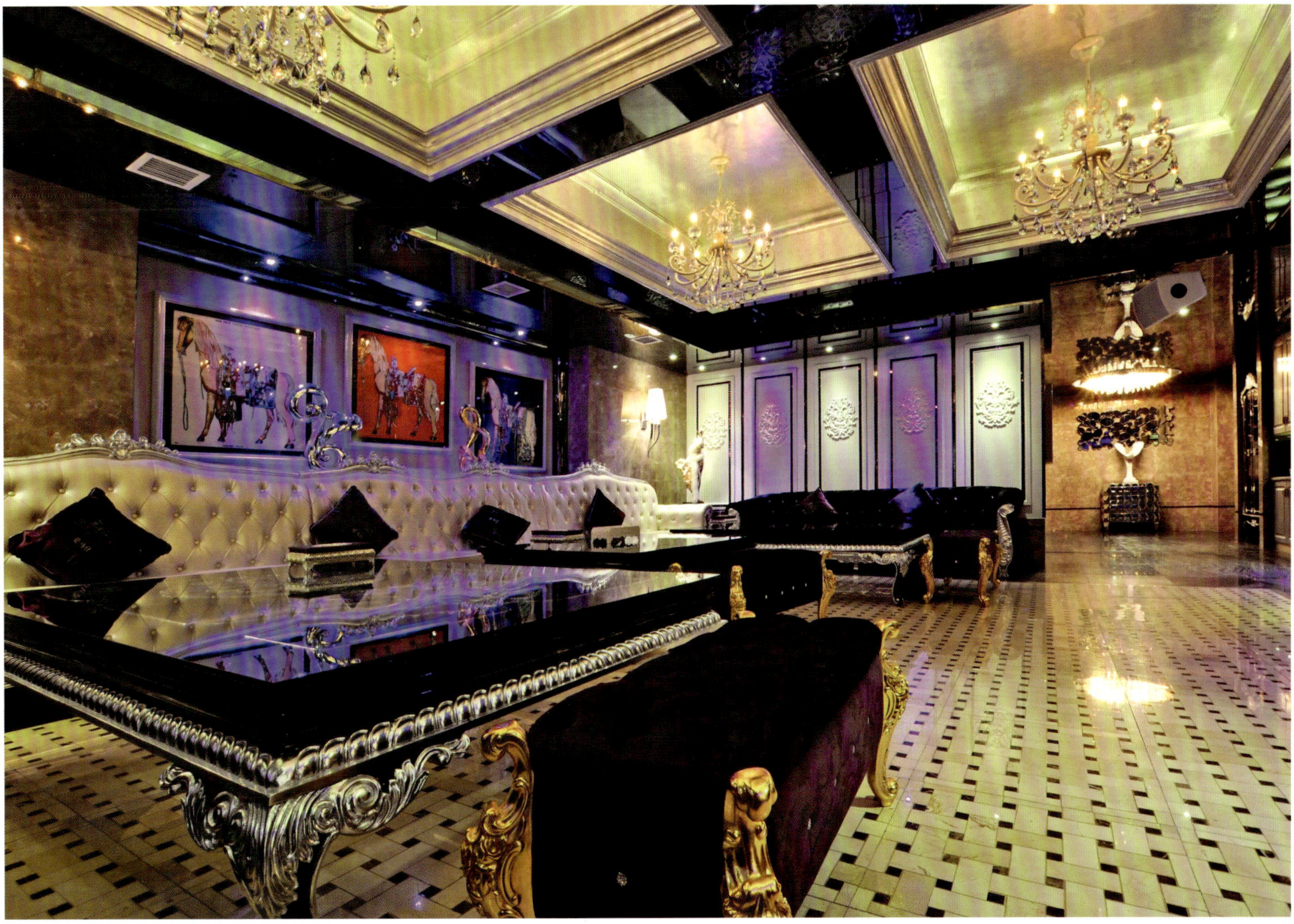

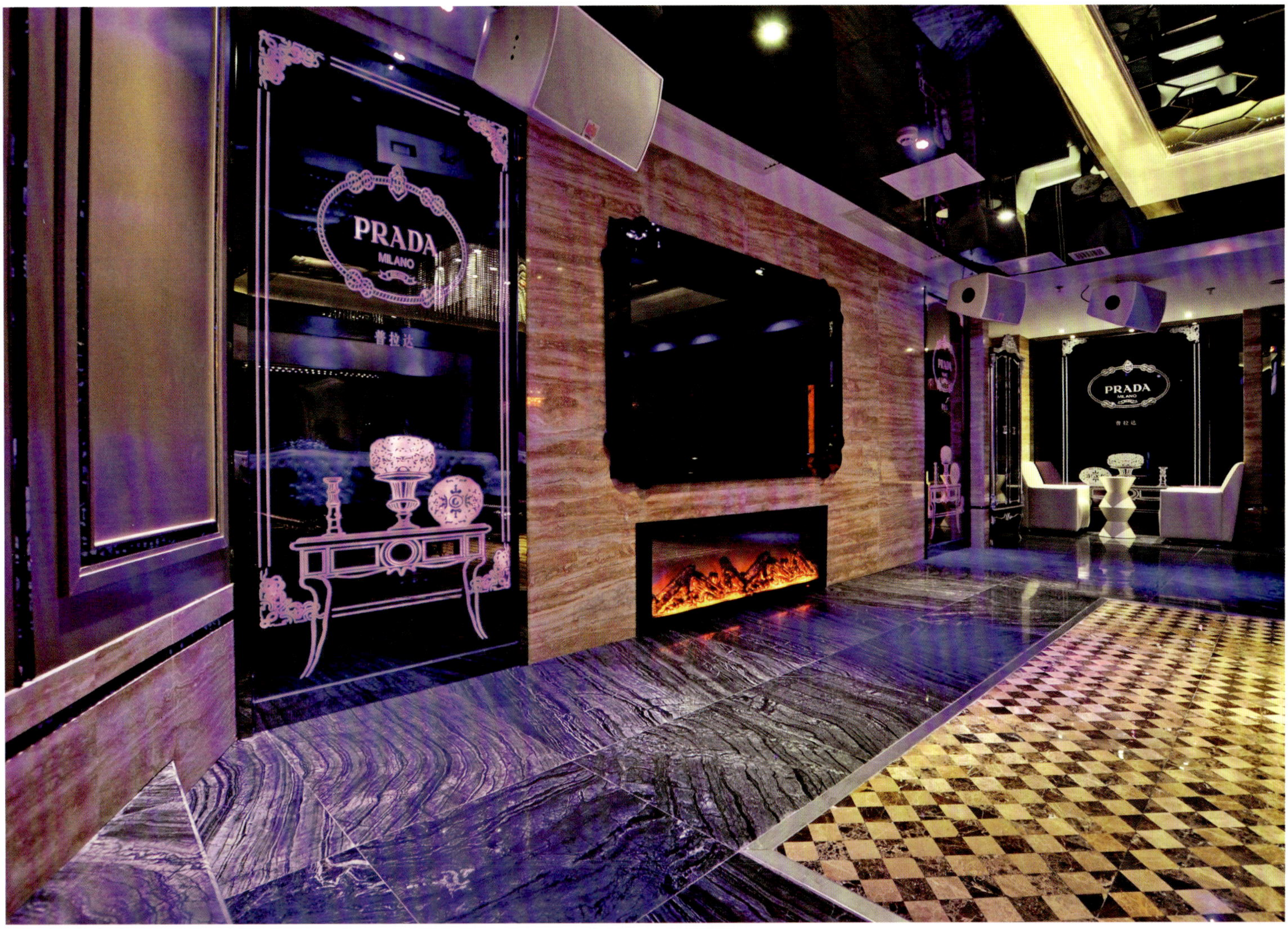
PRADA
MILANO
普拉达

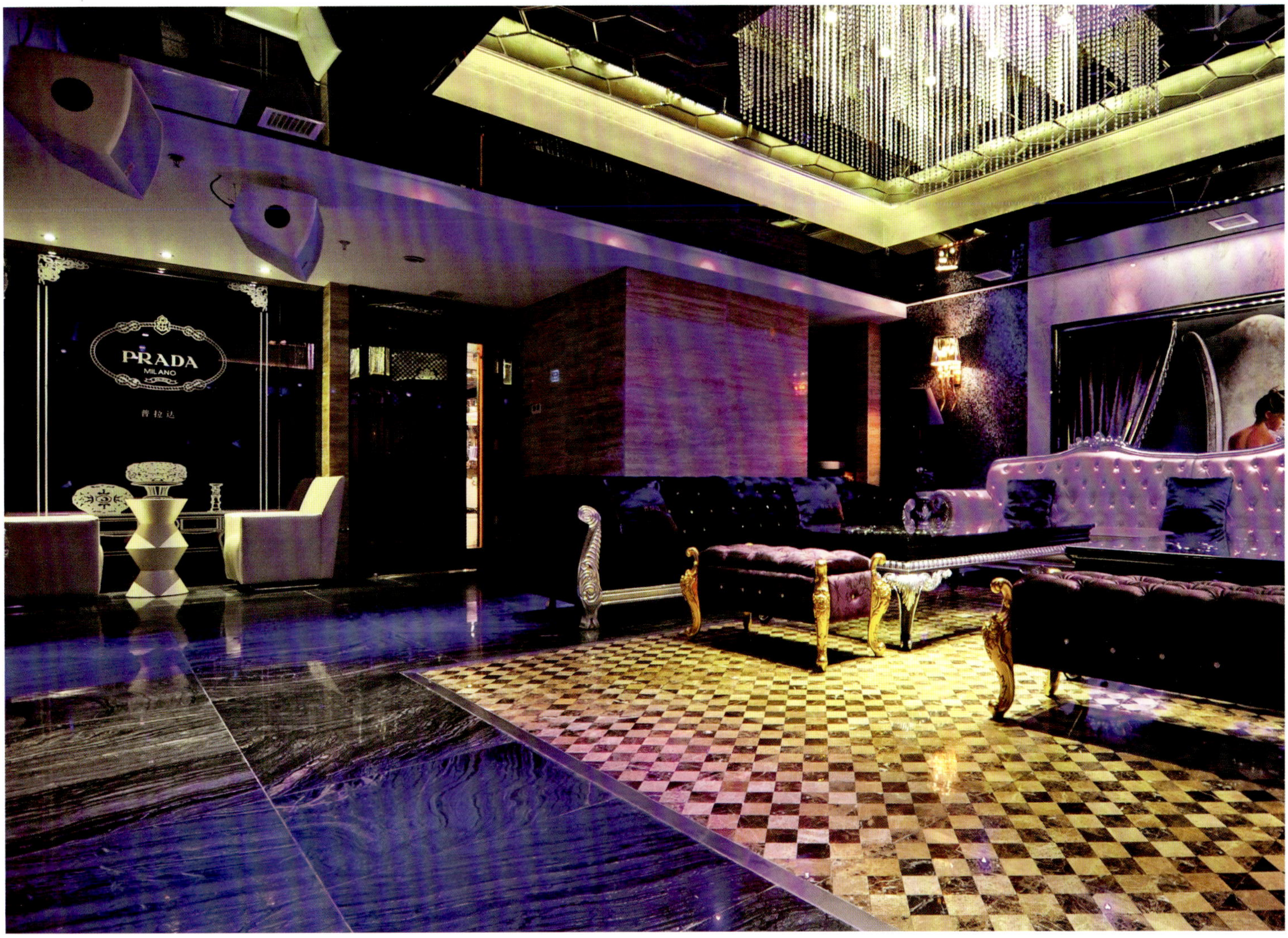
PRADA
MILANO
普拉达

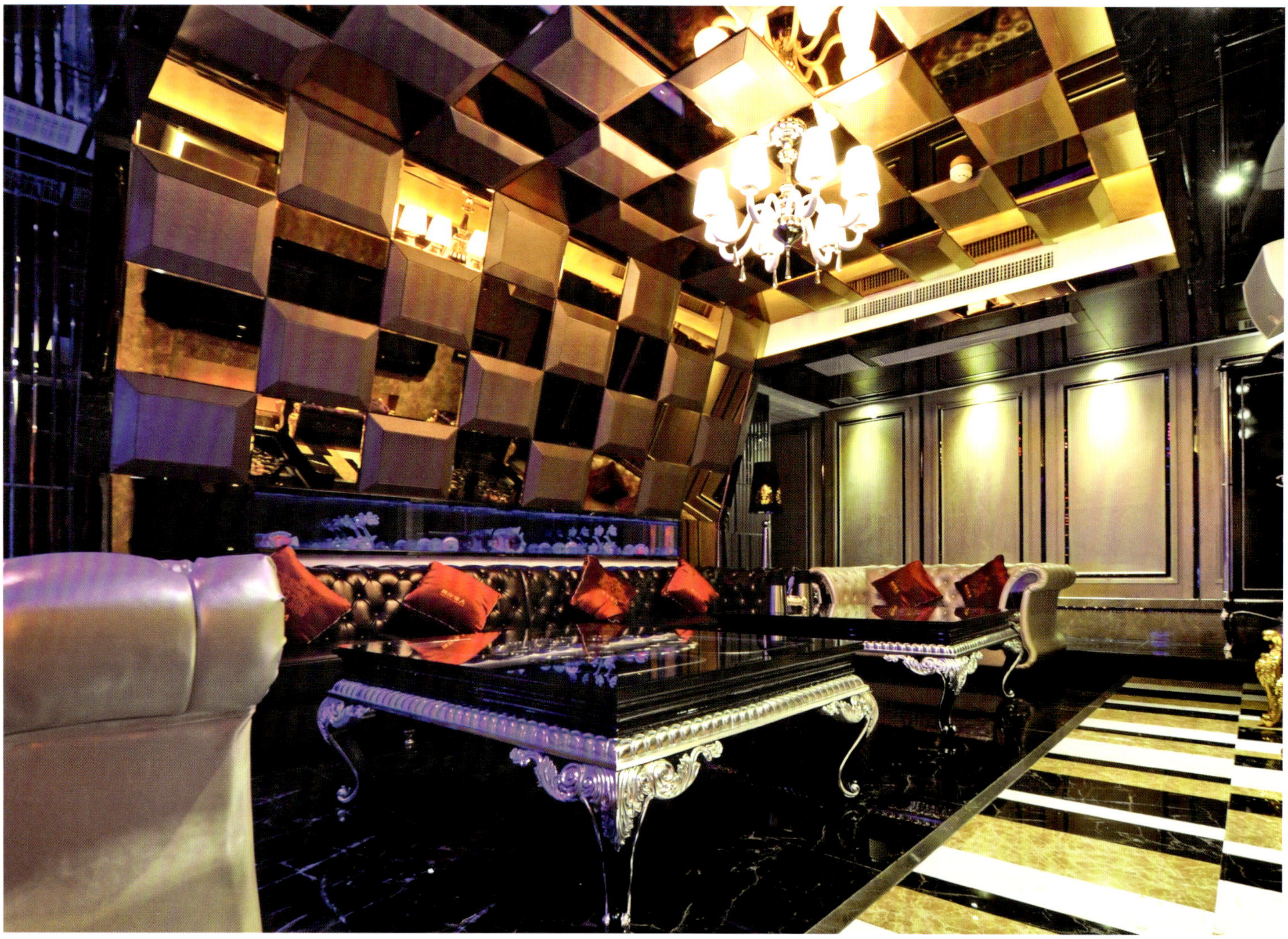

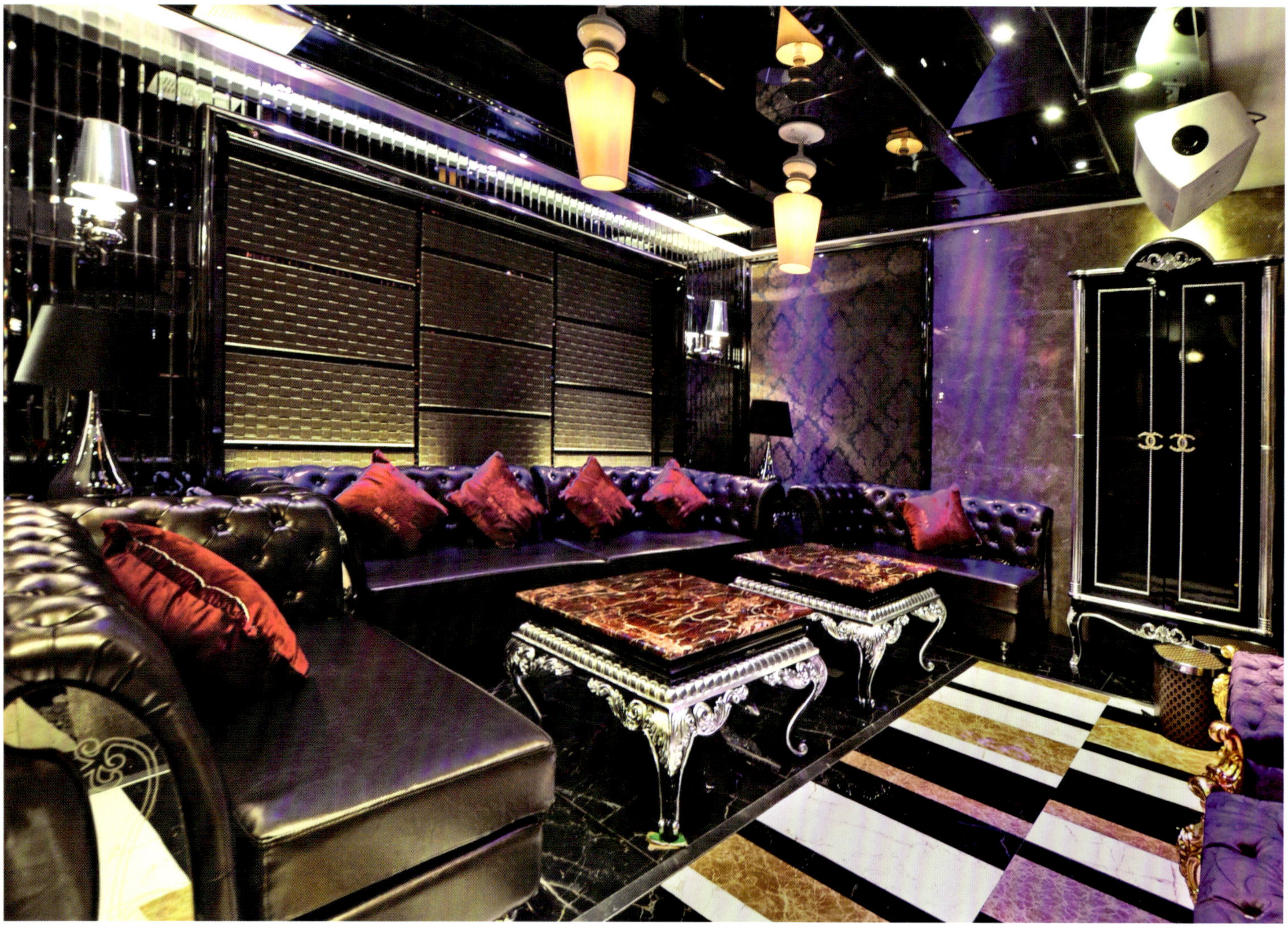

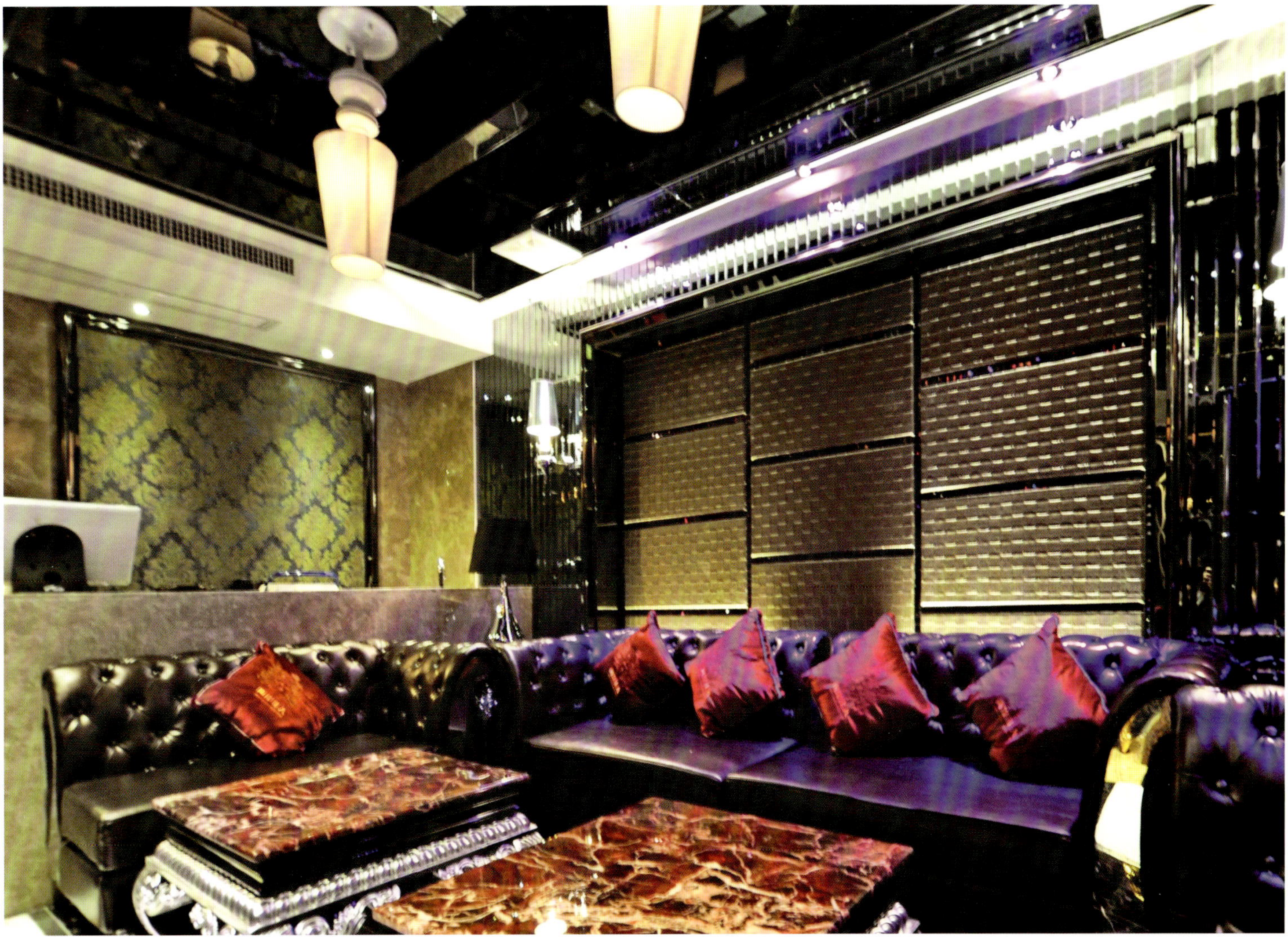

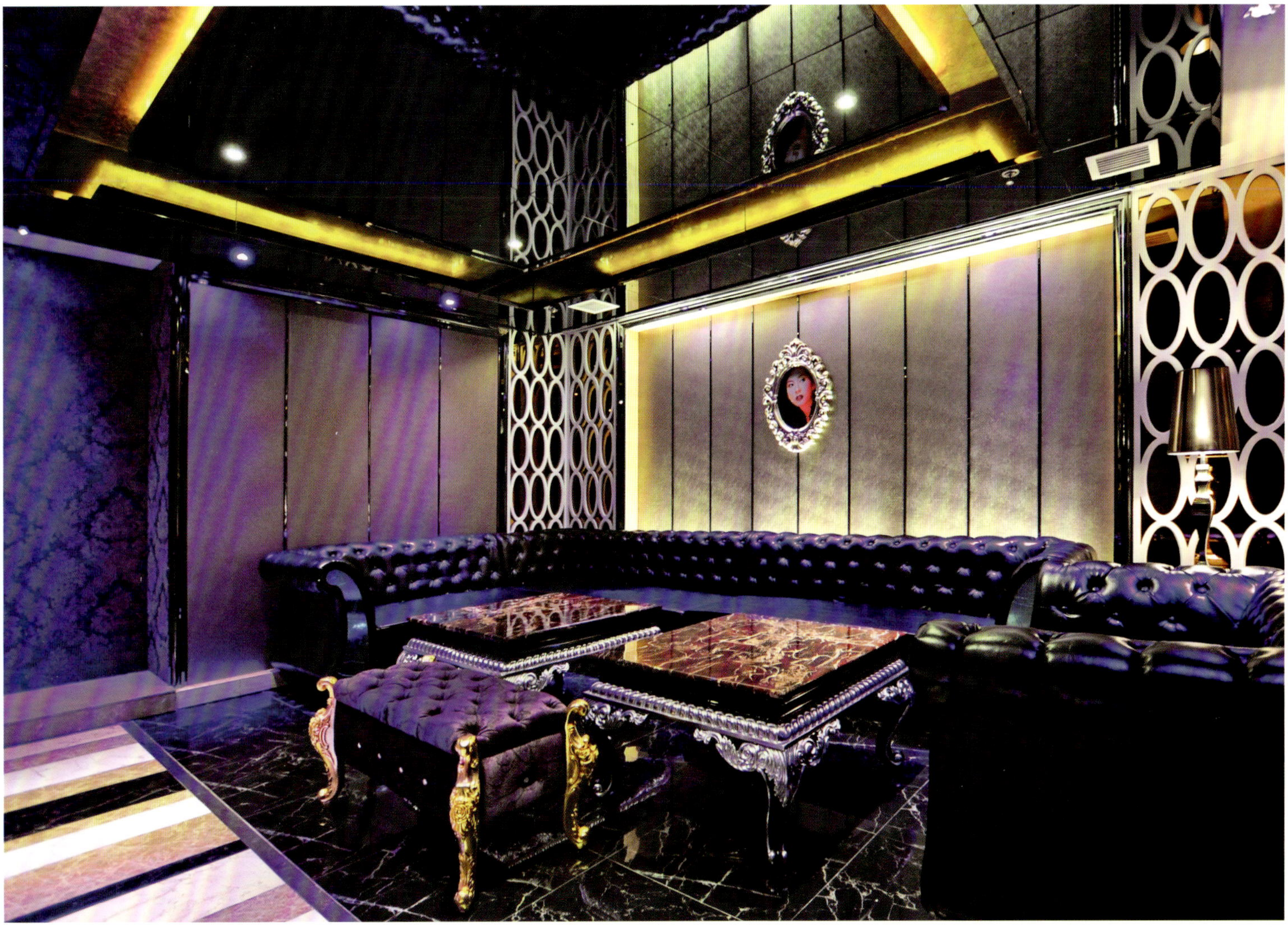

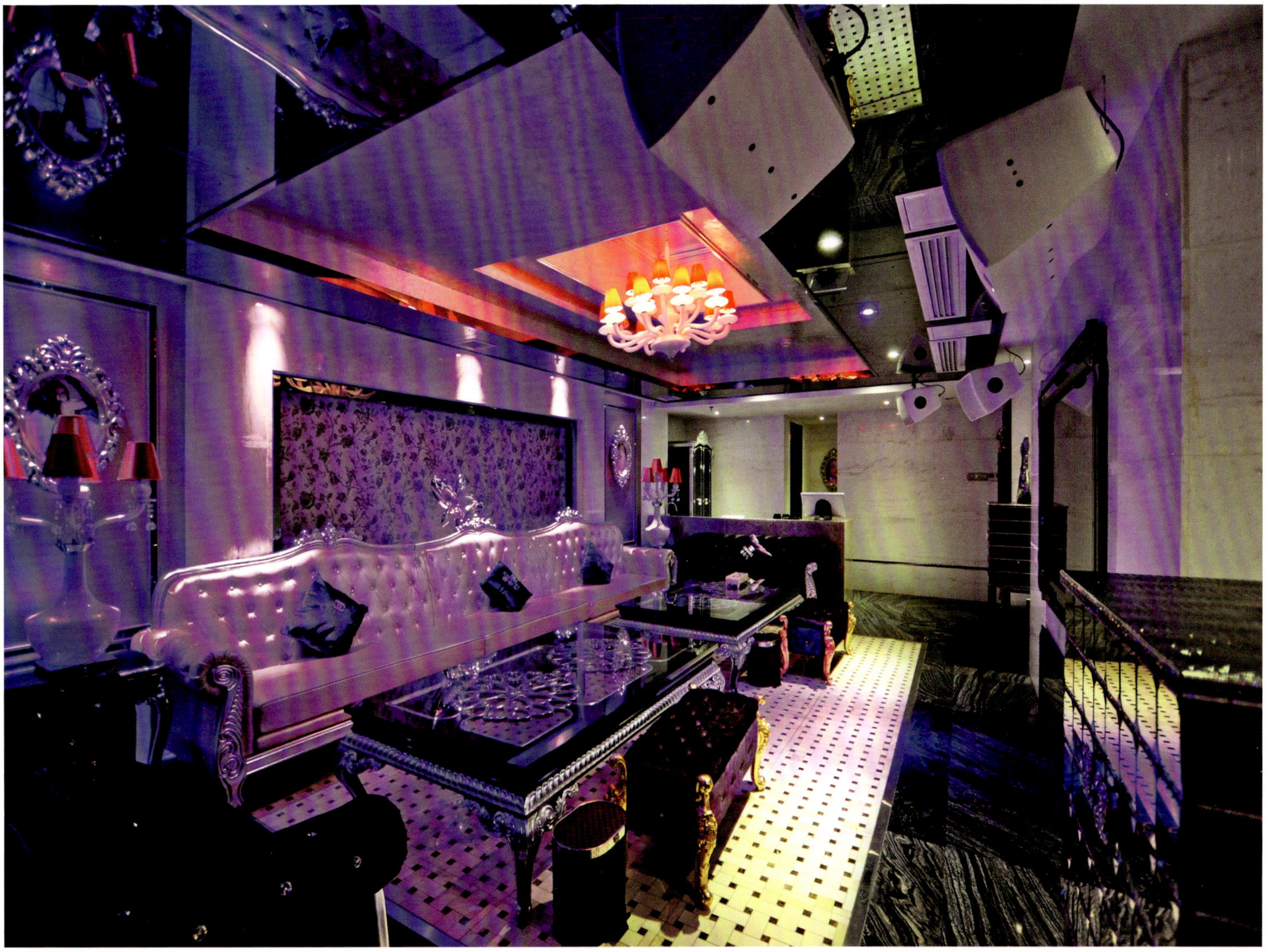

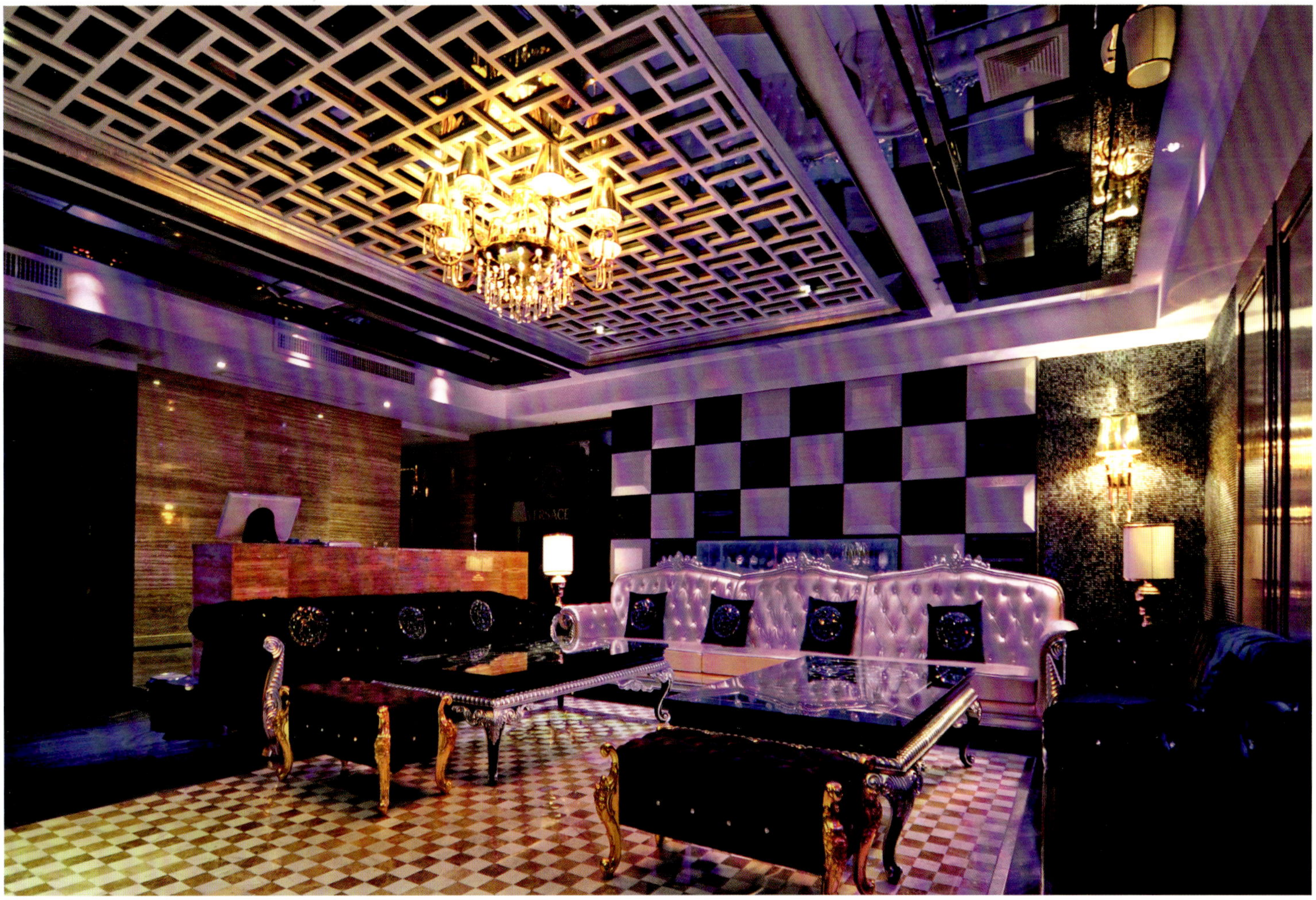

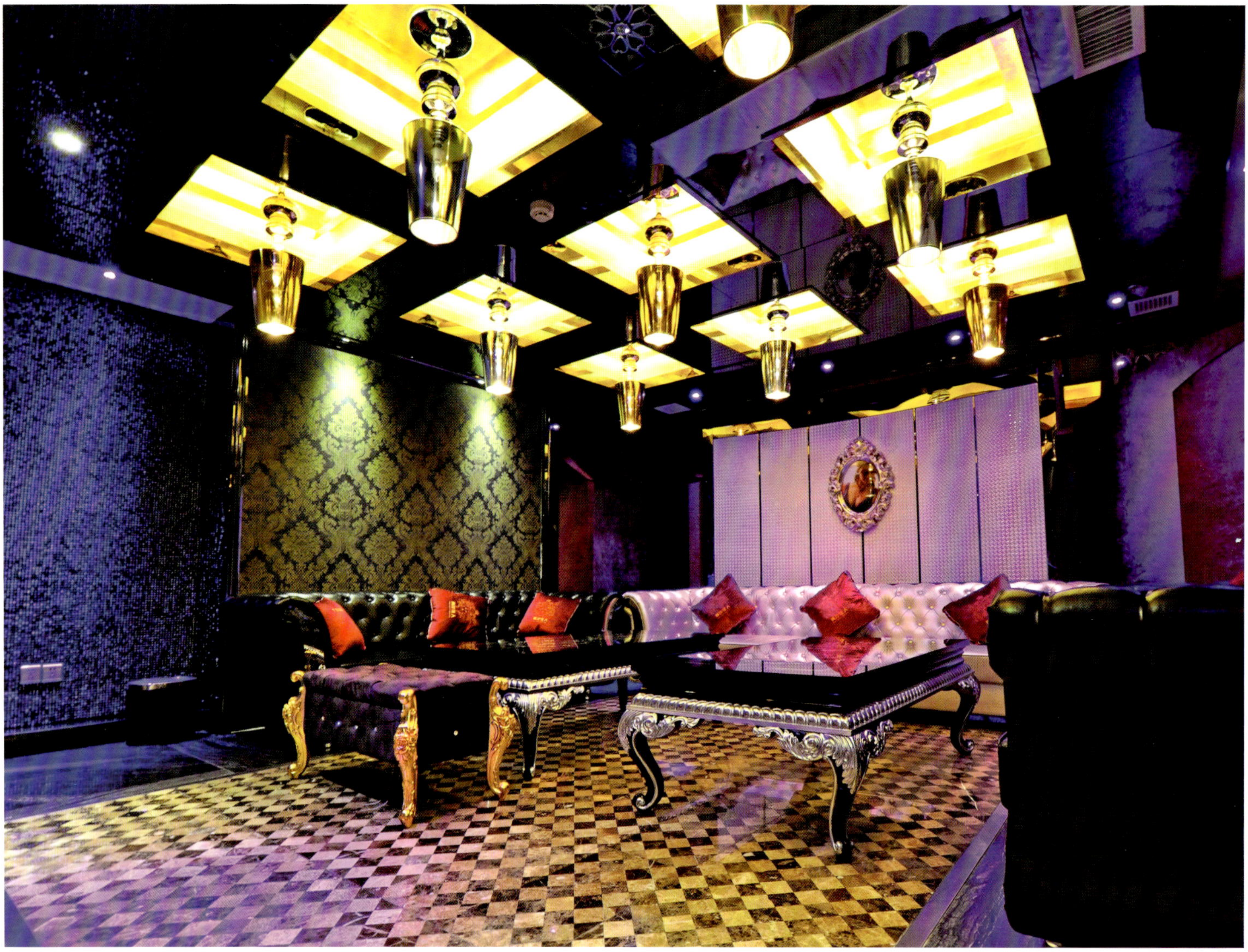

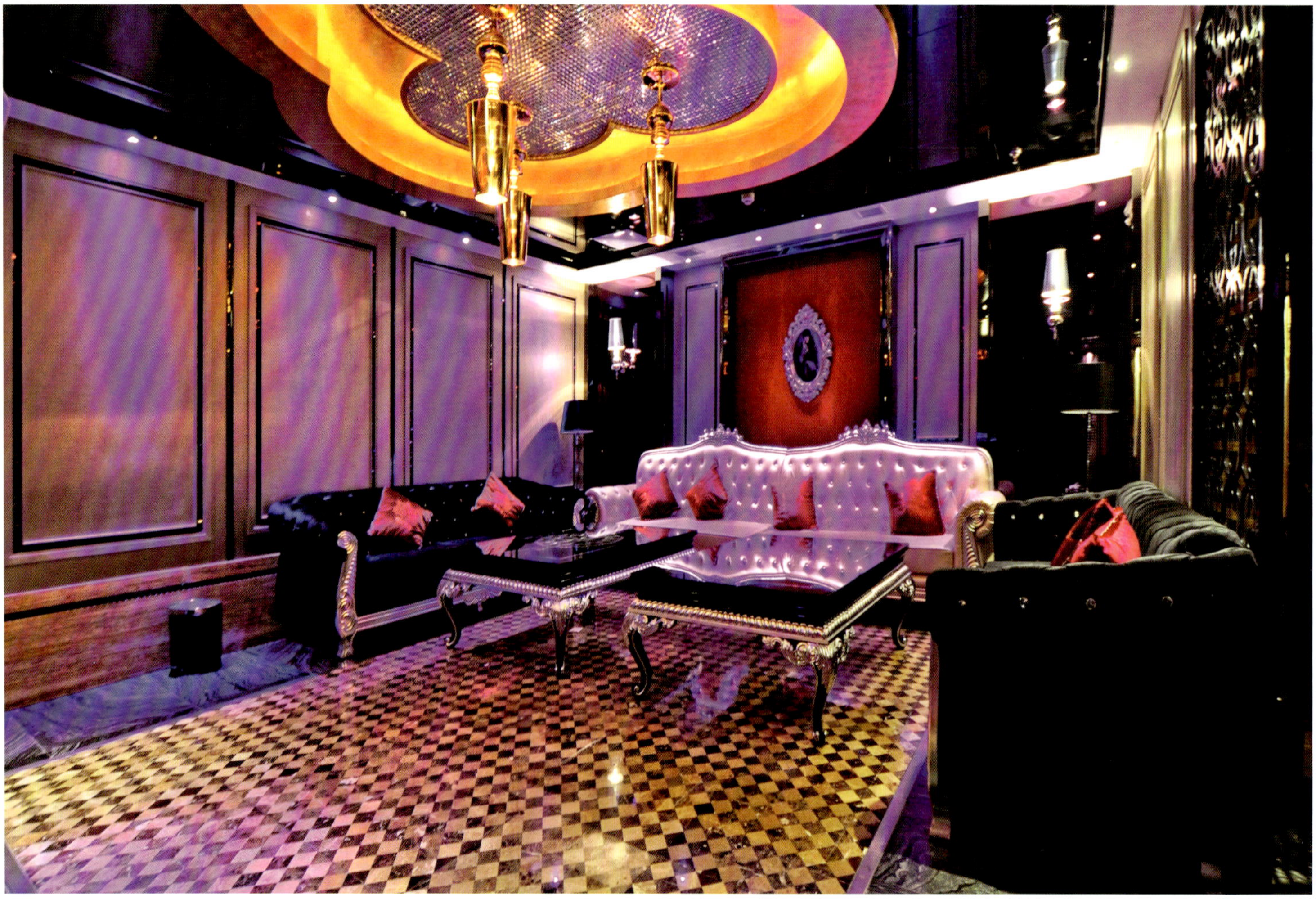

　　本案要打造的是面向都市高端消费人群的高档会所，因此设计师摒弃了金碧辉煌的装饰风格和迷离绚烂的夜场氛围，为消费者打造出一个明净高雅的休闲娱乐场所。

　　设计师以欧式新古典结合现代时尚风格来打造这个娱乐场所，没有绚丽的灯光烘托，也没有艳丽的色彩映衬，只有开阔明净的空间布局和优雅大气的家具组合，便将空间的气质和格调显示出来了。在空

间造型上，简化的罗马柱让空间更显开阔大气，结合简练的线条和恢弘华丽的吊灯，一个酷俏典雅的空间就此定型。家具方面，舒适华丽的沙发以金、银雕花镶边，在大理石地板和绒布墙纸的辉映下，流露出优雅气质，欧式壁画和图案的运用，更是将艺术气质隐于其间，让人感受高雅的夜店文化。

The World KTV

___ 深圳市大地KTV（歌唱中心）

The project is the first invested KTV place by Dadi Entertainment Company. The designer defines it as a good taste entertainment space with modern and fashionable artistic atmosphere. The space is mainly designed with black stone material, white carved panels, silver paint and plated mosaic, expressing a natural shape through materials of achromatic system, black, white and silver, such as the simplified tree shape and cell shape, easily meet the needs of modern fashion design.

The core concept of the designer and design team is effect-oriented, market-source and culture rooted, so the symbol of culture is indispensable. In the project, the designer selects the painting of Yue Minjun, a modernist and master, because his work is a sign of modernism, which is consistent with the modern and fashionable entertainment atmosphere pursued by the project, adding a little artistic charm into the space.

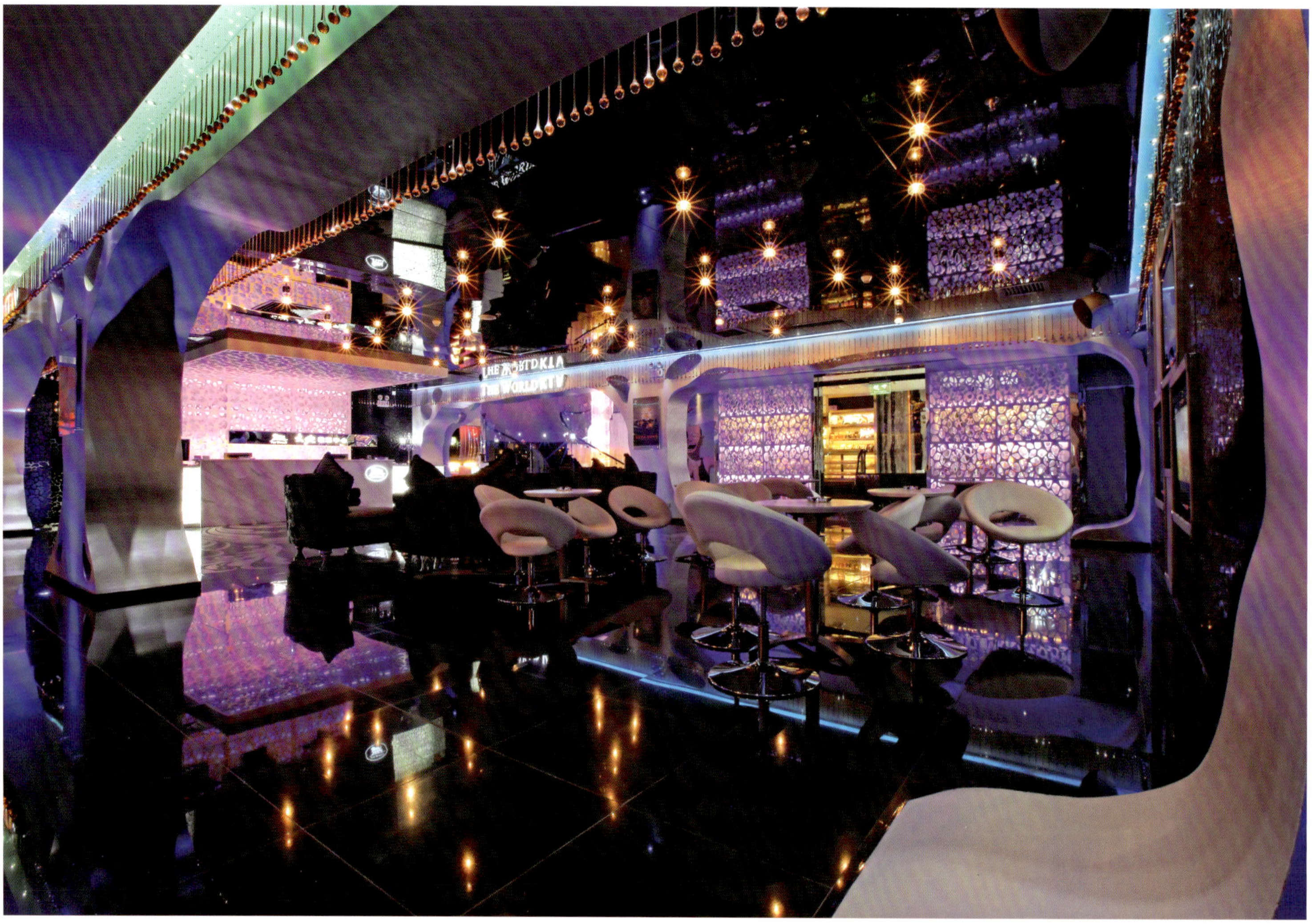

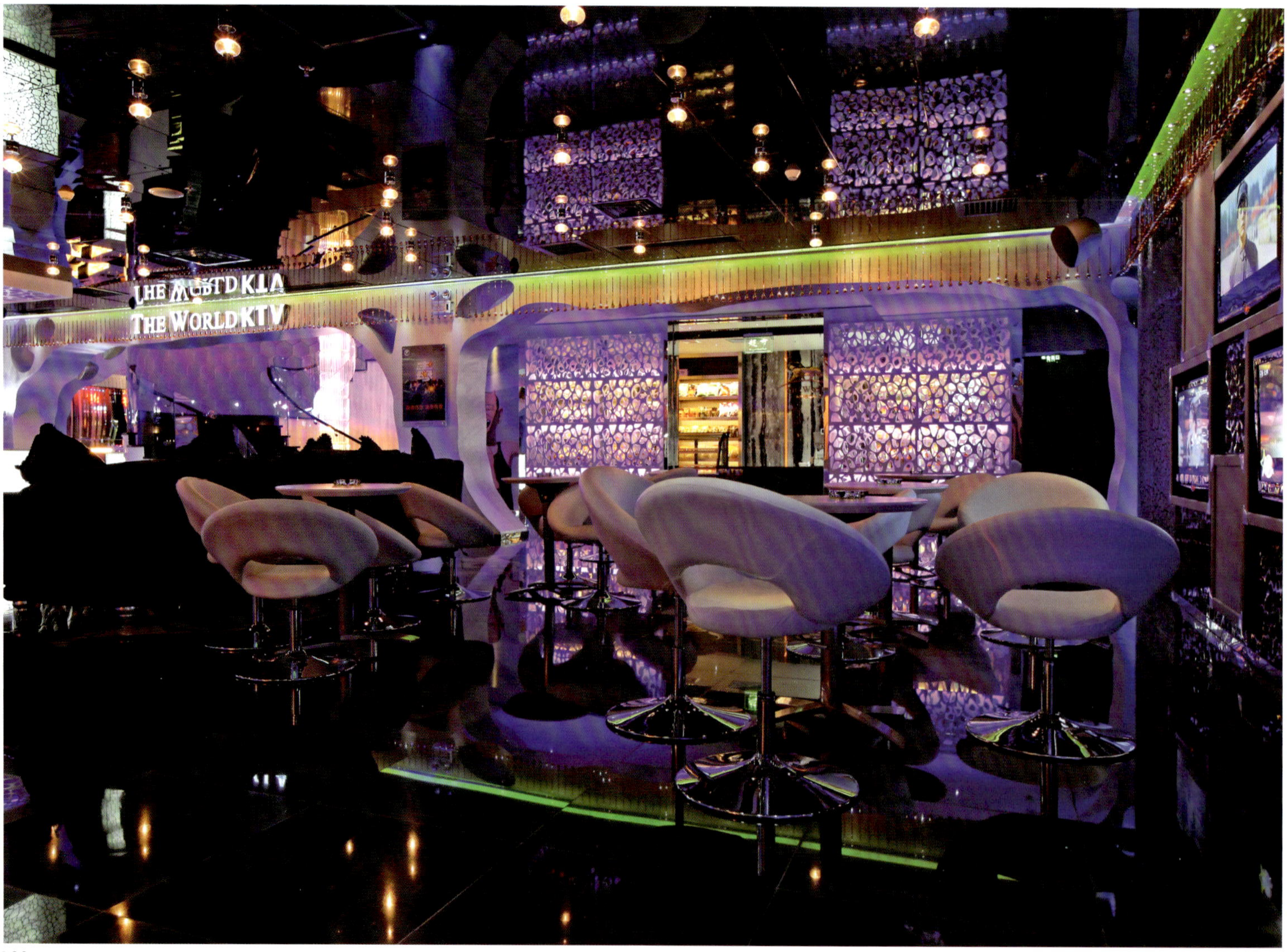

本案是大地娱乐公司首次投资的 KTV 场所。设计师对它的定位是：具有现代时尚艺术氛围的品位娱乐空间。该空间主要采用了黑色的石材、白色的雕花板、银色的油漆以及电镀的马赛克，通过黑、白、银等这些无色系的材料表现出自然的形态造型，比如简化树的形态和细胞的形态等，很轻松地完成了现代时尚的设计要求。

设计师及设计团队的核心理念为：效果为本，市场为源，文化为根，因此文化的符号是不可缺少的。在本案中，设计师选用了现代主义大师岳敏君先生的油画，因为他的作品是现代主义的标志，这与本案追求的现代时尚的娱乐氛围相吻合，同时也为空间增添了几分艺术韵味。

地点	面积	设计师	设计公司	主要材料
/ 广州	/ 1200m²	/ 吴恙	/ EMC（国际）室内设计公司	/ 瓷砖、墙纸、木、皮革、玻璃等

Music Country (M&C) KTV in Guangzhou

___ 广州市畅想国度KTV

In urban life, the fashionable entertainment space is the urban spectrum at night, pointing out the bustling area of the city. The project is a place like a vane of fashion, and is one of the essential chains in the urban entertainment life. As the design of KTV club, holding the principle of integrating elegance with worldliness, it enhances the environment to a higher level and creates an elegant and fashionable entertainment place with good taste. In the design of the entrance, the white columns in the wall are coupled with transparent glass to create a grand and generous atmosphere, and clear glass is matched with the gate design of white balls to make the interior space looming, but full of fashionable flavor. The reception hall is designed to be a central place in ring for customers, softening the space and making service convenient. In the design of the space, the use of mosaic, wallpaper, silver, leather, wood, glass and other materials allows the space to show out a concise and elegant temperament, coupled with the echo of light and colors, to bring some fascinate taste.

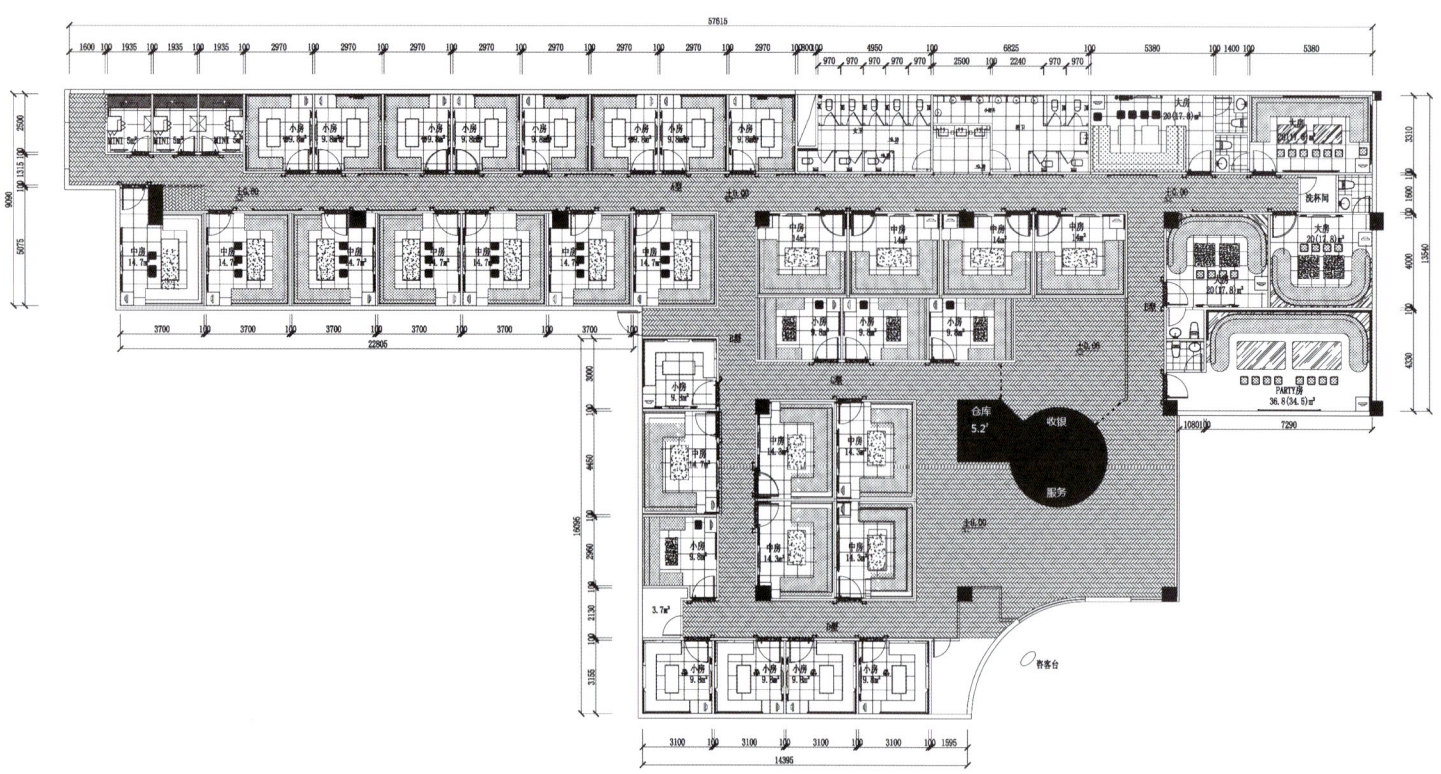

在都市生活中，时尚娱乐空间就是夜间的城市光谱，点出城市繁华地的所在。本案就是一个类似时尚风向标的所在，是都市娱乐生活必不可少的链条之一。

本案作为KTV场所的设计，秉着雅俗共赏的原则，将环境提升一个层次，打造一个优雅、时尚兼具品位的娱乐场所。入口处的设计，设计师运用白色墙柱结合透明玻璃营造出恢弘大气的氛围，清玻璃结合白色圆球的门庭设计让室内空间若隐若现，却又充满时尚气息。接待大厅以环形中岛设计面向顾客，既柔化了空间，又方便了服务。在空间的设计上，马赛克、墙纸、银箔、皮革、木、玻璃等材料的运用，让空间呈现出简约而高雅的气质，加之灯光与色彩的互相辉映，更添几分迷离情趣。

地点	面积	设计师	设计主管	设计公司	主要材料
/ 湖北	/ 3200m²	/ 郭友国、丁鹏	/ 程刚	/ 武汉逸臣装饰设计有限公司	/ PVC 雕花、黑镜、直木纹、亚克力灯片、黑钛钢、皮质硬包等

No.1 Mansion KTV

一号公馆KTV

It is a total of 3200 square meters, featuring four floors: the hall on the first floor, business area on the second floor in an elegant and refined style, VIP area on the third floor in a luxurious and distinguished style only to VIP customers, and entertainment area on the fourth floor with an emphasis on fashion and quality.

The large light column surrounded by patterns in spiral on the first floor extends to the atrium ceiling on the second floor and transforms to be a bloom of golden petals, leading to the second floor design with the theme of flowers. The huge golden peony on the aisle wall looks like a smiling face, greeting each distinguished guest in an elegant and refined style.

The third floor looks like a palace of art, where the murals, dome, exquisite chandeliers, and the light wall carved with gorgeous crystal show out the exquisitely arranged artistic sculpture, allowing the space full of luxurious and noble temperament, so the VIP customers can experience an elegant service.

On the fourth floor, the space is divided with the introduction of curves and irregular way, and a lot of glass and translucent wall refract mutually, setting off the fiber-optic light source on the top, to create a fashionable and fantasy entertainment space and allow everyone hanging around enjoy themselves freely.

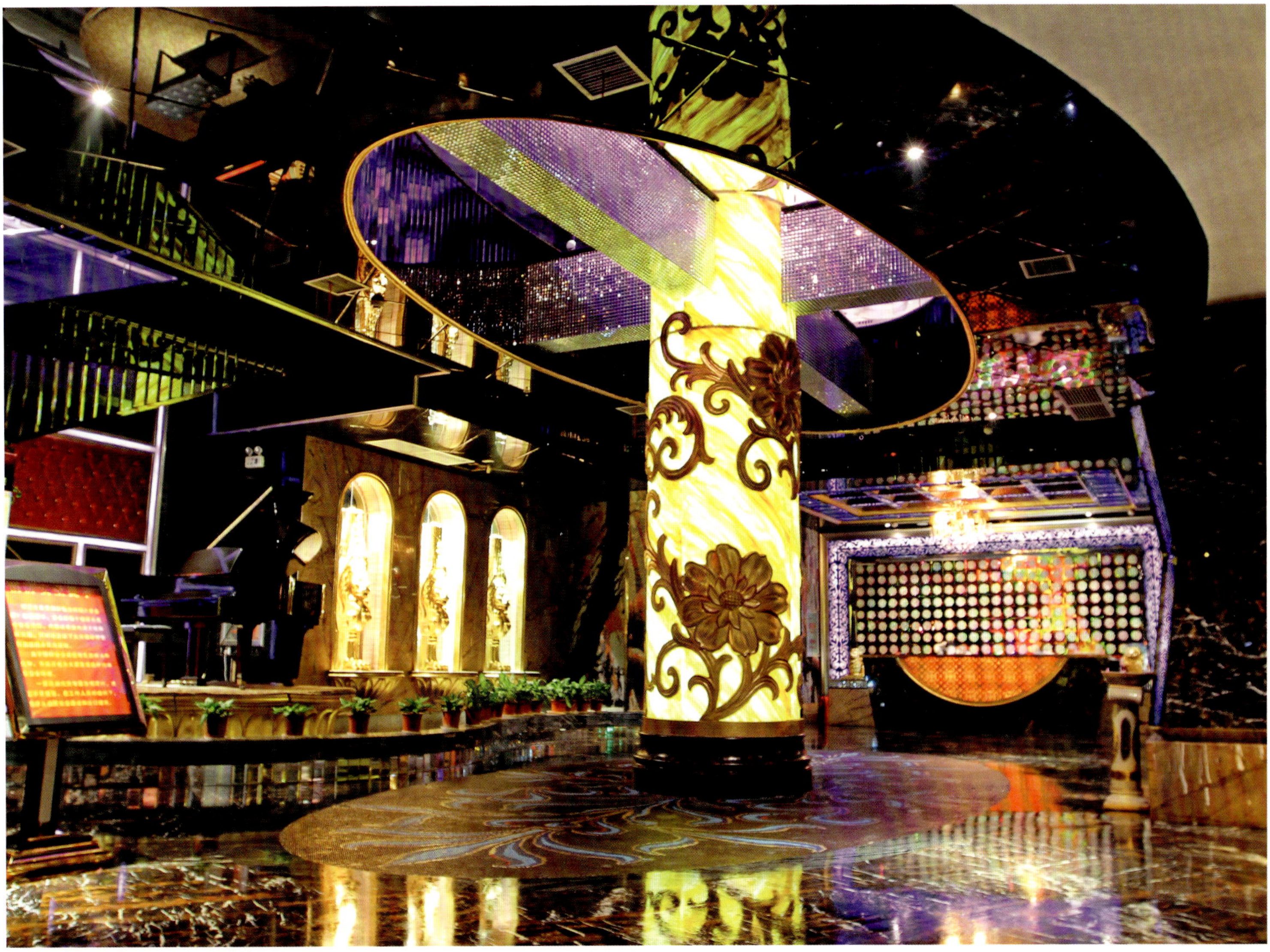

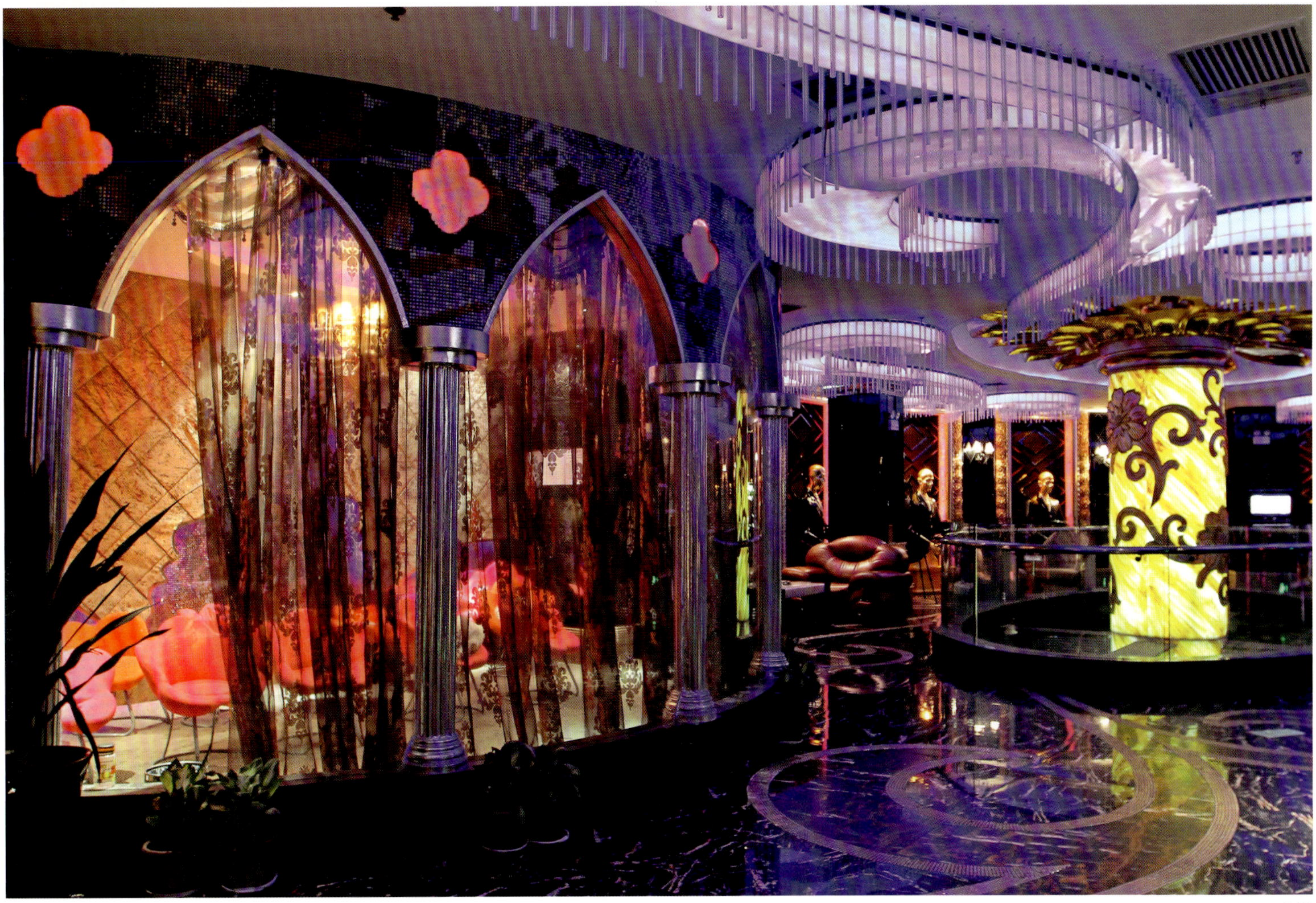

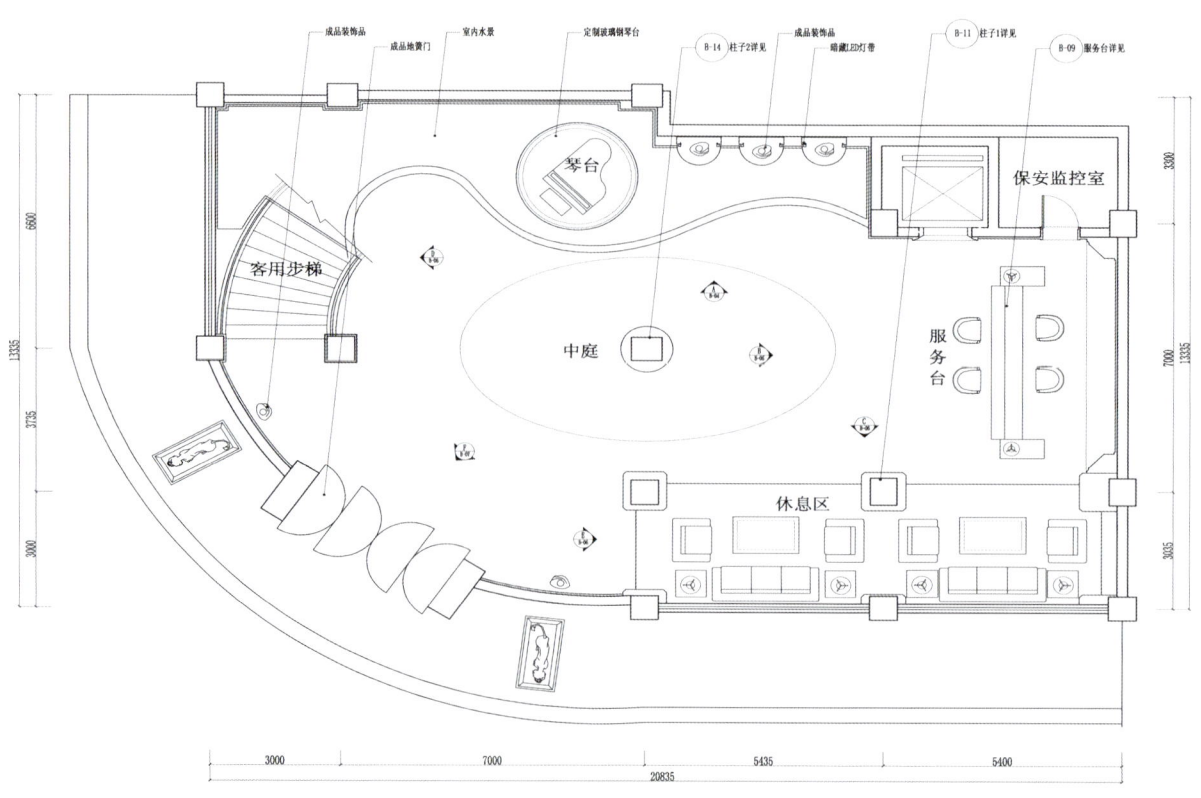

成品装饰品　成品地弹门　室内水景　定制玻璃钢琴台　B-14 柱子2详见　成品装饰品　暗藏LED灯带　B-11 柱子1详见　B-09 服务台详见

琴台

保安监控室

客用步梯

中庭

服务台

休息区

6600　13335　3735　3000

3300　7000　13335　3035

3000　7000　5435　5400
20835

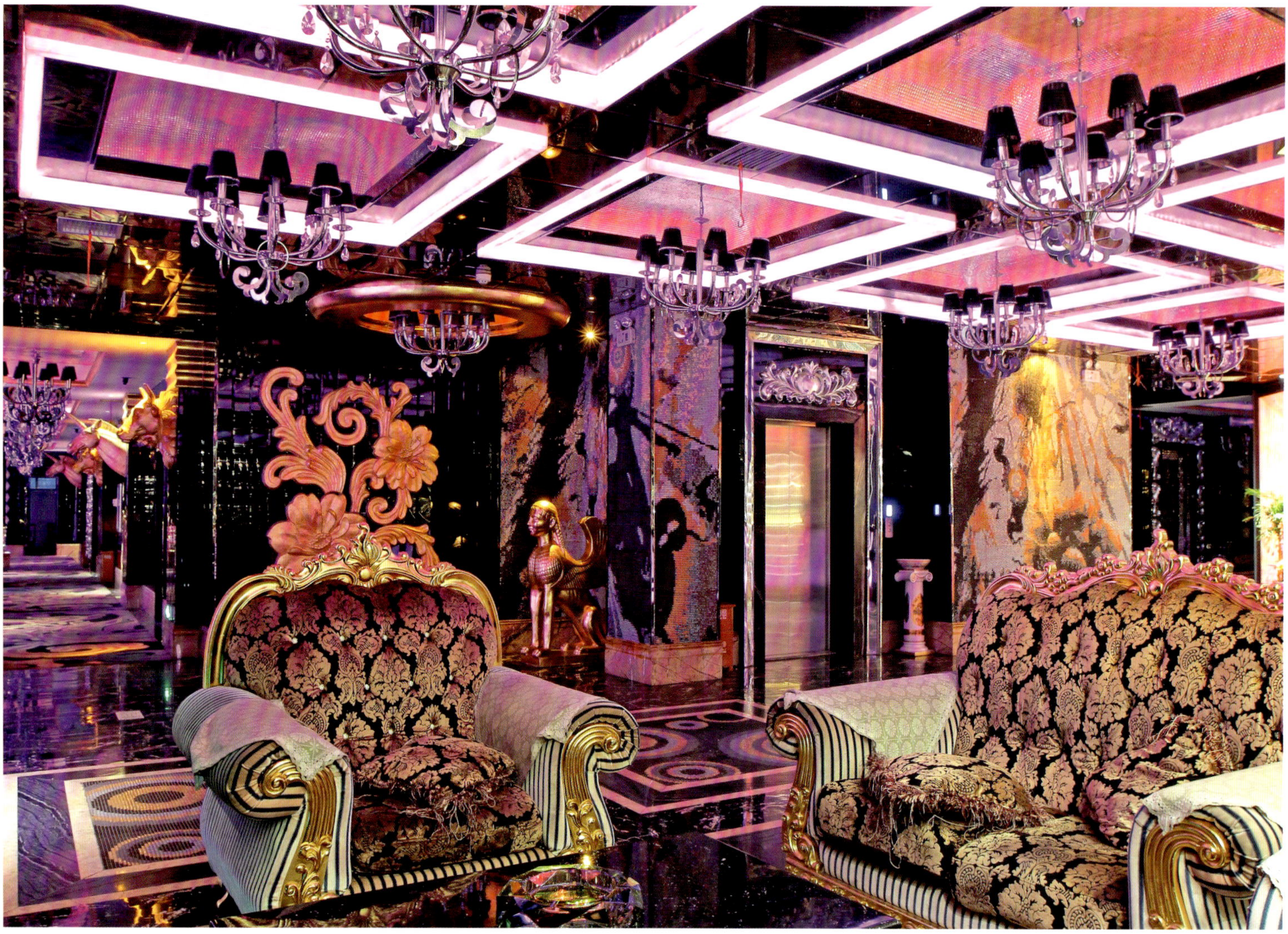

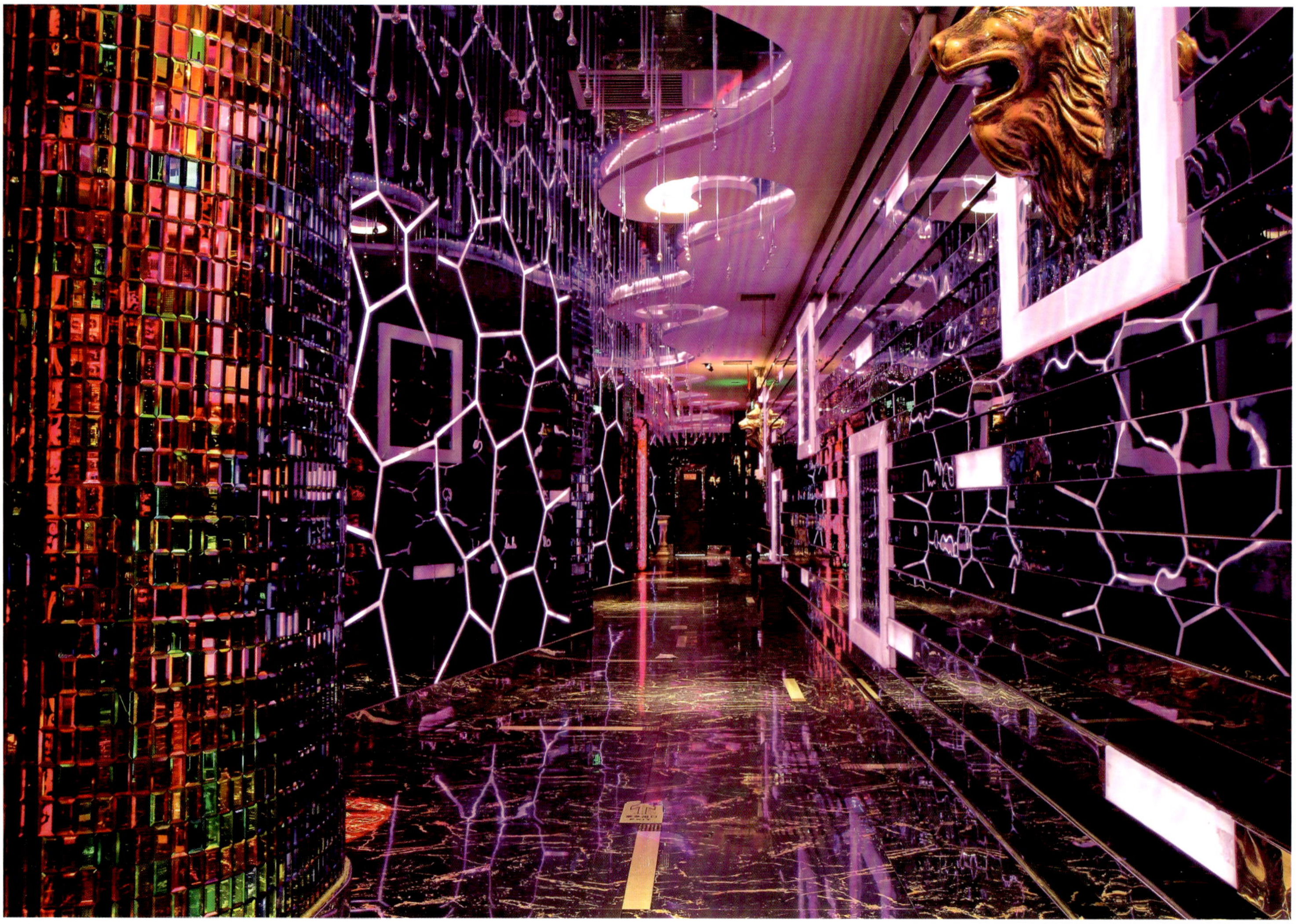

包房2
36.2 M²
包房3
37.1 M²
包房4
41.5 M²
包房5
包房6
33.2M²
包房7
33.2M²
技师房
14.6M²
技师房
22.7M²
下
包房8
56.3M²
包房10
42.8M²
包房11
44.7M²
包房12
31.3 M²
包房13
41.7 M²
包房14
51.7 M²
2100
5700
6600
5400
5400
6600
3735
1565
37100
3000
7000
2400
7000
1600
21000
1565
7035
5400
5400
6600
5700
6000
3600
1450
42750

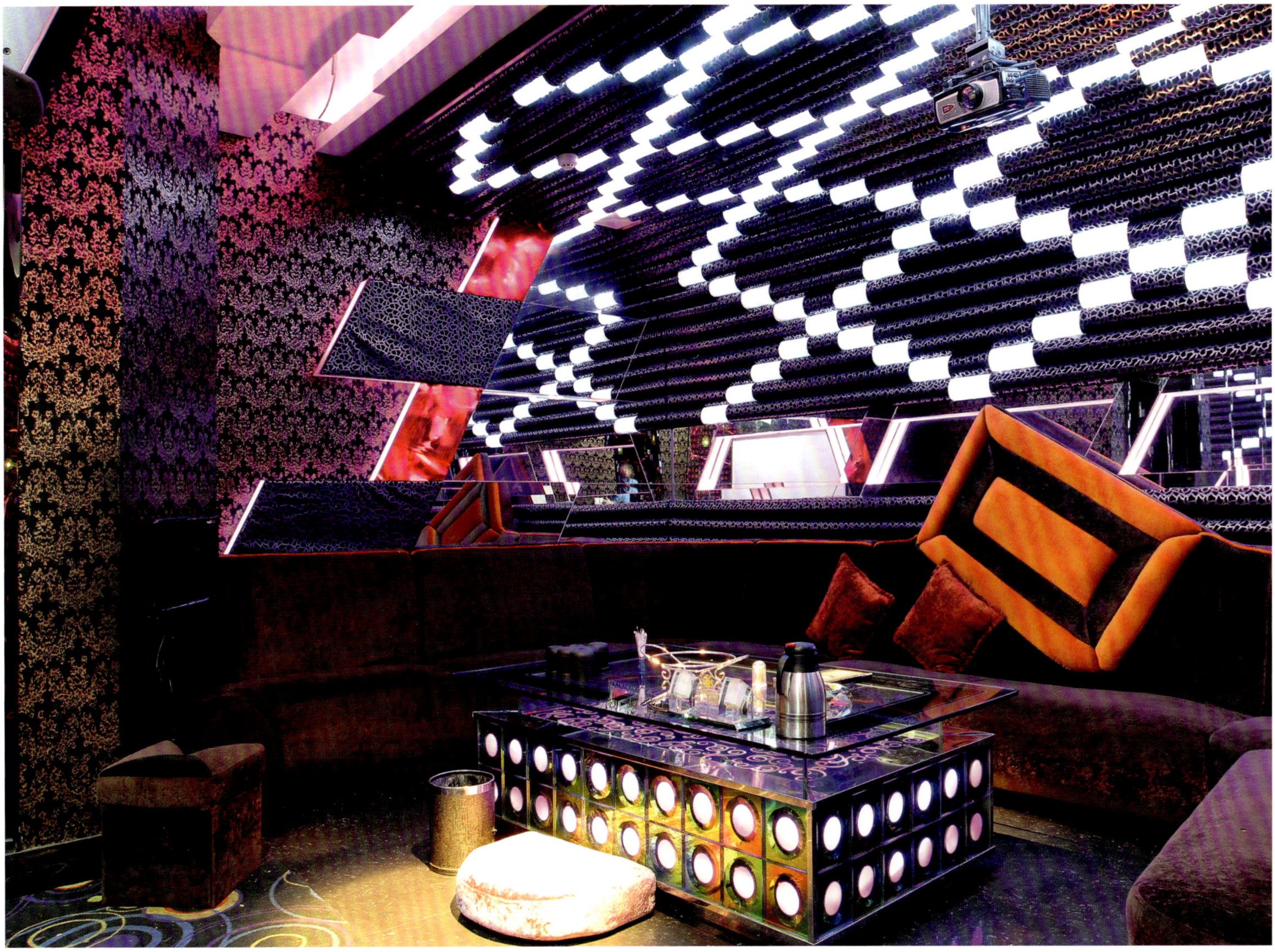

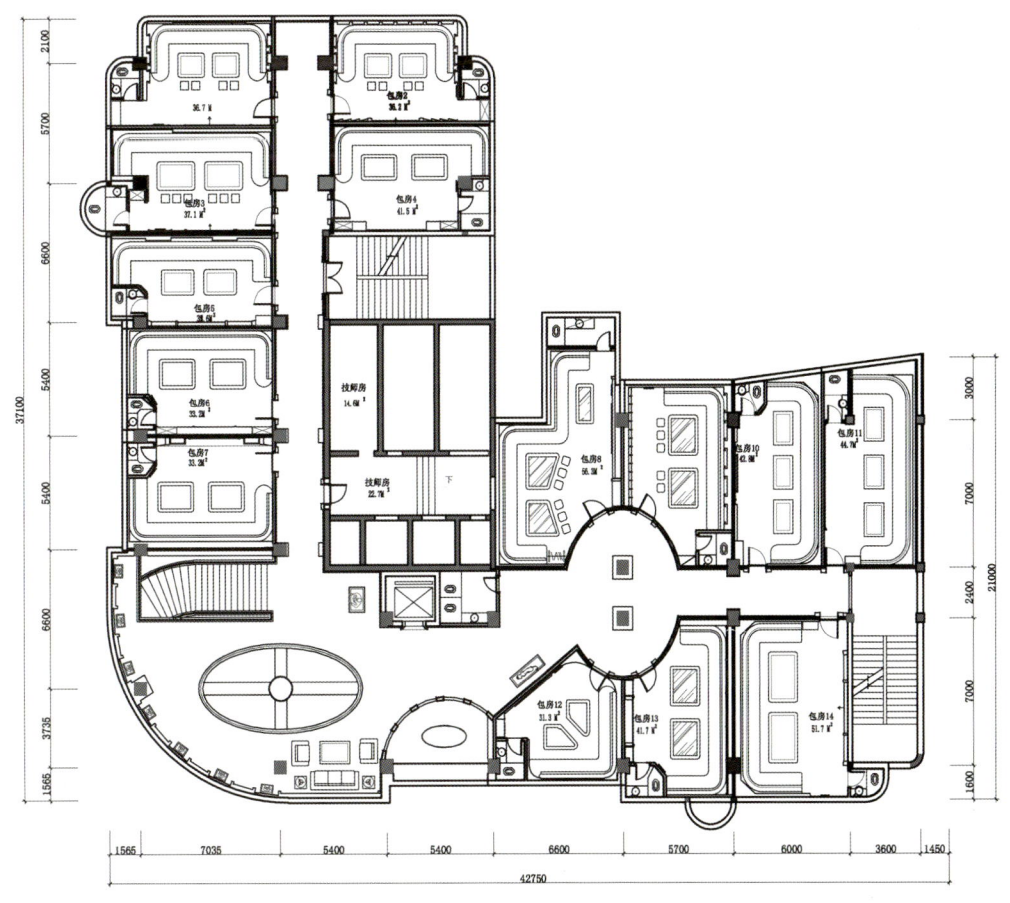

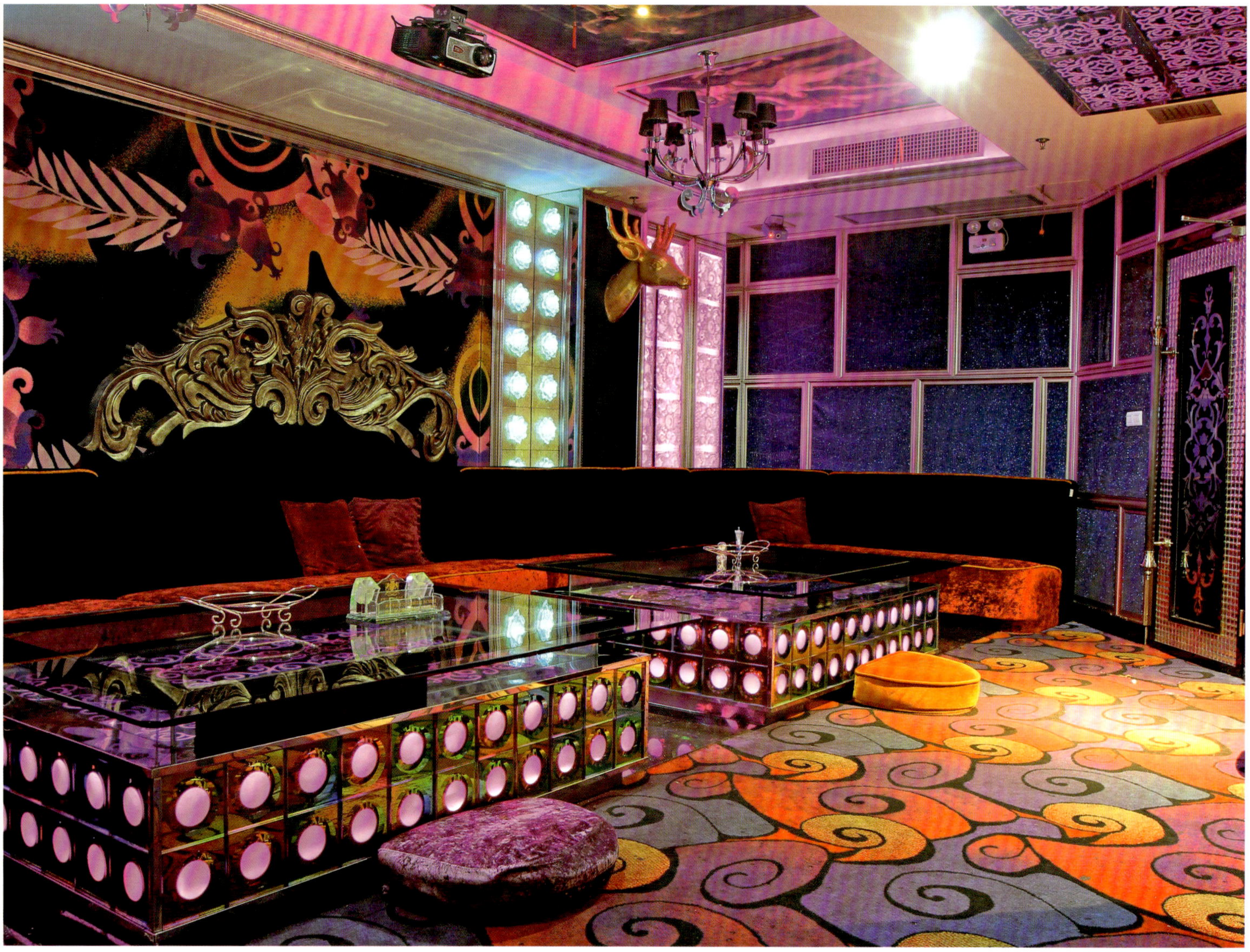

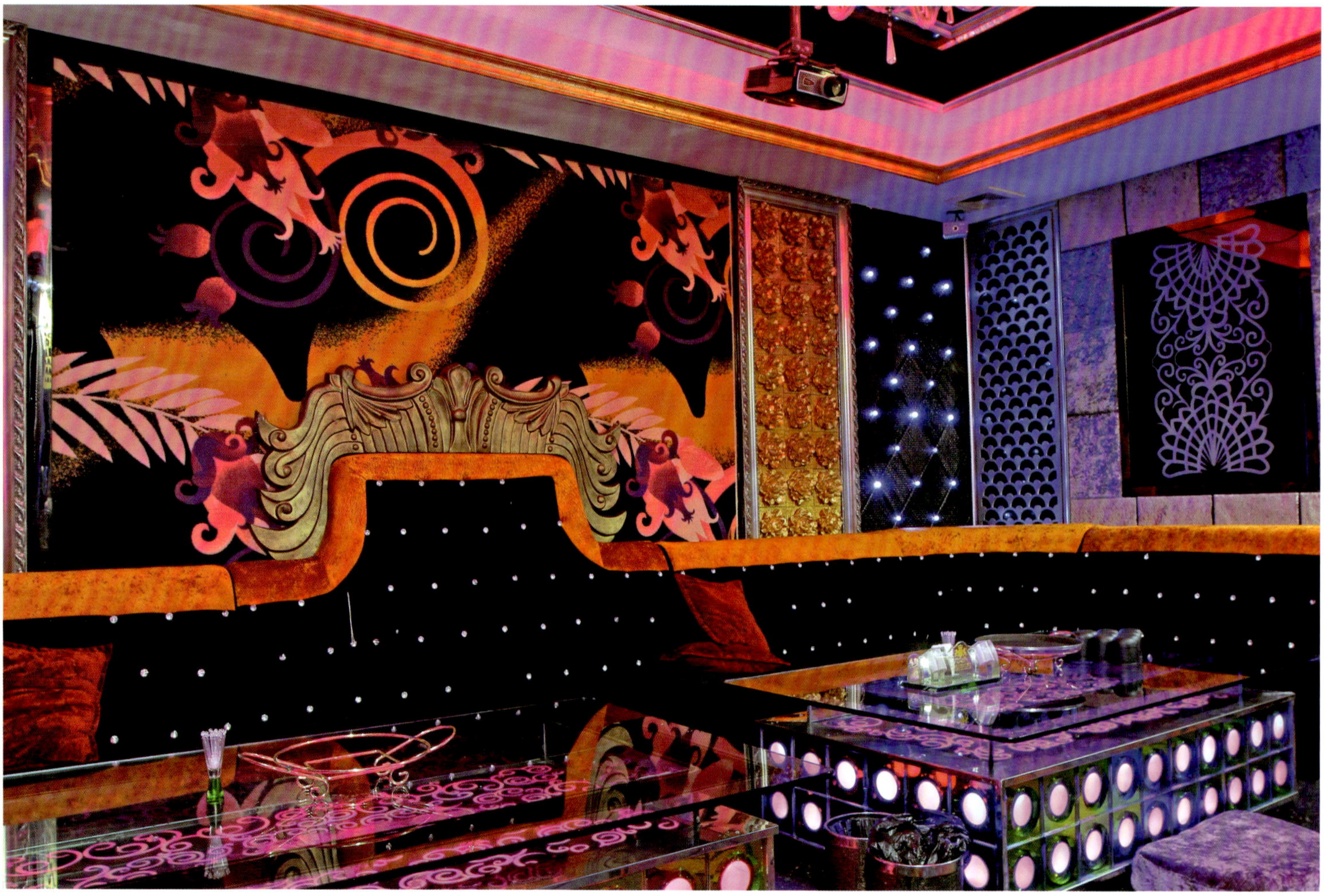

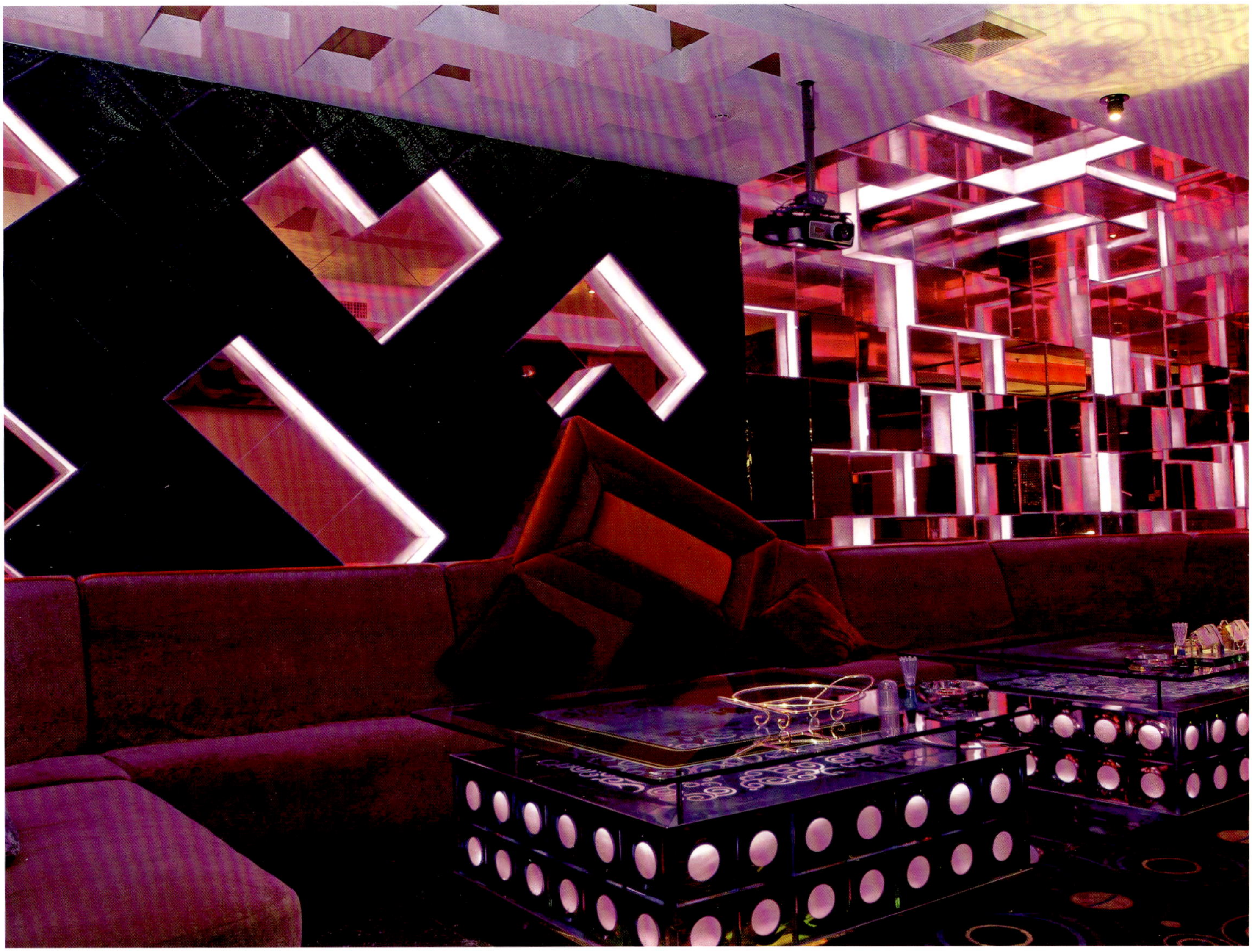

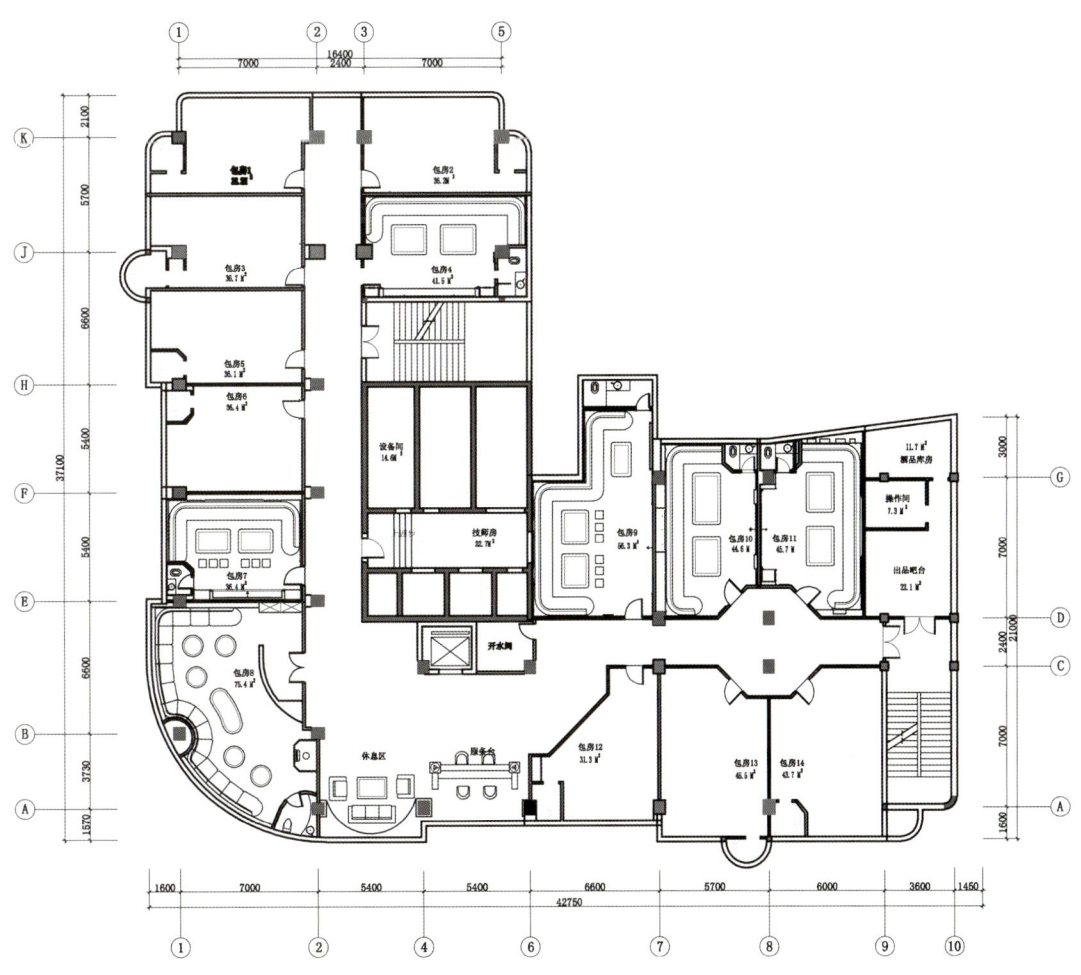

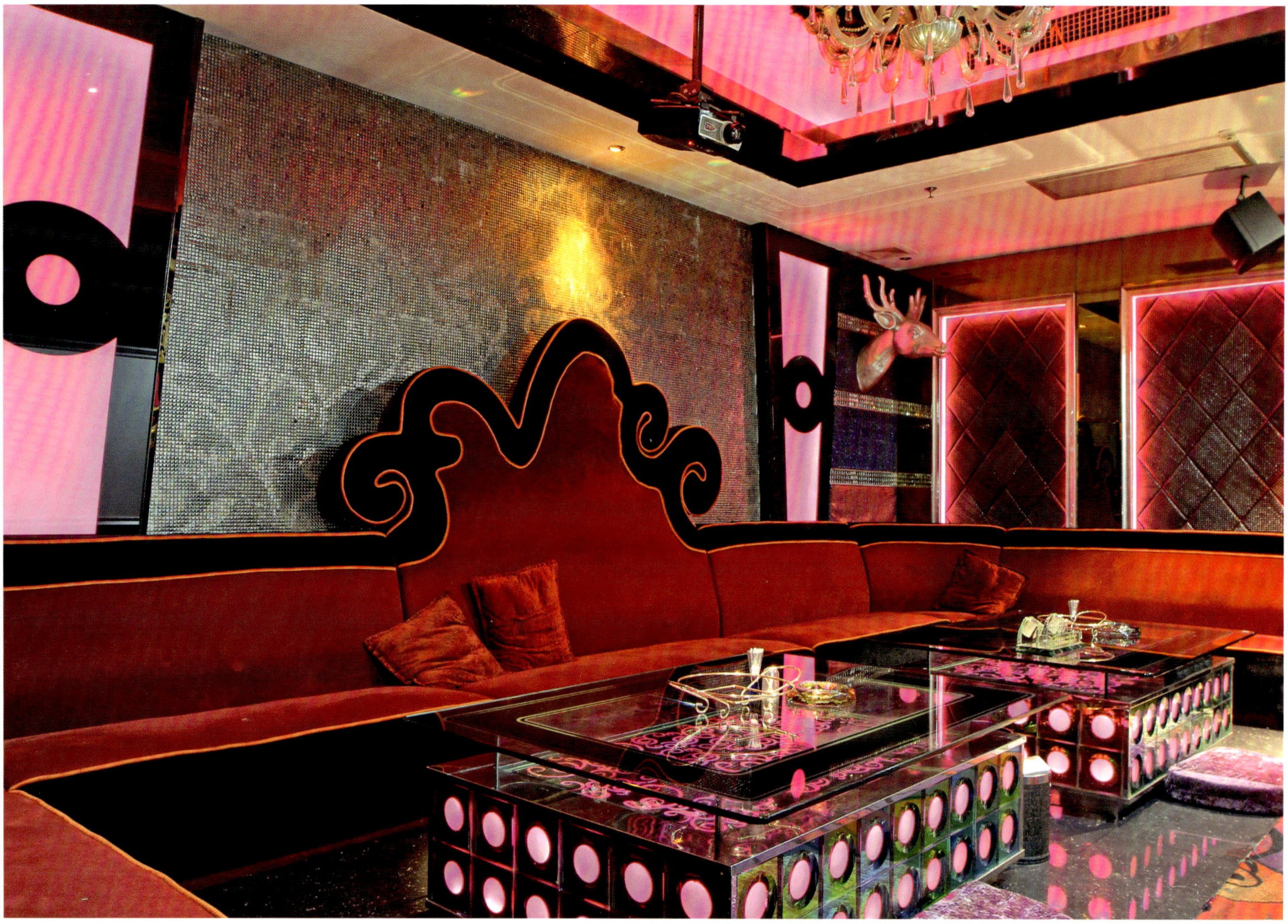

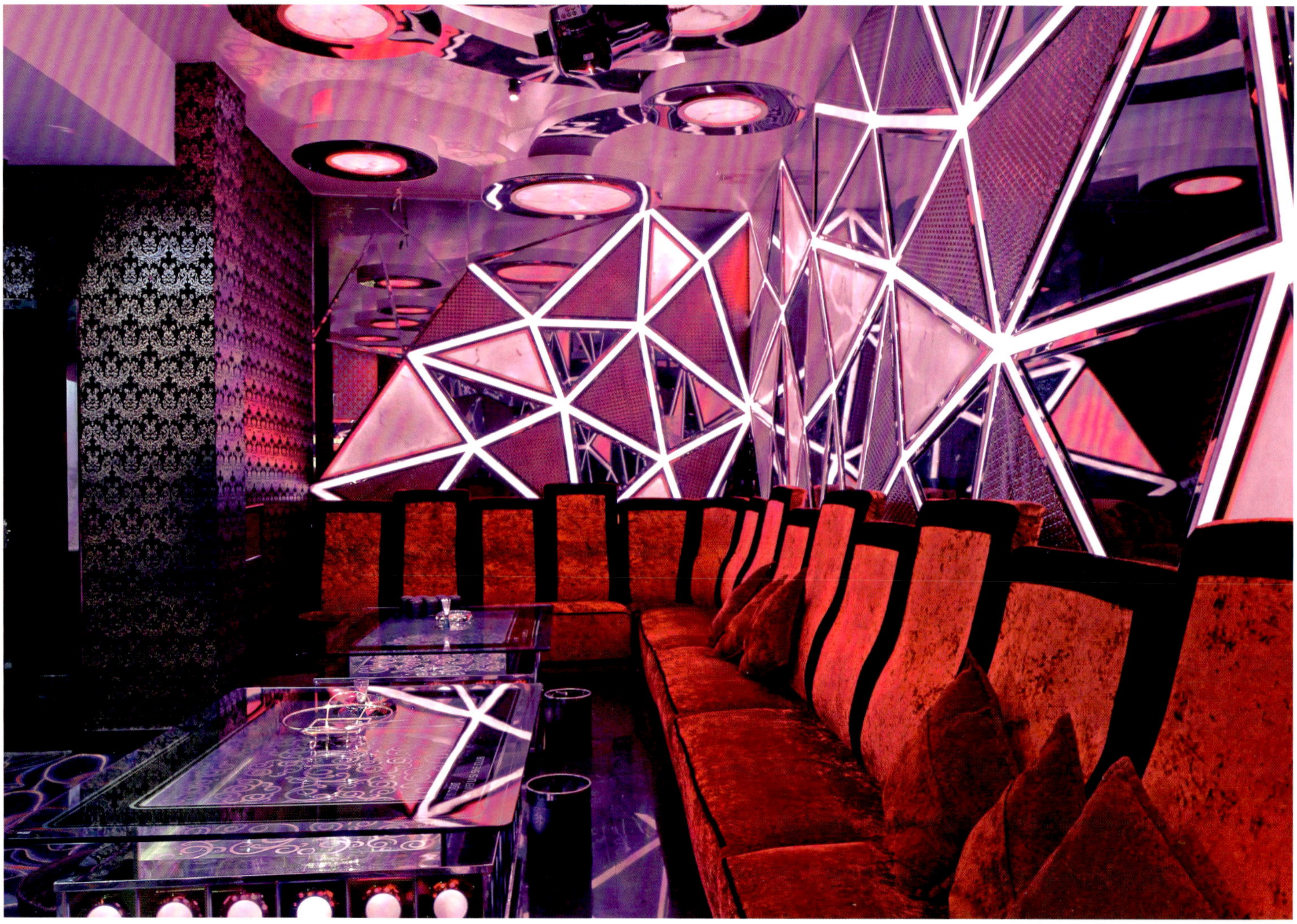

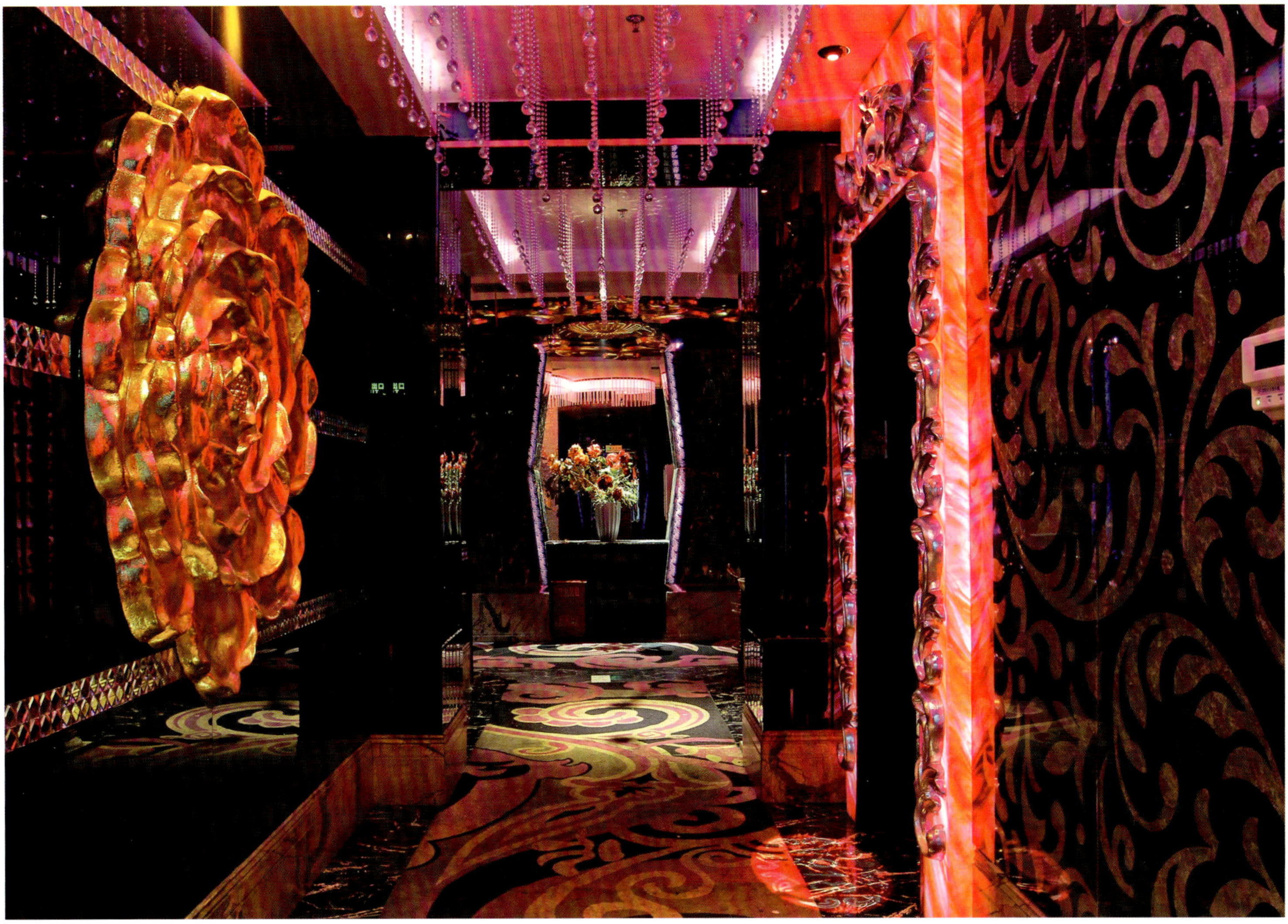

1
2
3
5
16400
1600
7000
2400
7000
1400
H
J
H
F
E
B
A
2100
2850
5700
2850
2400
2400
4200
4200
2700
5400
2700
33400
2700
5400
2700
3300
6600
3300
3700
3700
1600
1200
G
D
C
A
3000
7000
2400
16400
7000
1600
7000
5400
5400
6600
3300
2400
6000
3600
1
2
4
6
7
8
9
10

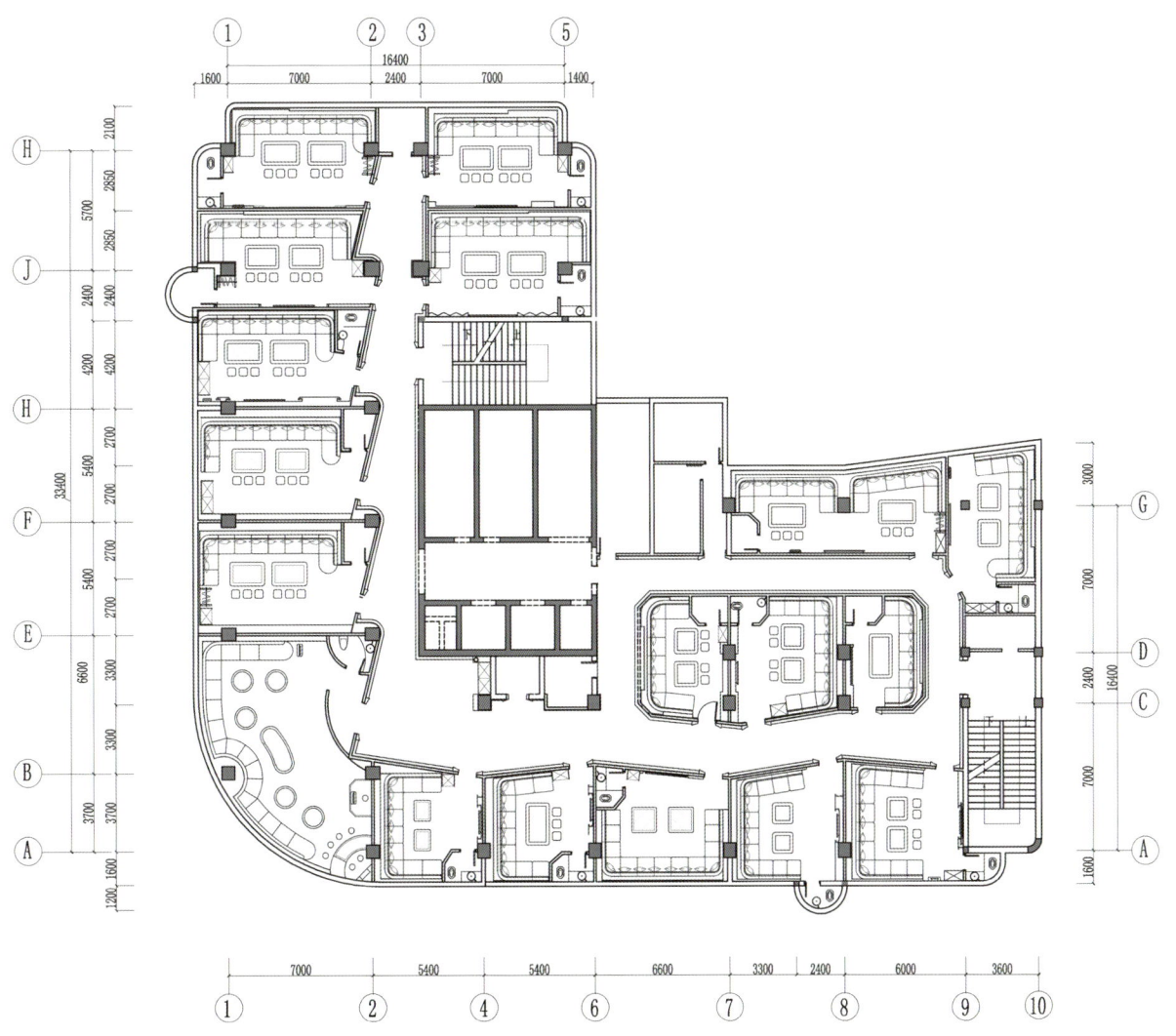

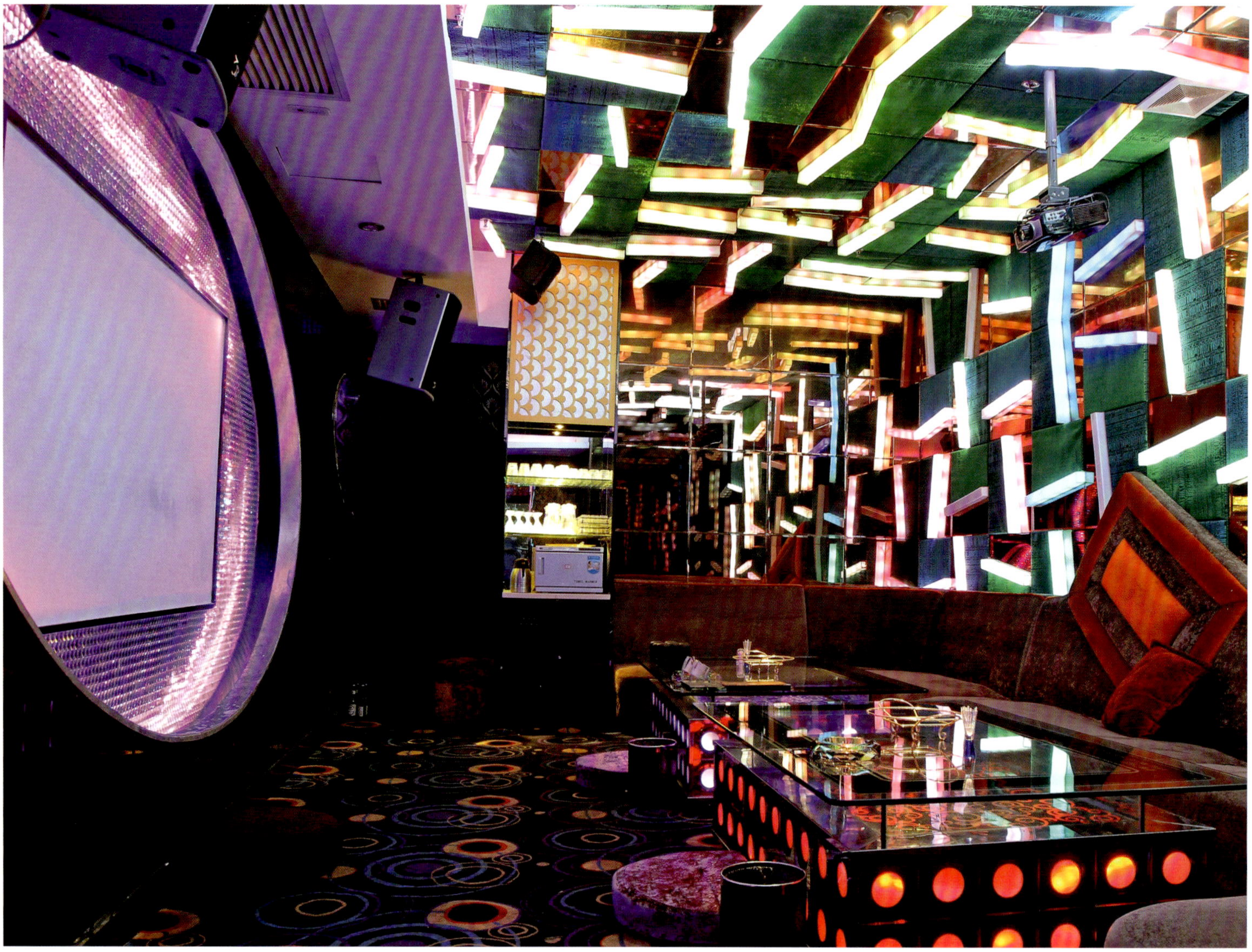

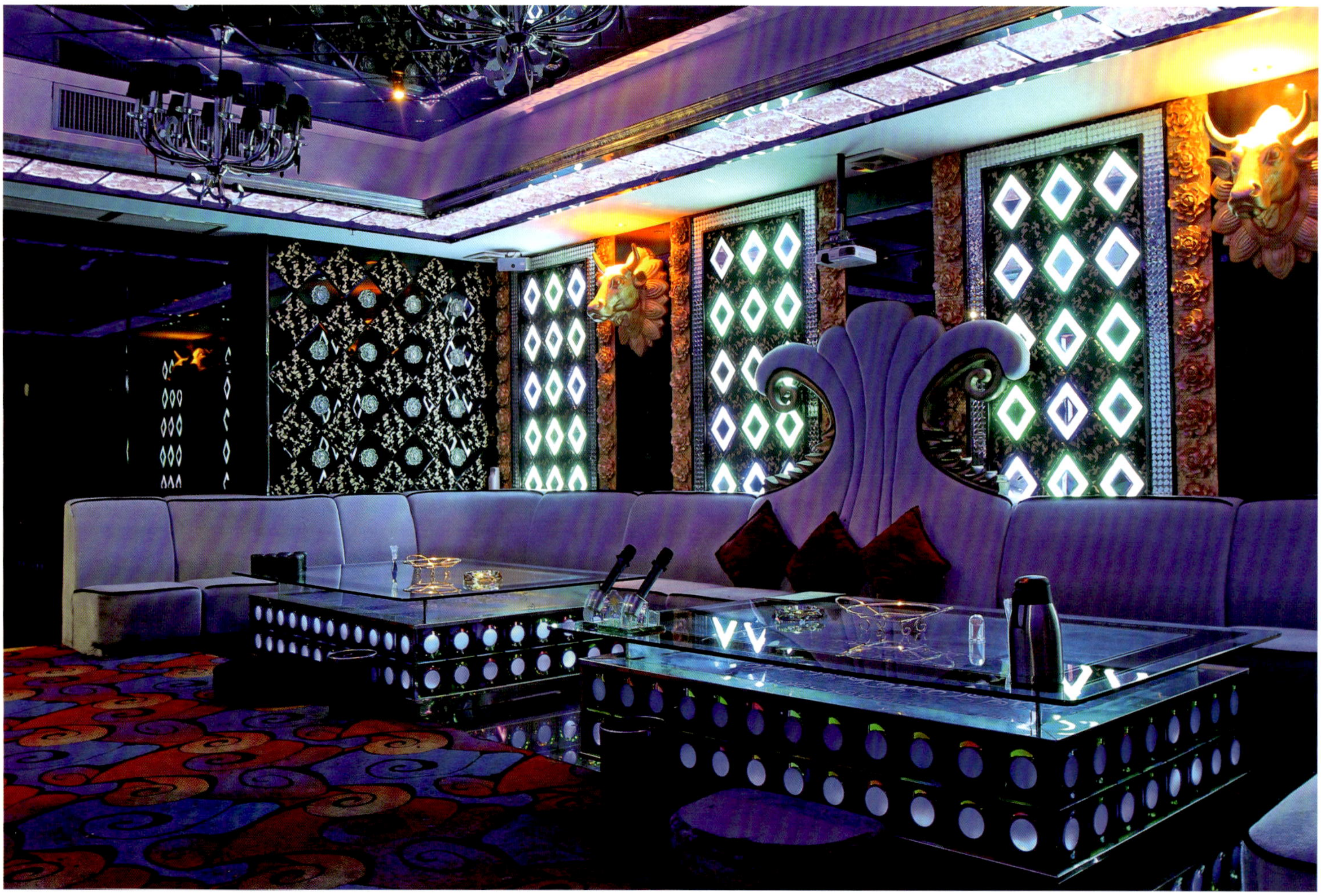

本案共 3200 平方米，有四个楼层：一楼人厅，二楼商务层（风格高雅精致），三楼 VIP 层（风格奢华尊贵，只对 VIP 客户开放）和四楼娱乐层（注重时尚与品质）。

一楼的发光盘花大柱延伸至二楼中庭天花变换出盛开的金色花瓣，引出二楼花卉为主题的设计。走道墙面巨大的金色牡丹仿佛灿烂的笑脸迎接每一位尊贵的客人，风格高雅而精致。

三楼仿佛是一座艺术殿堂，壁画穹顶、精致吊灯以及华丽的水晶雕刻的发光墙面，映衬着精心布置的艺术雕塑，使空间弥漫着奢华尊贵的气质，让 VIP 客户感受到高雅的服务。

四楼运用曲线和不规则的手法来划分空间，大量的琉璃及透光墙面相互折射，映衬着顶面光纤光源，营造出一个时尚梦幻的娱乐空间，让每一个流连其中的人尽情挥洒热情。

地 点	面 积	设计师	设计公司	摄 影
/ 东莞	/ 3800m²	/ 叶福宇、蔡秉才	/ 菲尚装饰	/ 朱亚照

Dongguan Golden Palace Nightclub

—— 东莞金座夜总会

The project takes black as a key tone, supplemented with purple, red, silver gray and other calm tones, integrating European-style elements into the luxurious modern style to create a low-key and luxurious place for entertainment. Unlike common nightclubs, the project presents a high-quality leisure place. The designers plan the space with skilled approach, create an elegant temperament with calm and noble colors, render a quiet atmosphere with calm and pure lighting, create a comfortable environment with materials of rich texture, and decorate the space with elegant patterns...the smooth and bright floor tiles reflecting light, like beacon; the golden patterns within reach are combined with metal frame, like carefully framed murals; the star-like lights in private rooms are shining like galaxy; the lobby and the silver theme walls in hallways are shiny in the reflection of lighting, floor and ceiling, like a flowing water curtain, which makes people unconsciously stop and listen to the sound of water. Surprise and delight are just here.

155

57210
2345
6190
8990
8540
8400
8415
8105
6230
4170
3575
5300
4340
8375
8400
8380
8480
6465
9280
66765
7765
7080
7055
5390
2300
29540
7755
3880
4490
5630
7690
29435

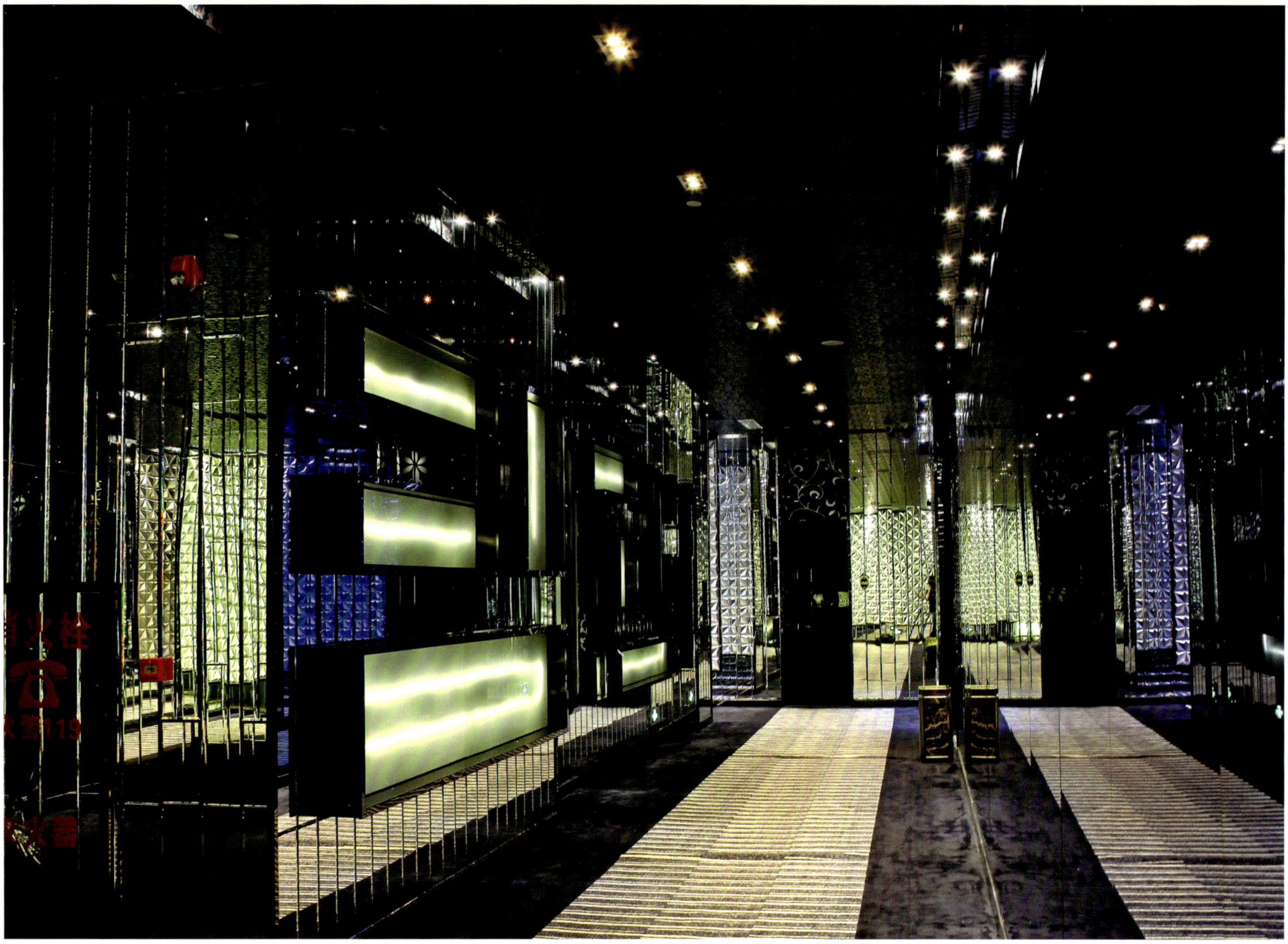

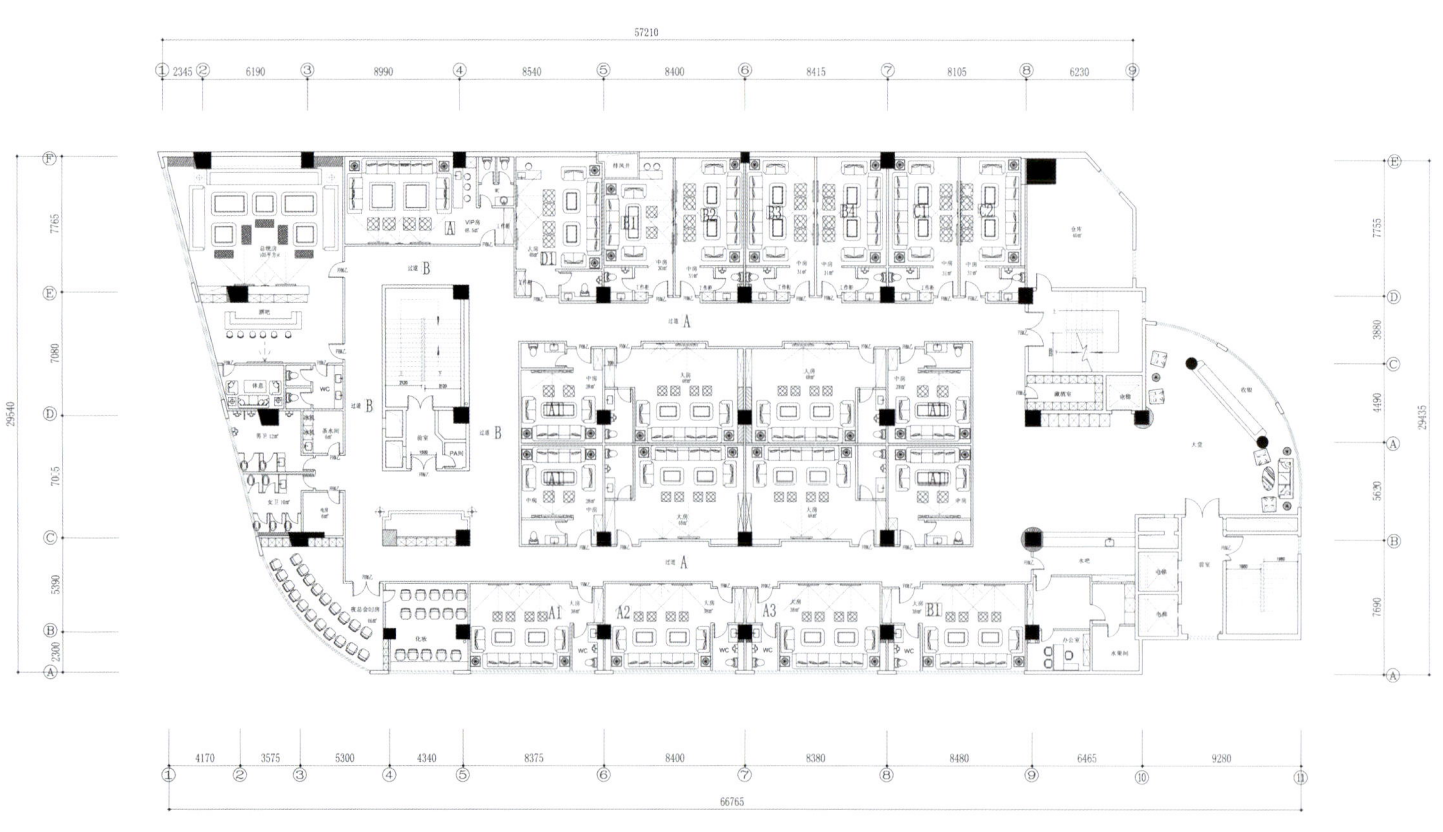

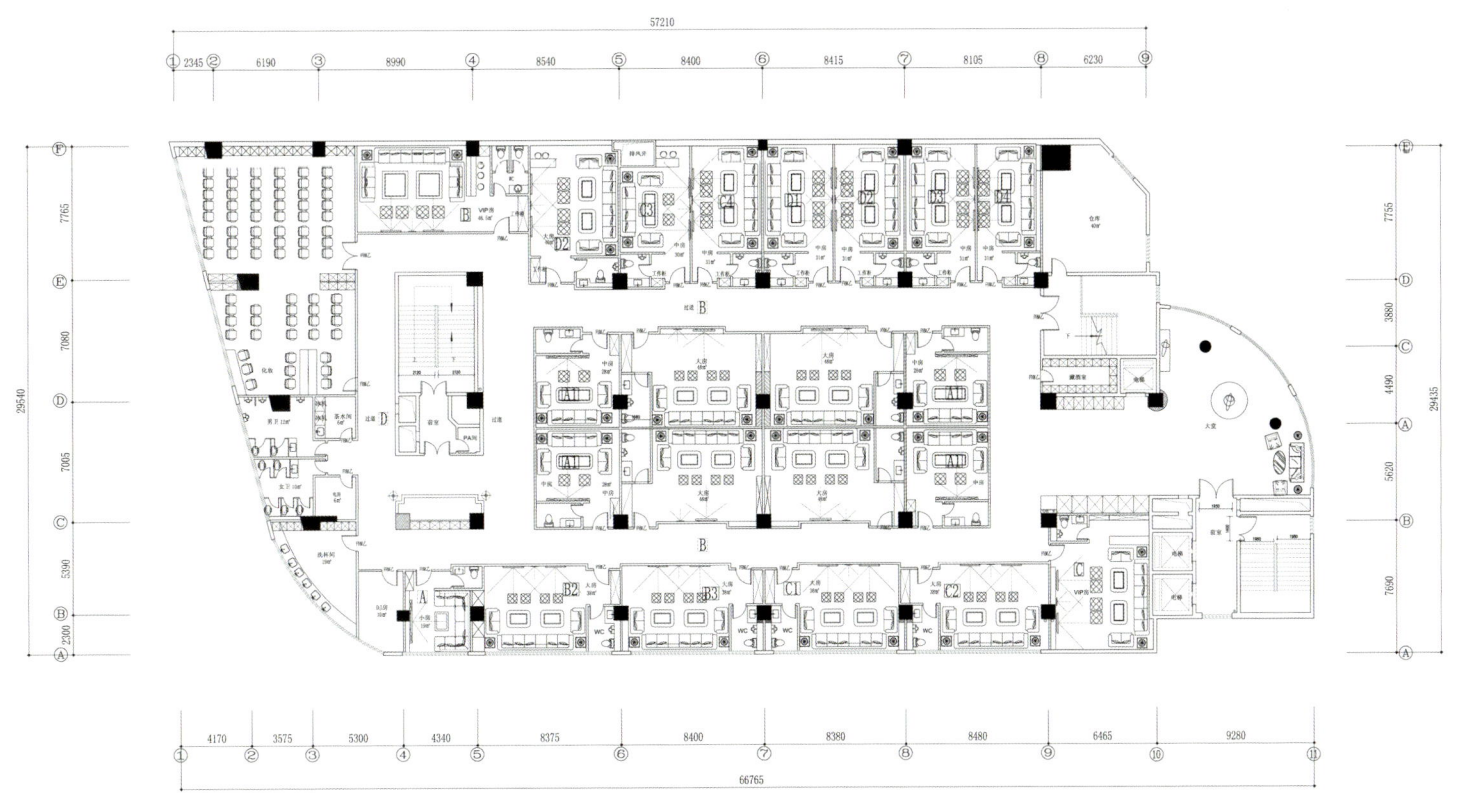

本案以黑色为主色调，辅以紫色、红色、银灰等沉稳的色调，在现代奢华风格中融入欧式元素，打造出一个低调奢华的娱乐场所。

不同于常见的夜店，本案呈现的是一个高品质的休闲场所。设计师以纯熟的手法规划空间，以沉稳且高贵的色彩营造高雅气质，以冷静而纯粹的灯光渲染清静的氛围，以富有质感的材料打造舒适的环境，以典雅的图案装点空间……只见那光滑明亮的地砖映着灯光，仿佛指明灯；触手可及的金色花纹与金属框架的结合，仿佛精心装裱的壁画；包房里星星点般的灯光，仿若星河闪烁；大堂和走道里银色的主题墙在灯光、地板、天花的反射下熠熠生辉，仿佛流动的水幕墙，让人不自觉地驻足聆听水的声音。惊奇和欣喜就在这里。

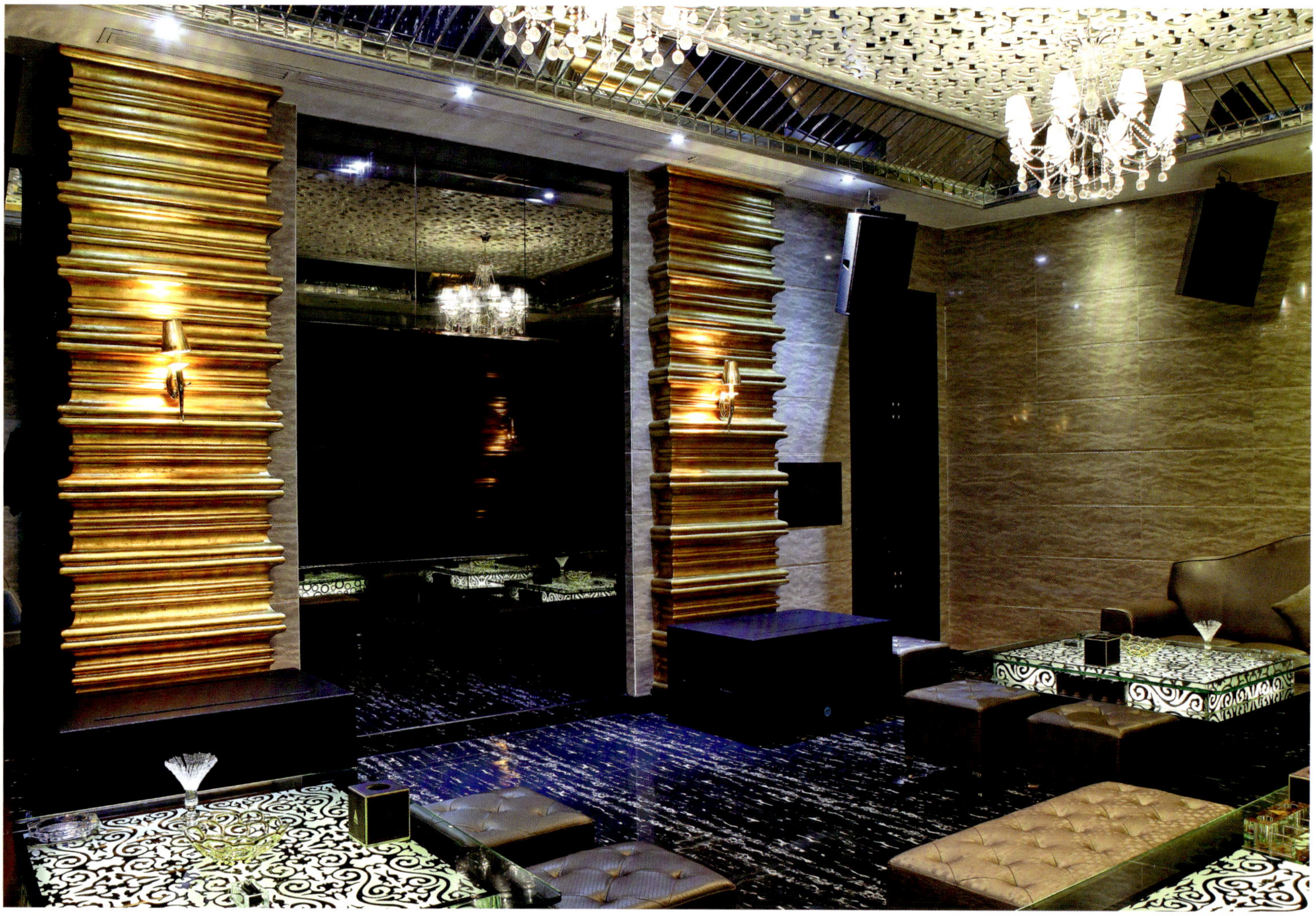

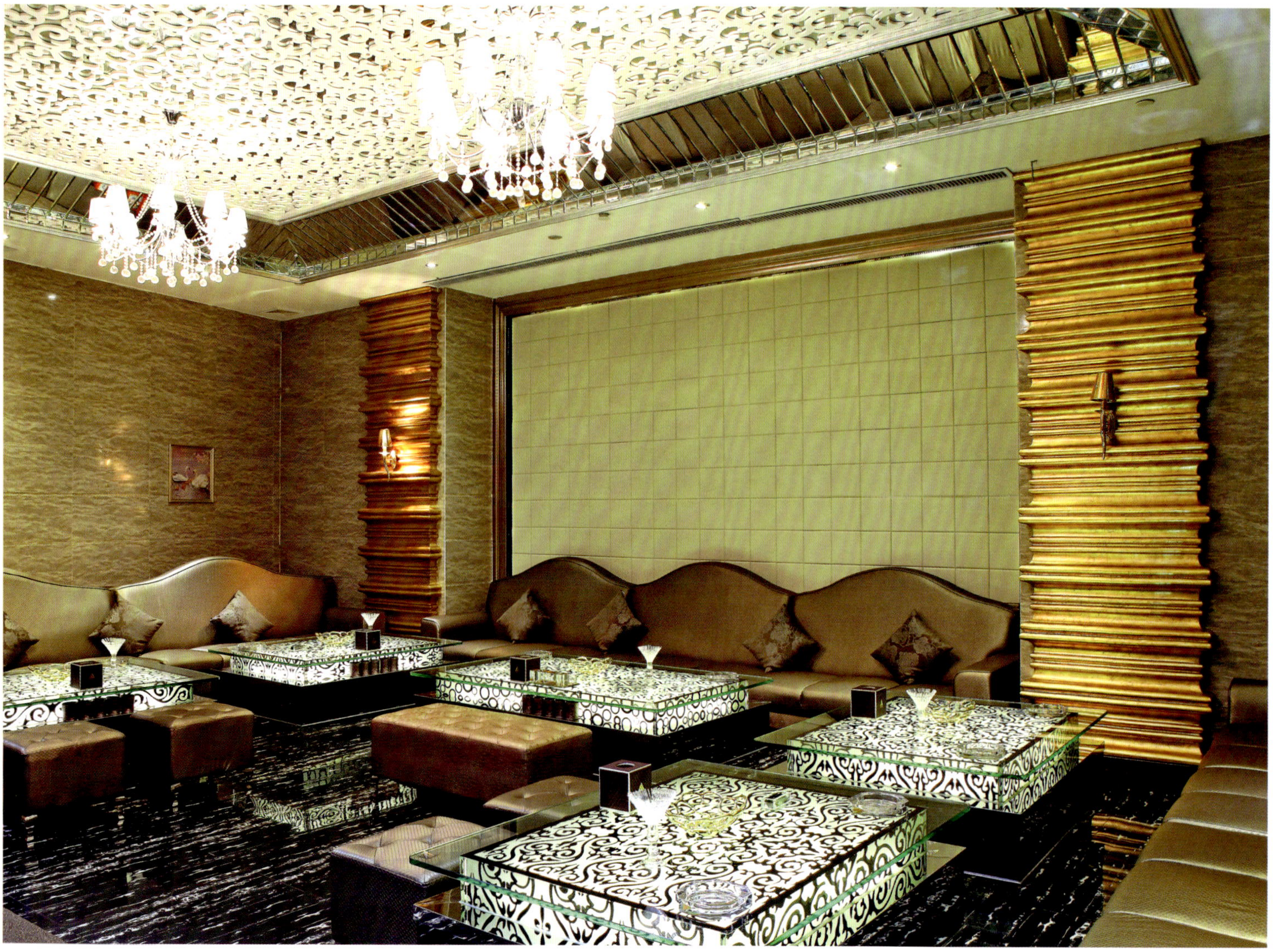

地点	面积	参与设计	设计公司	主要材料	摄影
/ 惠州	/ 3200m²	/ 叶福宇、蔡秉才	/ 菲尚装饰	黑金沙大理石、白色人造石、黑钛金、不锈钢、茶色镜、银镜、玻璃钢、墙纸	/ 朱亚照

Times Wholesale KTV in Huizhou

___ 惠州时代量贩式KTV

Times Wholesale KTV is an ideal place for business meetings, singing and gathering, and it is also the first entertainment club introducing the technique of producing oxygen in Huizhou. The designers define the space with the concept of "music oxygen bar", breaking the design concept and model of the previous wholesale KTV.

The project features more than 80 private rooms, and its design is simple rather than luxurious with its own unique charm. The hall is bright and spacious, reflecting the generous, bright and fashionable style. In terms of materials, the designers make use of black, white and gray tone and veneers easy to clean. The ground is paved with patterned marble to enhance the stereoscopic effect of the space, at the same time it takes into account the problem of reducing noise pollution.

In KTV, a too narrow corridor makes people feel a sense of uneasiness, but a spacious one features a quiet and warm feeling. A well-designed corridor can sweep away a boring feeling and become a beautiful landscape. In the private rooms, the glass carved with patterns and silver mirror endow the space with a broader vision and increase the charm of the space.

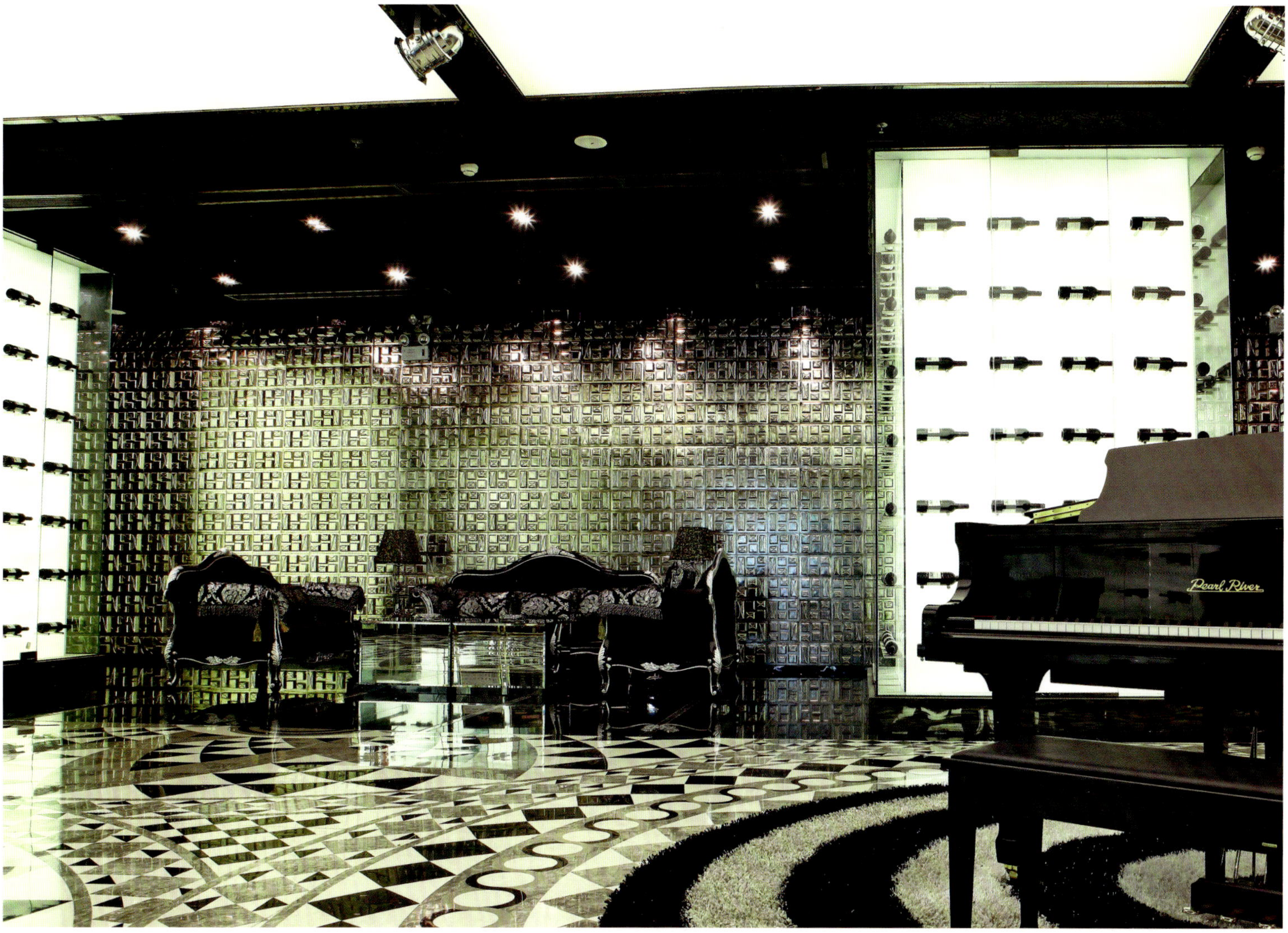
Pearl River

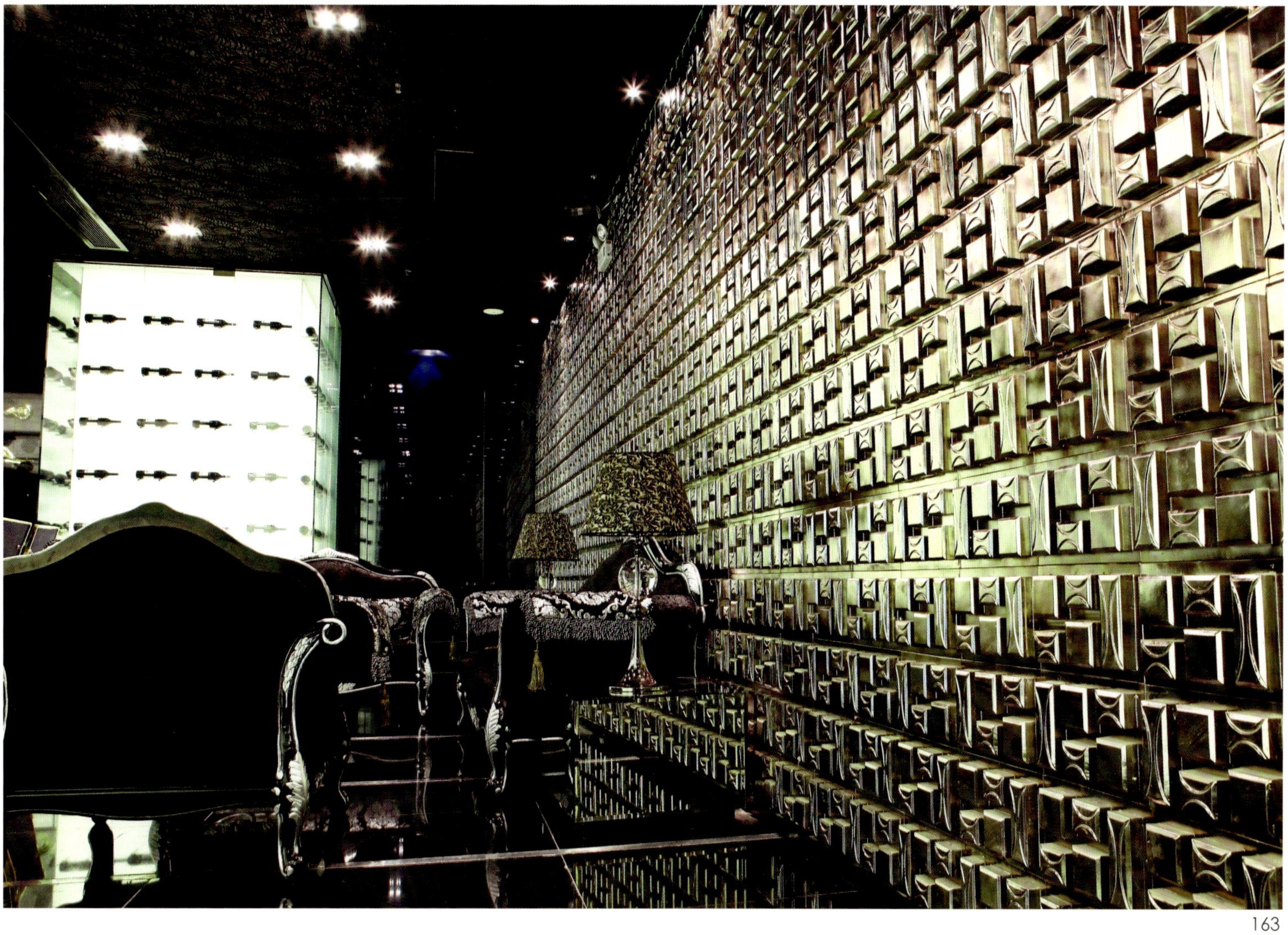

惠州时代量贩式KTV是商务洽谈、欢歌、聚会的理想去处，也是惠州第一家引进独立制氧技术的娱乐场所。设计师以"音乐氧吧"的概念定义空间，打破了以往量贩式KTV的设计概念和模式。

本案设置了80余间包厢，包厢设计简约而不奢华，有自己的独特魅力。大厅明亮宽敞，体现出大气、明朗、时尚的风格。材质上，选用黑、白、灰色调和易清洁的饰面为材料。地面采用有图案的大理石，以加强空间立体感，同时也考虑到了减少噪音污染的问题。

KTV走廊太窄会给人局促感，而宽敞的走道则给人安静、温馨的感觉。精心设计的走廊可以使过道的沉闷一扫而空，成为一道亮丽的风景线。进入包厢，雕花图案的玻璃和银镜的运用让空间视野更为开阔，同时也增添了空间的魅力。

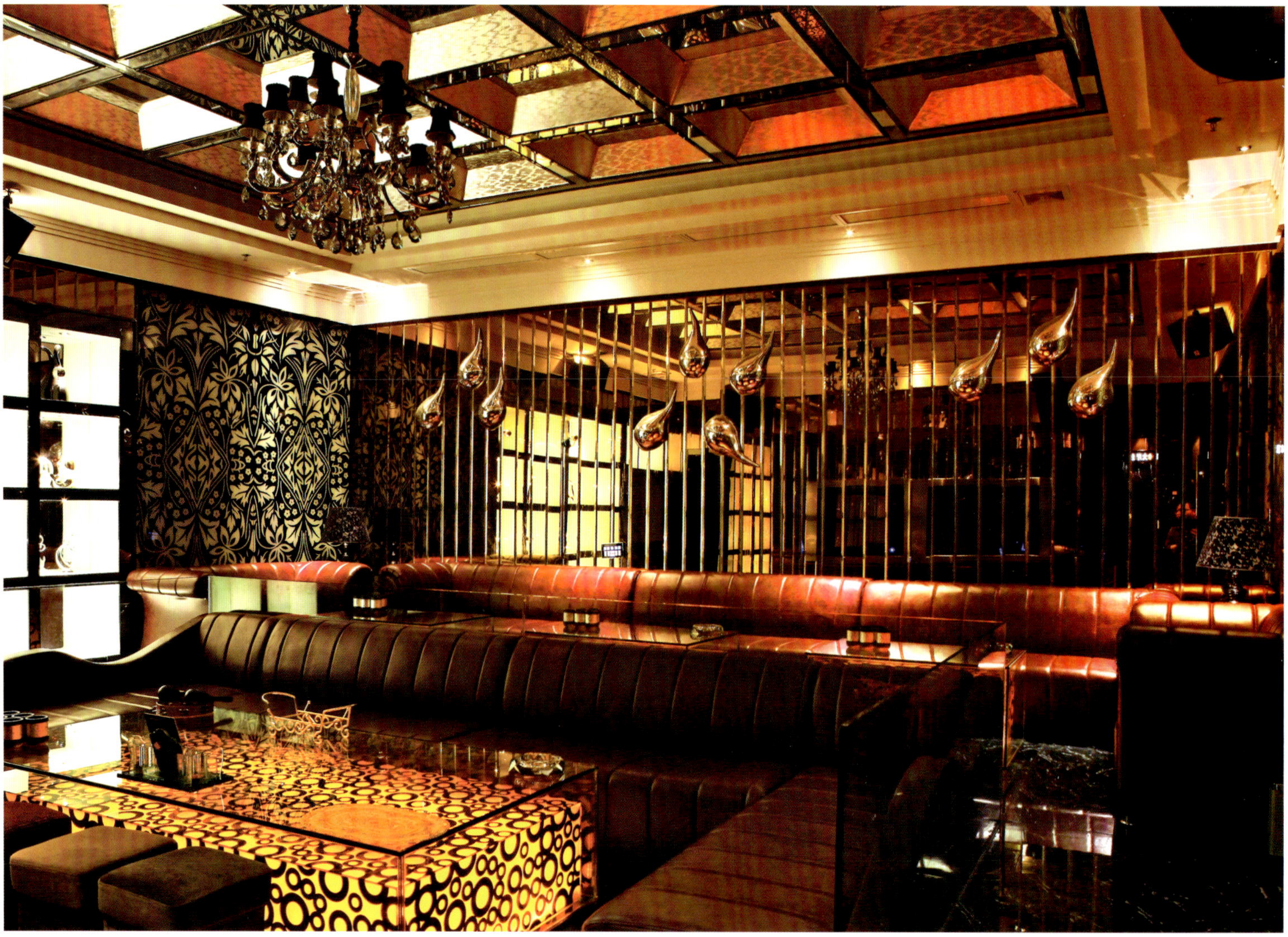

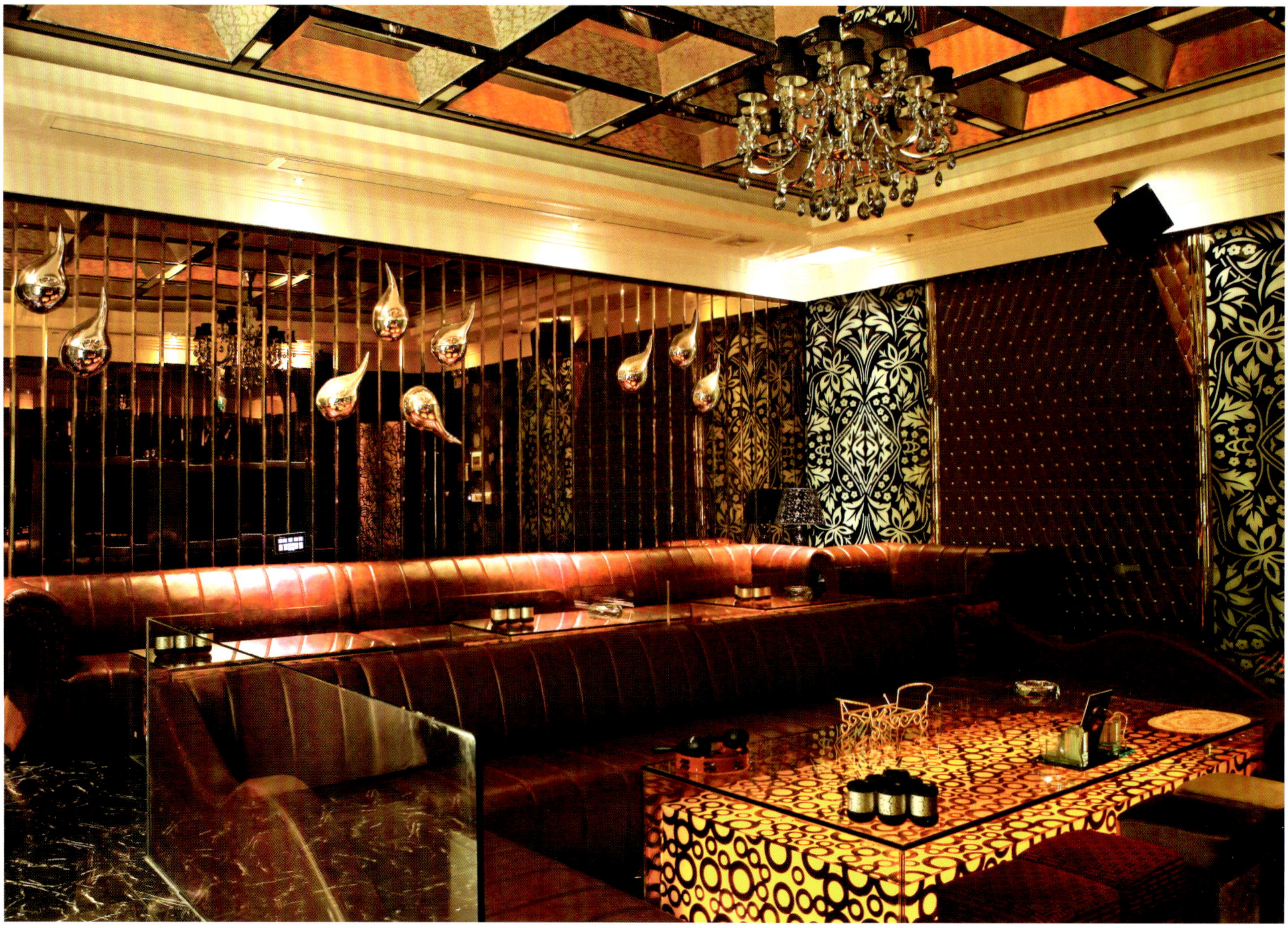

地点	面积	设计师	参与设计	设计公司	主要材料	摄影
/ 珠海	/ 2800m²	/ 吕道伟	/ 赖科、赵剑	/ 珠海空间印象建筑装饰设计有限公司	/ 灰镜、镜面不锈钢、烤漆玻璃、吸音板、天然大理石、人造石、木纹砖、抛光砖、墙纸、收口铝型材、乳胶漆	/ 谭继福

Kigo Wholesale KTV

—— 凯歌量贩式KTV

Space Impression Design Company creates eighty-eight KTV rooms with chic design and exquisite decoration for Kigo Dynasty, and it is a second joint project by Space Impression and Kigo Dynasty, showing the "non-entertainment and real joy" brand proposition of Kigo Dynasty to the end.

Kigo Wholesale KTV is Kigo Dynasty's first step of chain strategy, where the name is updated, and the new visual image invites master to do the design. Space Impression and Kigo Dynasty develop seven series of entertainment product plans, and more than thirty tailor-made themed rooms are produced for the target consumer groups of each product. At the same time, in order to define the space impression, they design a main image and a supporting logo "little wheat doll" for new Kigo; different themed rooms are designed with different exclusive graphics to form a unique space with Kigo characteristic, which can not be surpassed or copied.

In addition, the special "six-plus-one" mode of operation and 360° visual management design endow the operator with more peace of mind, and the definition of joy is broader, with unique taste of their own experience.

　　空间印象为凯歌王朝打造了88间设计新颖、装修别致的KTV房，本案是空间印象与凯歌王朝再度合作的项目，并将凯歌王朝"非娱乐，真欢乐"的品牌主张进行到底。

　　凯歌量贩式KTV是凯歌王朝连锁战略的第一步，不仅案名更新，新的视觉形象更是力邀大师担纲设计。空间印象与凯歌一同制定了七大系列娱乐产品规划，又为每个产品对应的目标消费群量身定做了30

多个不同的主题房间。同时，为了定性空间的印象，还为新凯歌设计了主形象和辅助图形"小麦人"；不同的主题房间也开发了不同的专属图形，以此形成别具凯歌特色、无法超越和复制的独特空间。

　　此外，非常"6+1"的作业模式和360°视觉管理设计，也让经营者更省心，欢乐的定义也就更宽泛，个中滋味，各自体验。

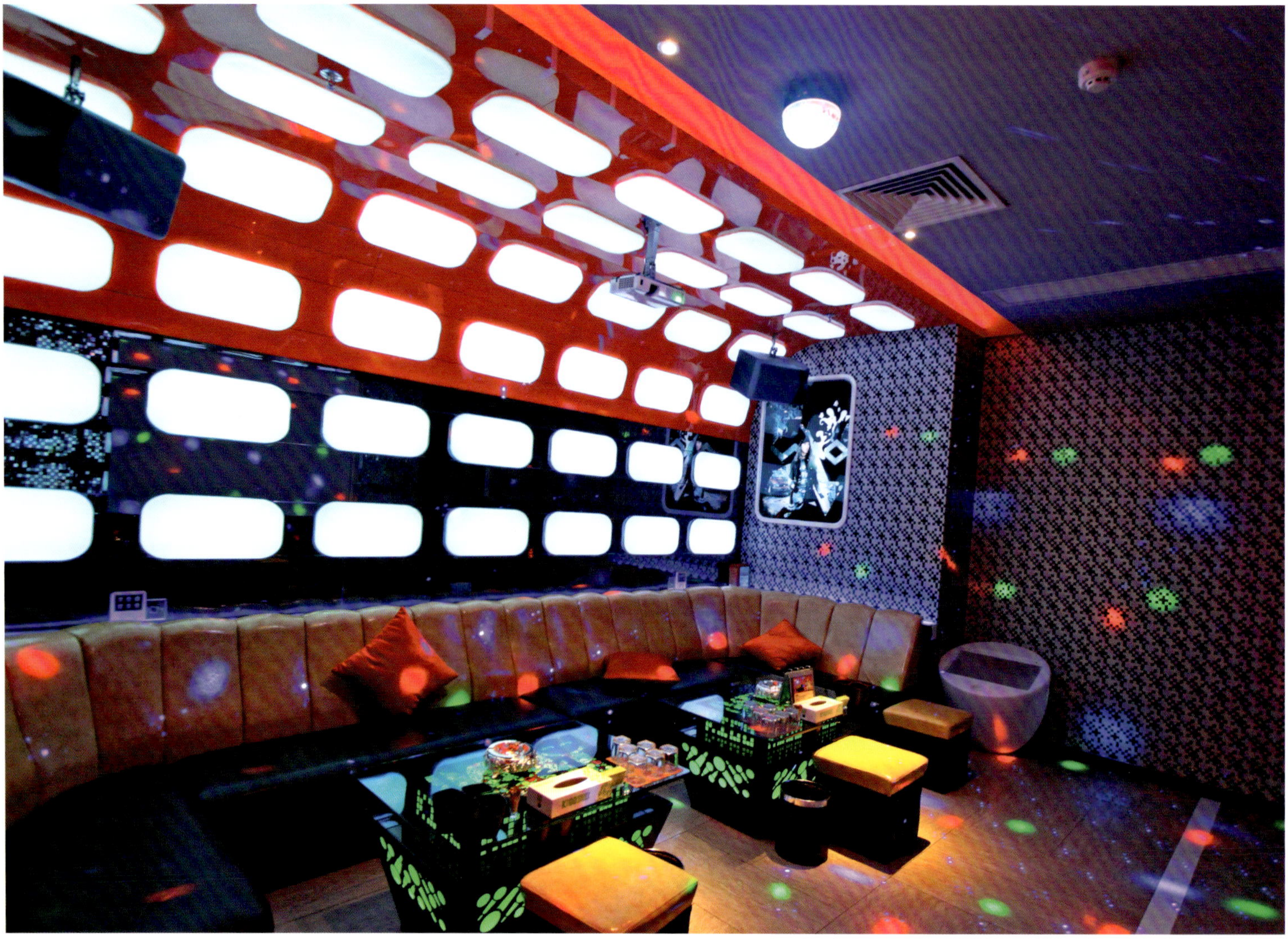

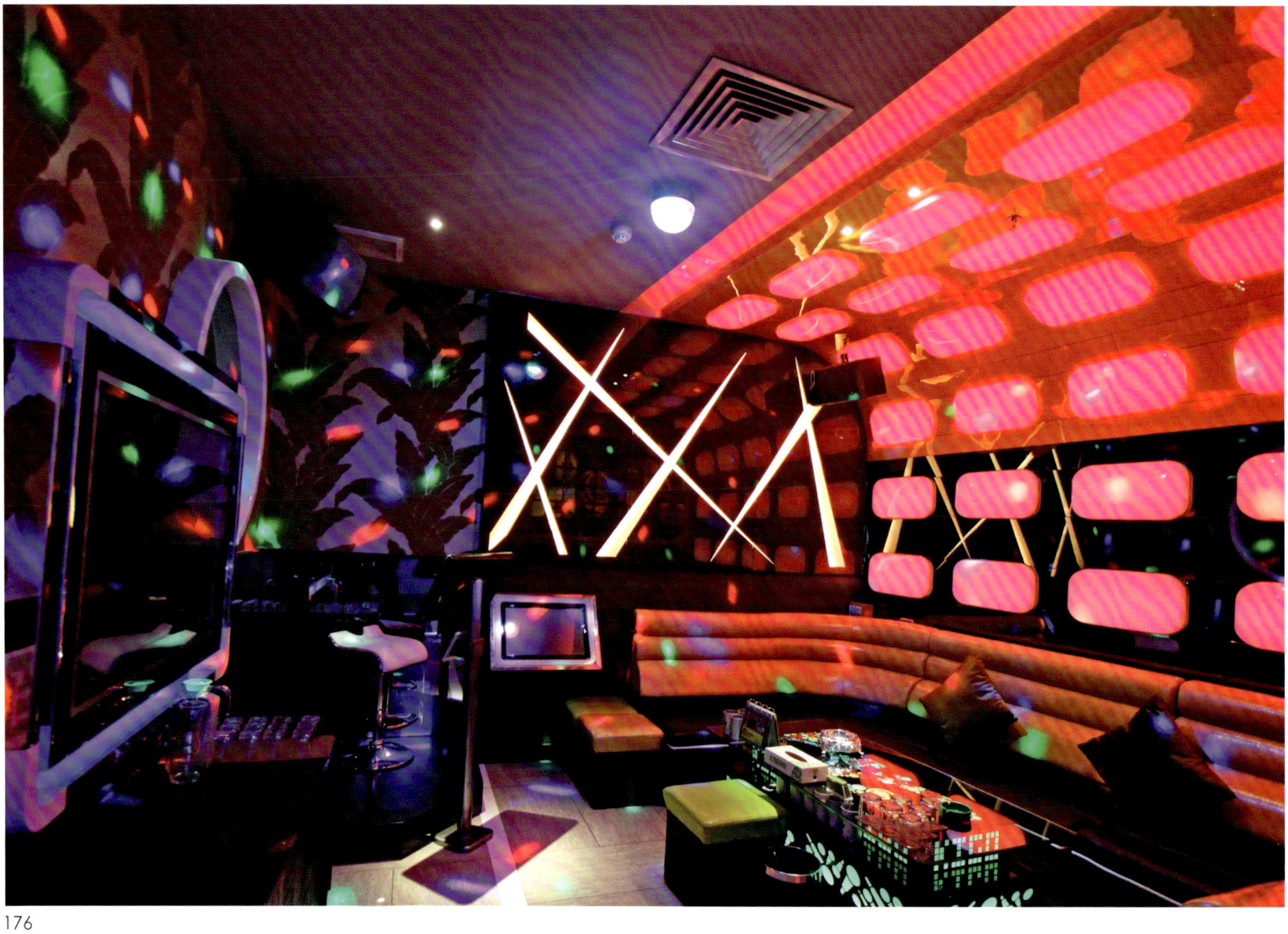

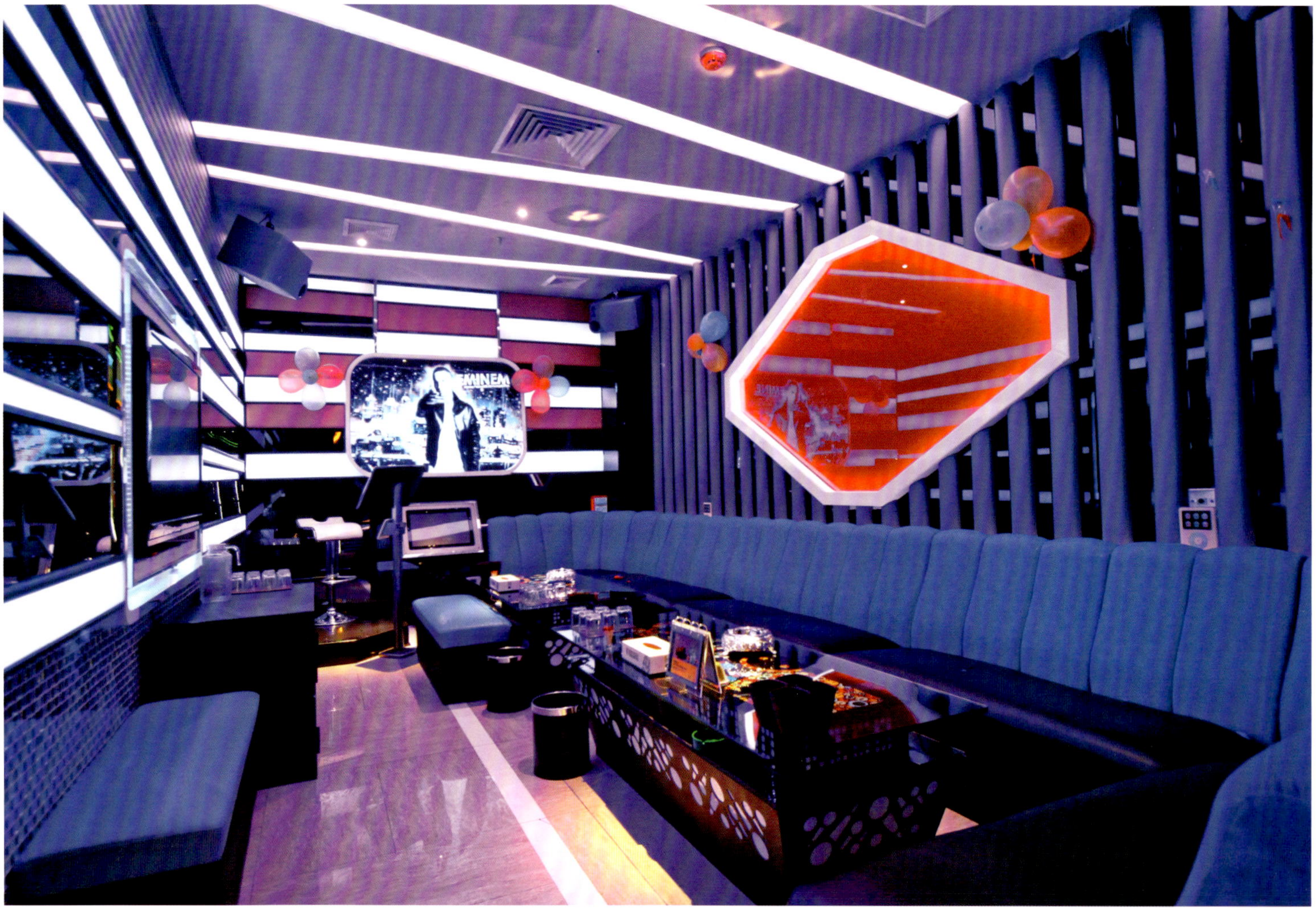

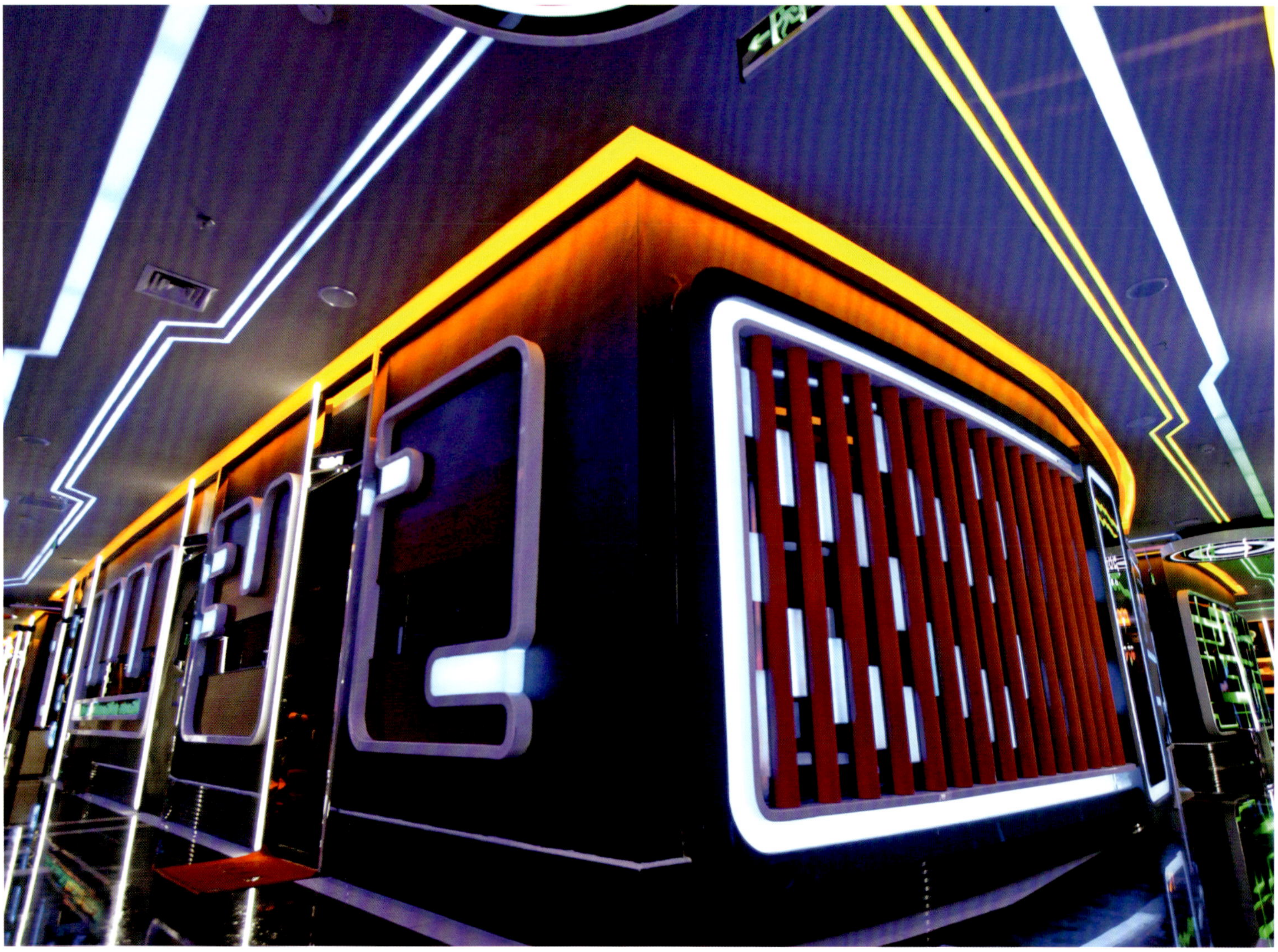

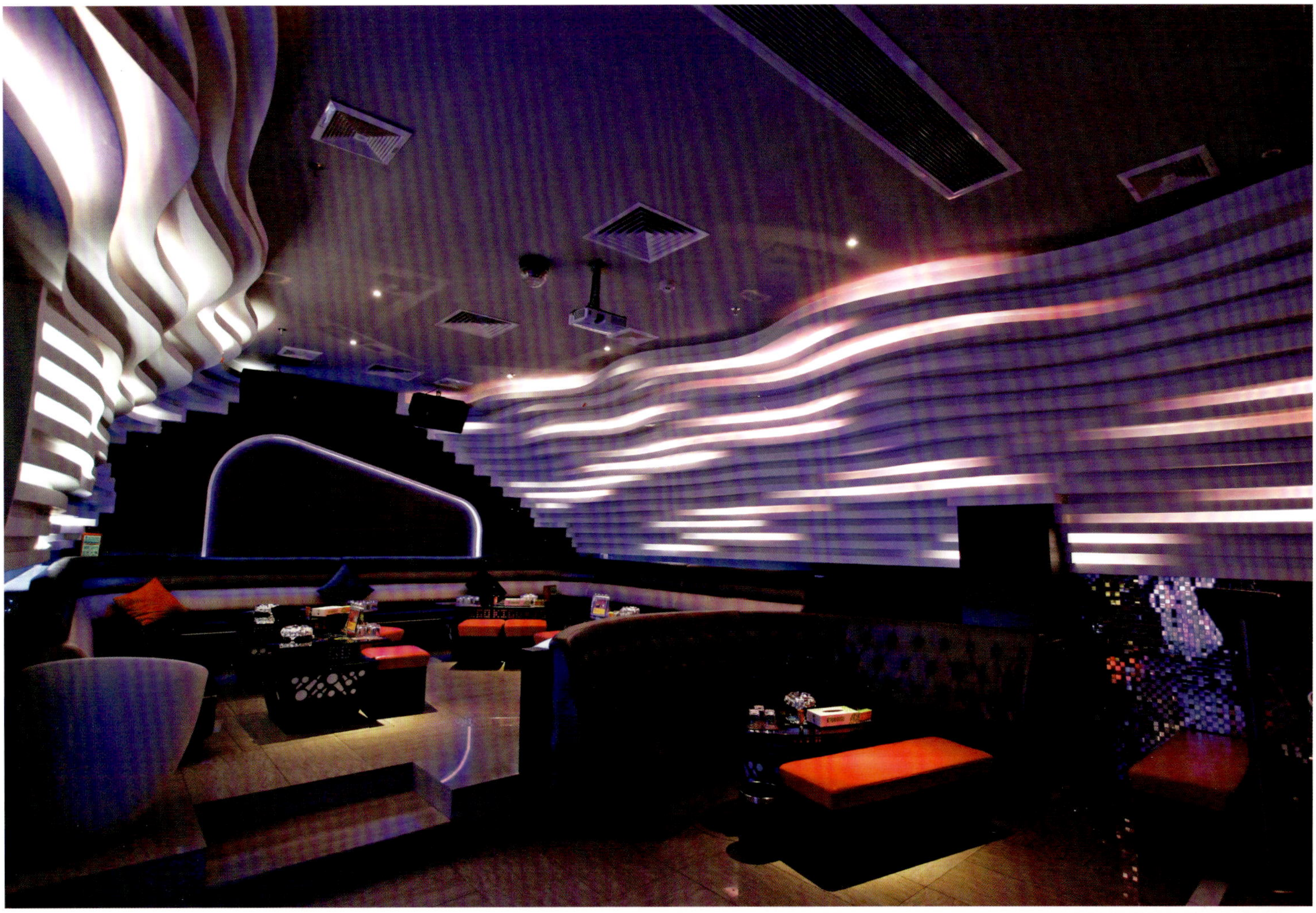

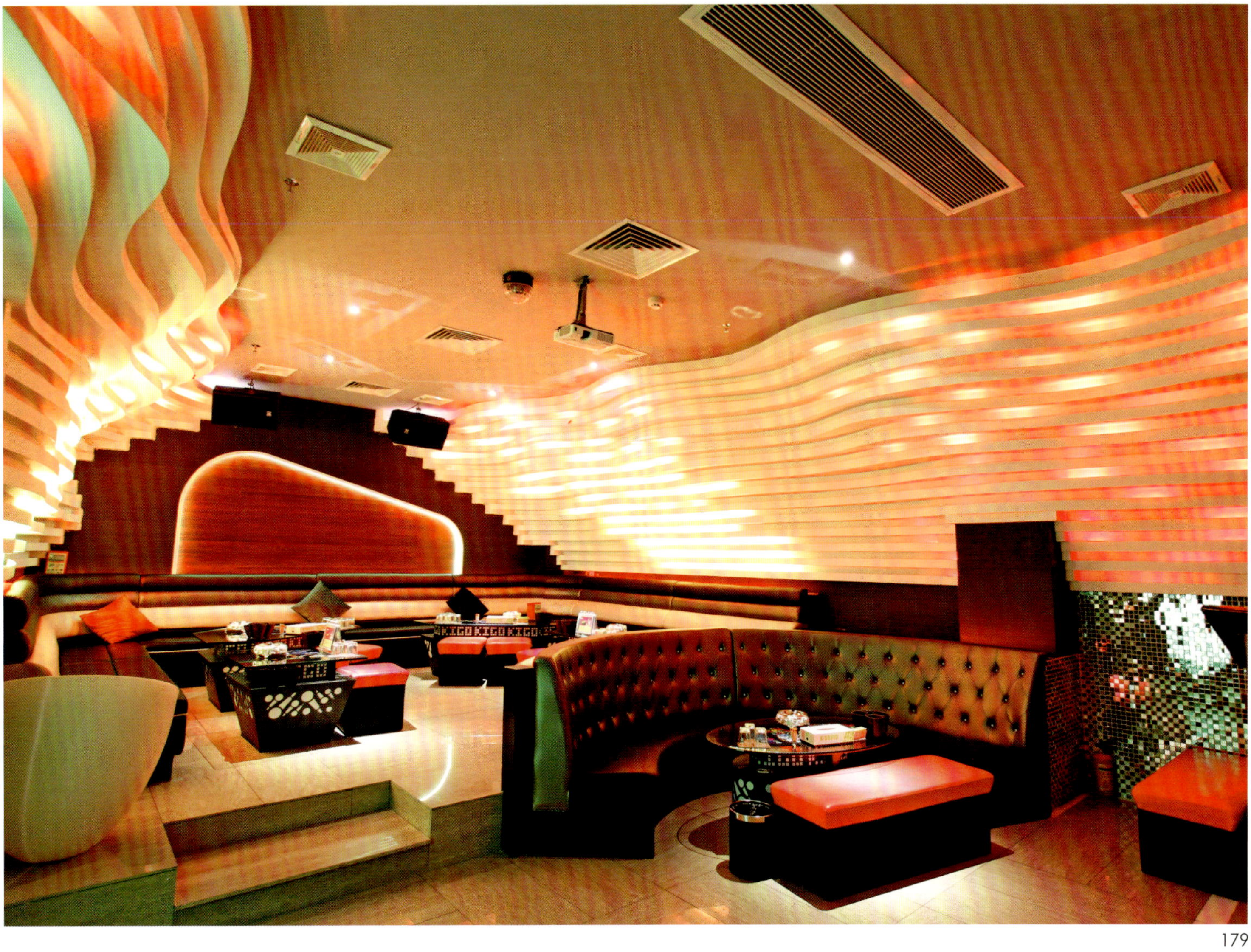

地点	面积	设计师	设计公司
/ 郑州	/ 3500m²	/ 罗国春	/ 罗一博装饰设计有限公司

Chant Song Phase II

___咏歌汇二期

It is a project of Chant Song Phase II located in a prosperous district with main customer groups of urban white-collar workers and businessmen, so the designer chooses bird's nest culture as the theme to create a trendy and fashionable atmosphere.

The star-grade lobby highlights the honor and wealth, even from the aisle, we can enjoy little ingenuity and some luxury. The project features sixty-eight KTV private rooms in different styles, including mini rooms, small, medium and large rooms, VIP rooms, and presidential rooms, to meet different needs of customers. Each private room features its unique characteristic, where the fashionable design, brilliant lighting and professional KTV audio create a comfortable, healthy, fashionable and happy image, and create an elegant and warm environment for customers, so they can relax and enjoy the happiness from entertainment.

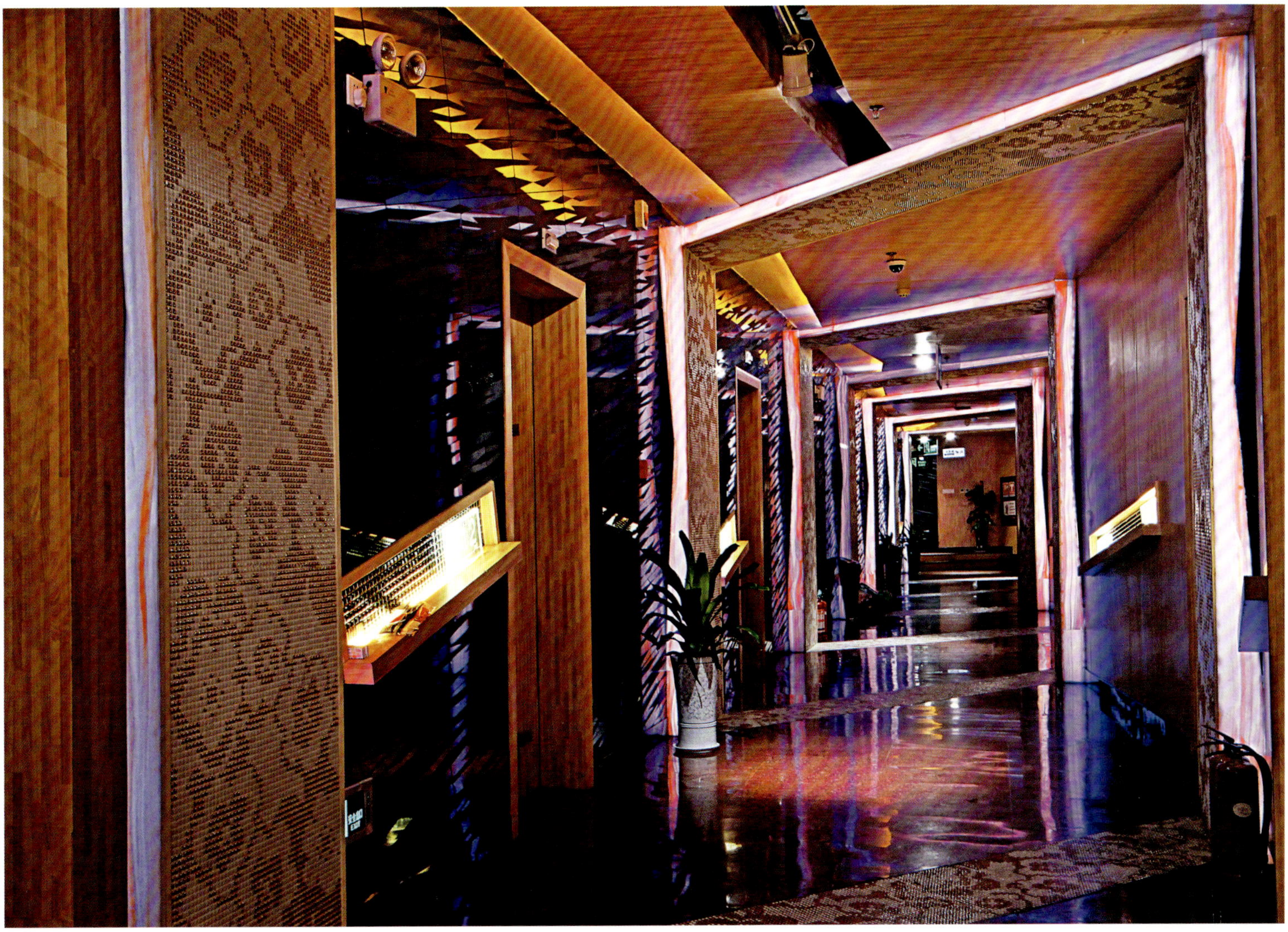

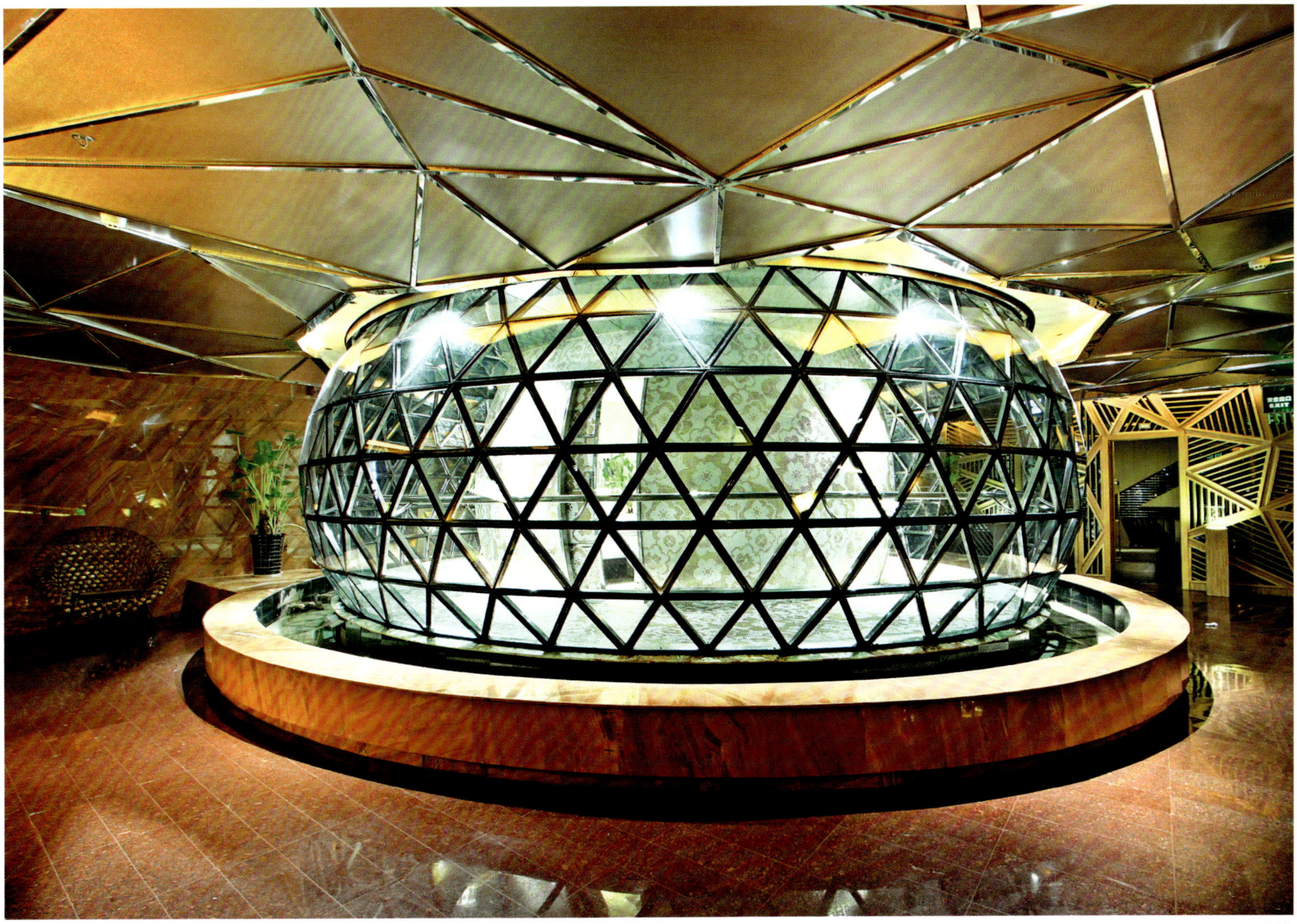

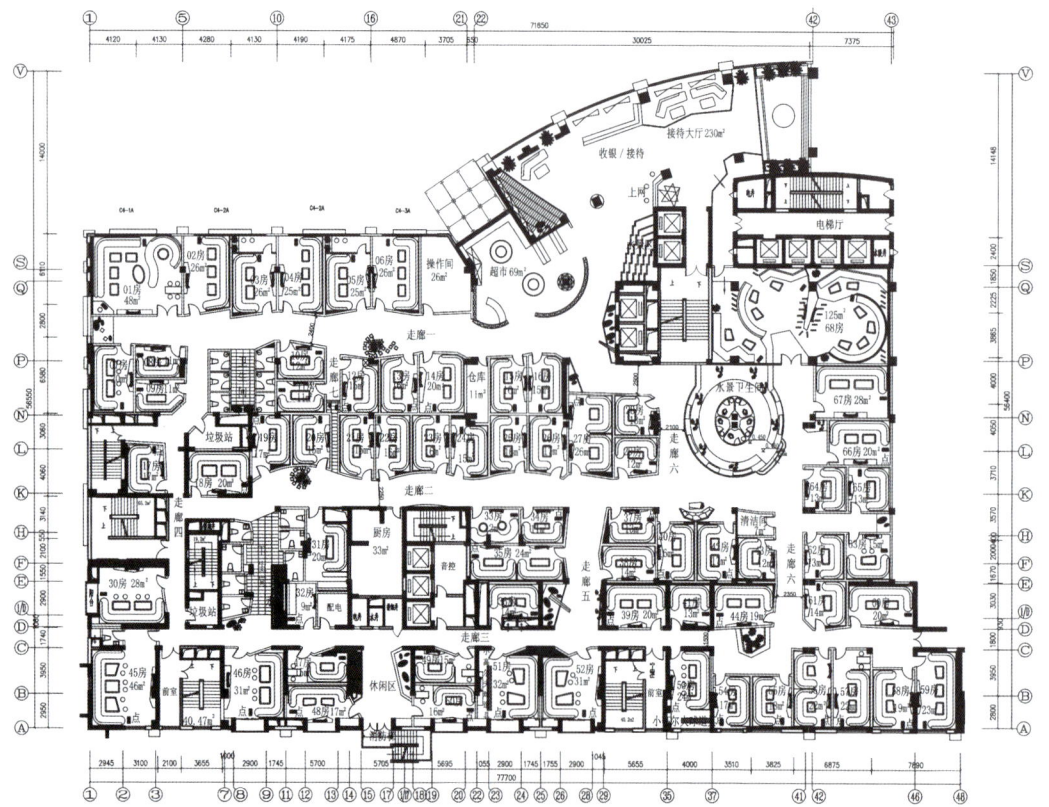

本案是咏歌汇二期的项目，地处繁华地带，客户群以都市白领和商务人士为主，因此，设计师以鸟巢文化为主题，营造出前卫时尚的氛围。

星级的大堂彰显出尊荣和富贵，即便是过道，也能读到一种匠心、三分奢华。本案共有风格迥异的 KTV 包厢 68 间，囊括了迷你房、大中小房、VIP 房、总统房等各种房型，以满足顾客的不同需求。每个包厢都各有特色，时尚的设计、绚丽的灯光以及专业的 KTV 音响塑造出舒适、健康、时尚、欢乐的形象，为顾客营造出优雅、温馨的环境，让其放松心情，尽情享受娱乐带来的欢乐。

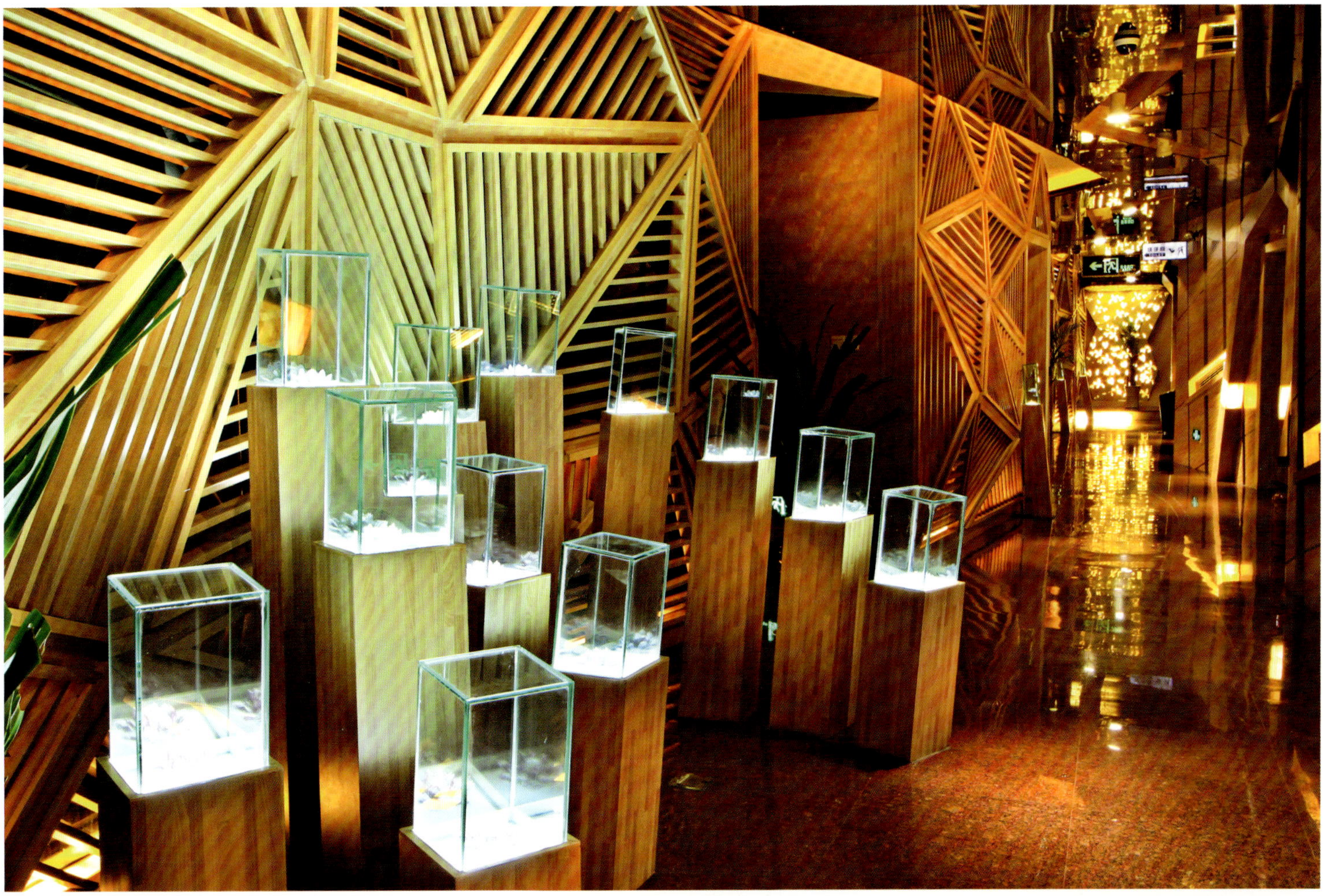

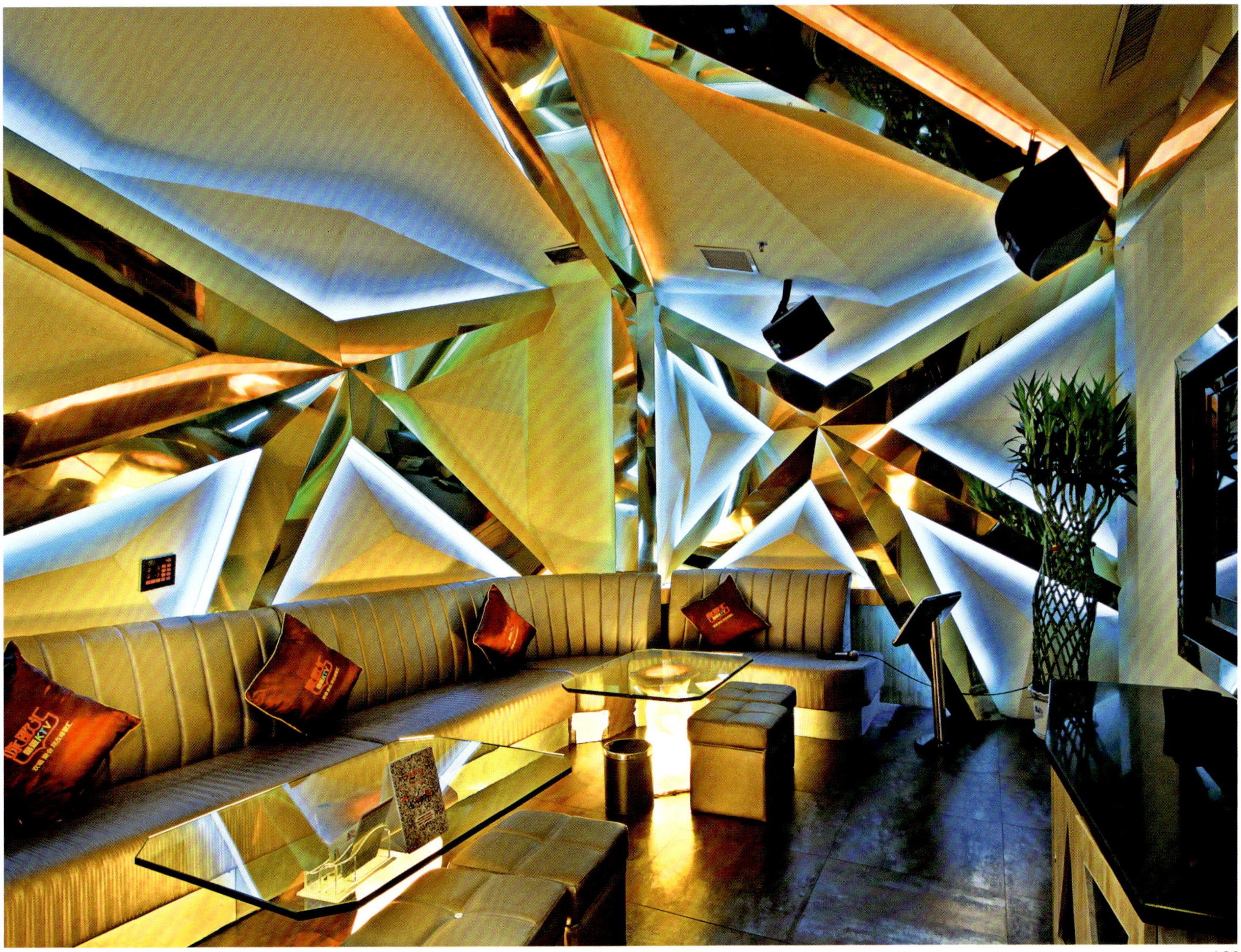

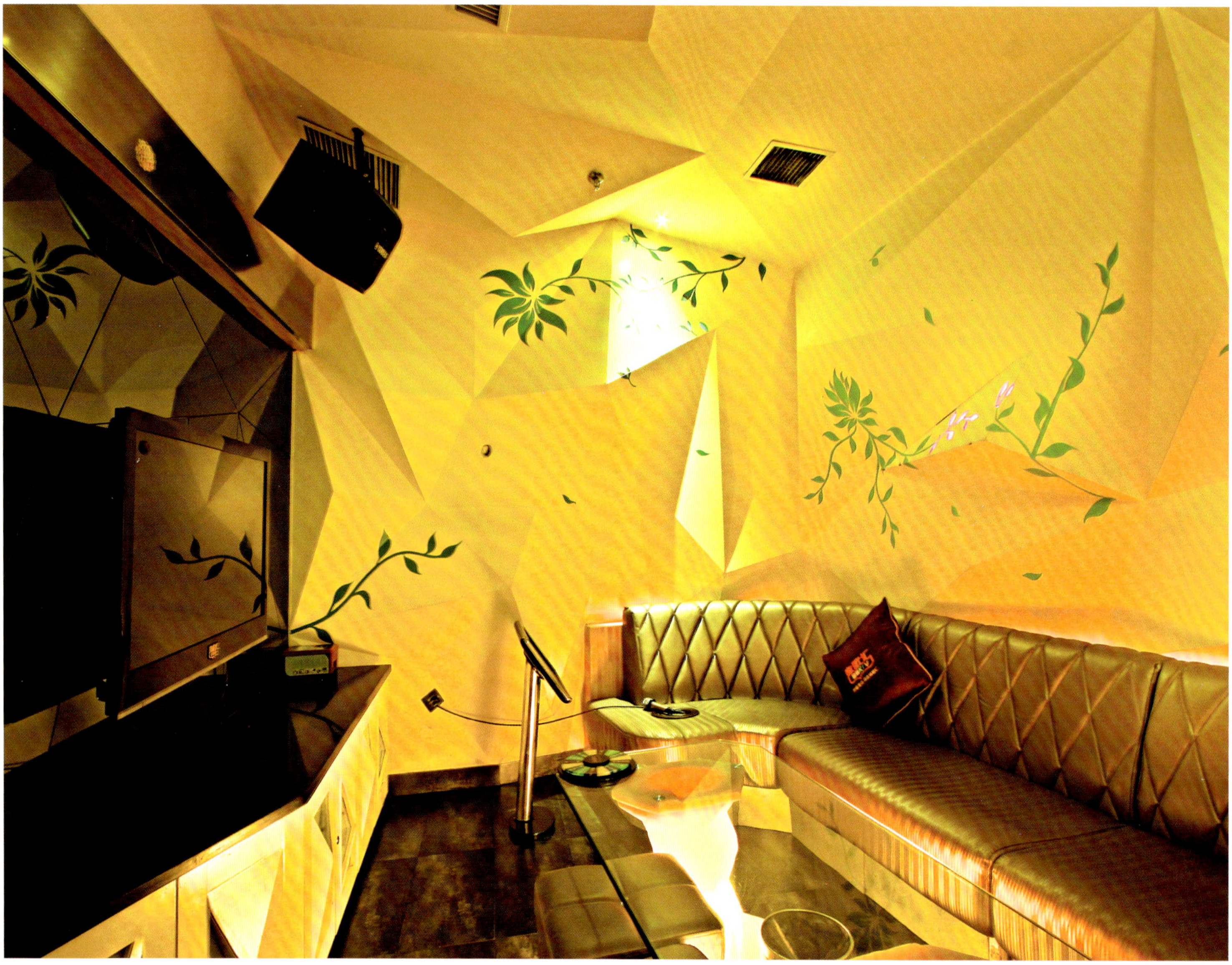

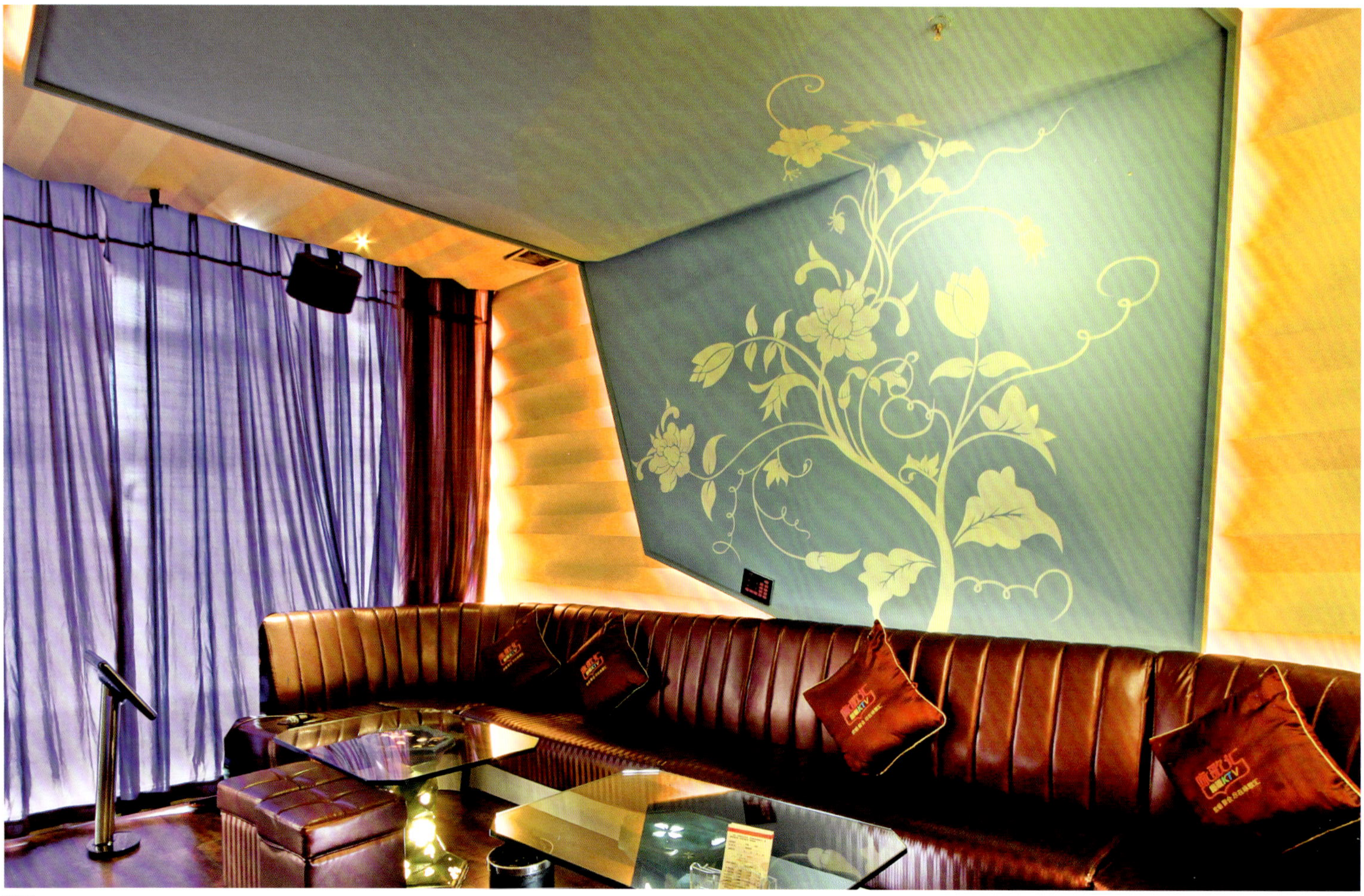

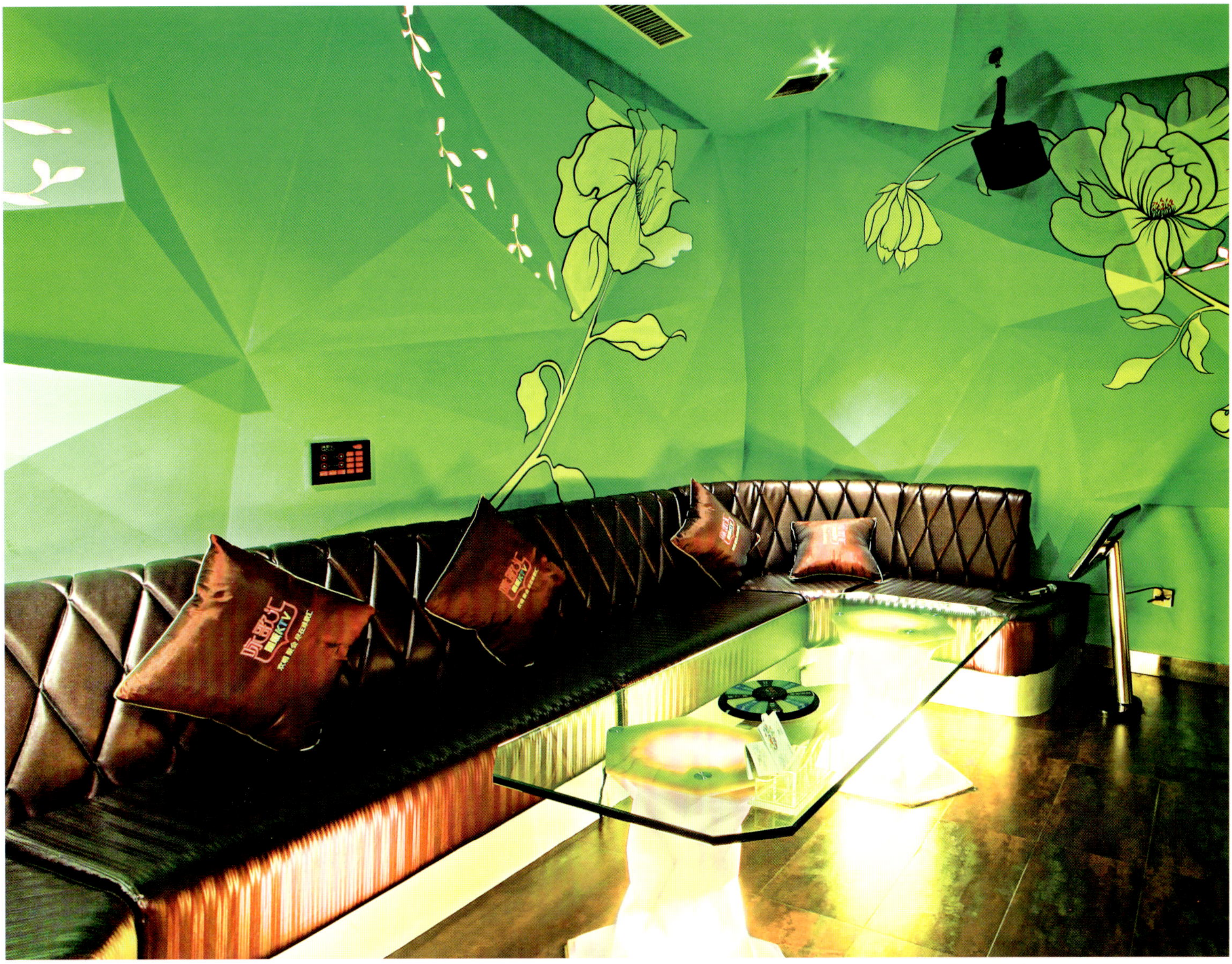

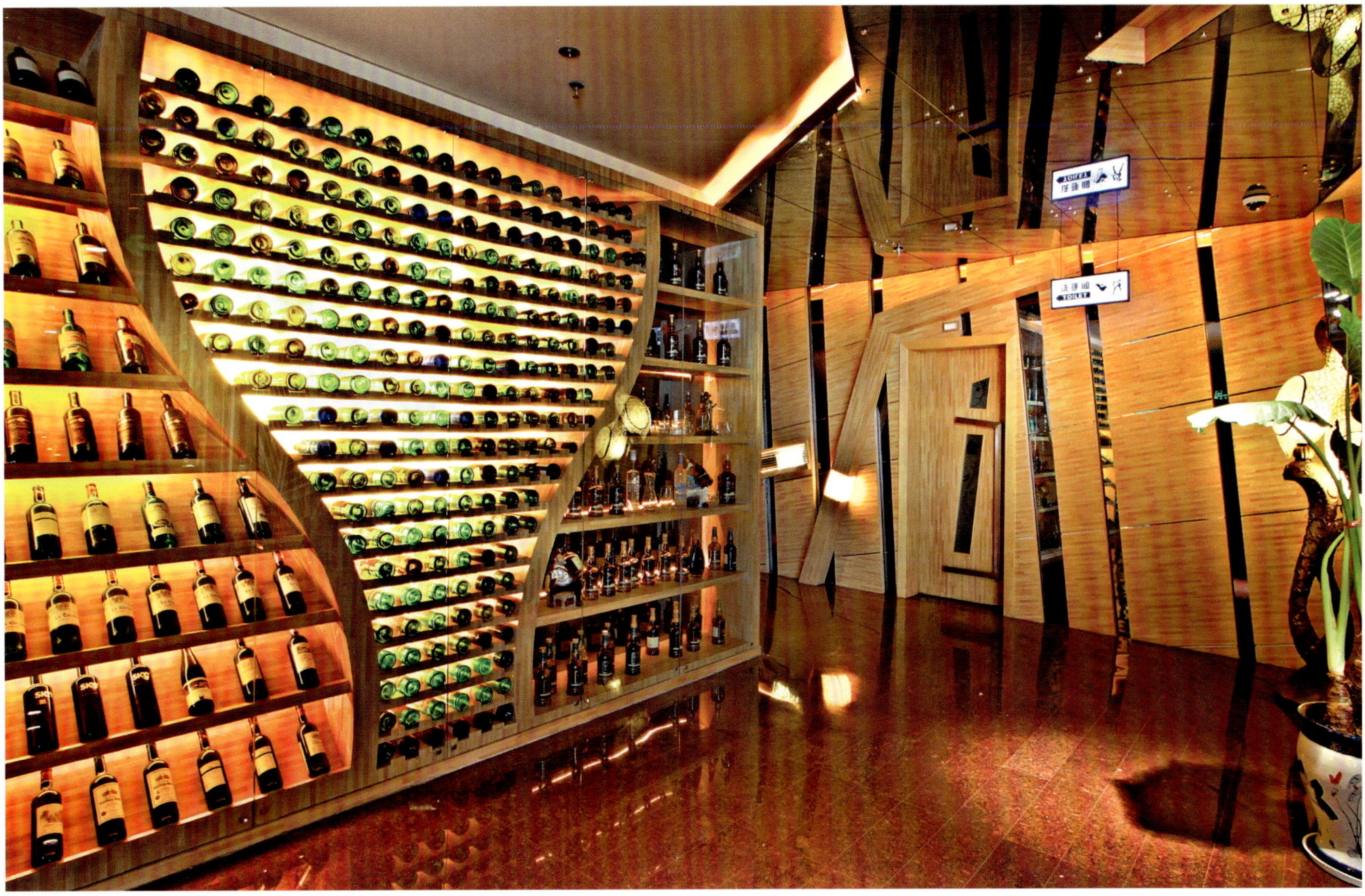
洗手间
TOILET

地点	面积	设计师	设计公司	主要材料
/ 南充	/ 2324m²	/ 杨彦	/ 深圳品彦室内设计有限公司	/ 进口羊绒地毯、仿景泰蓝砖、梵高金石材、玻璃马赛克、仿帕斯高灰砖、抛光砖、亚克力透光灯片、马赛克、墙纸、布艺软包

Martha Rhea Club in Nanchong

——南充玛莎·瑞亚会所

It is a KTV project, an entertainment place for stylish and trendy people, so the club space is defined to be bright and lively. As a design of night club, the lighting effect is crucial in the design, which must catches people's eyes and firmly grasps the minds of consumers to create its own feature. The designer makes use of a variety of bricks combined with mosaic, coupled with the coordination and contrast of lighting, to highlight the momentum of the space and the artistic atmosphere in the gorgeous night-club. Such a design can impress us, then attract more people to come and consume.

The first is the design of the reception hall, where the petal-like ceiling design echoes with the patterns of the floor tiles, combined with the design of half-close curved walls around, highly setting off the momentum of integrating the sky with the earth, and filling the space with the overall sense. Various kinds of bricks and lights are perfectly combined and integrated, and the private rooms and walkways are designed with different approaches, endowing customers with a different feeling and charm.

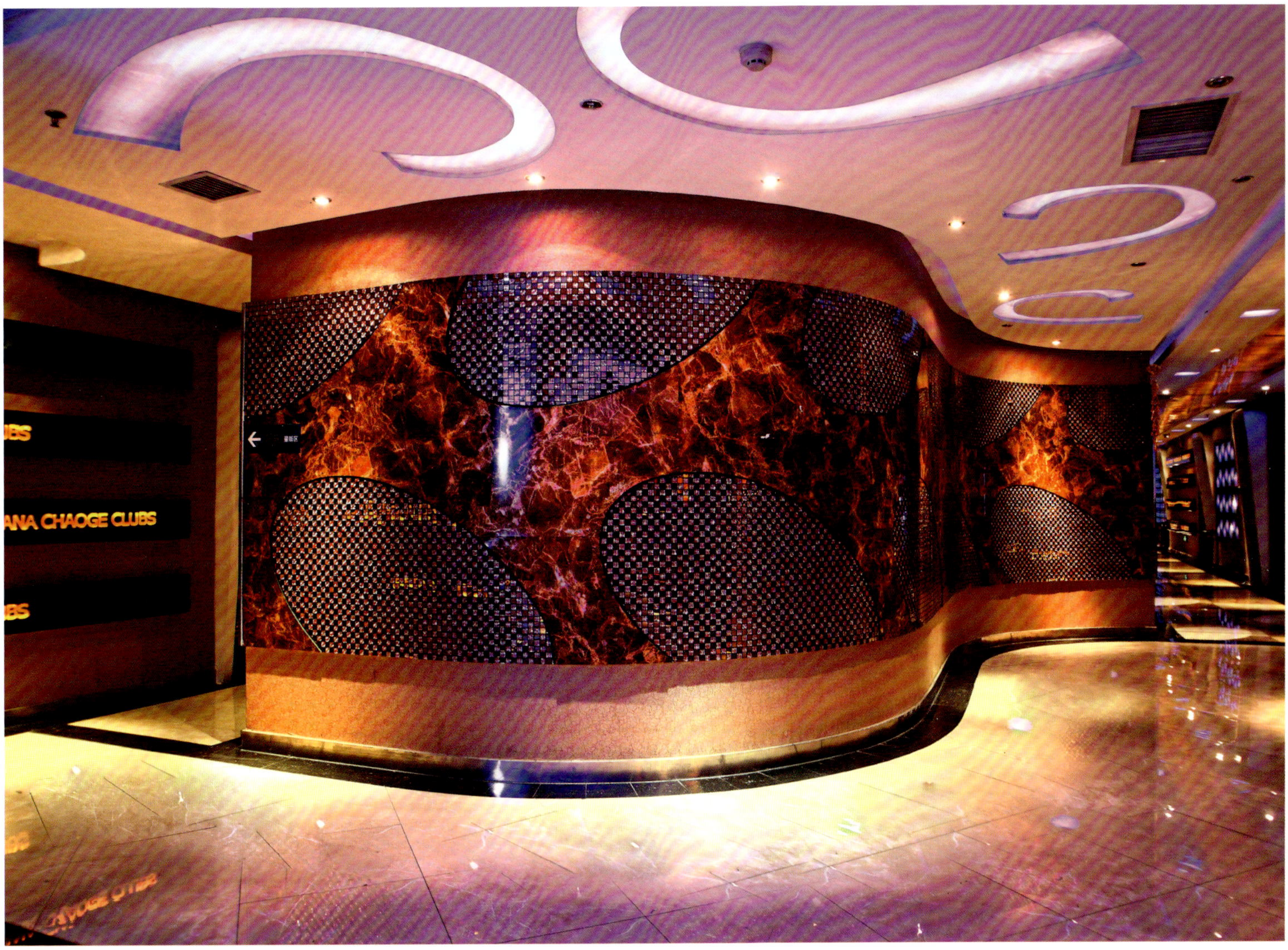
ANA CHAOGE CLUBS

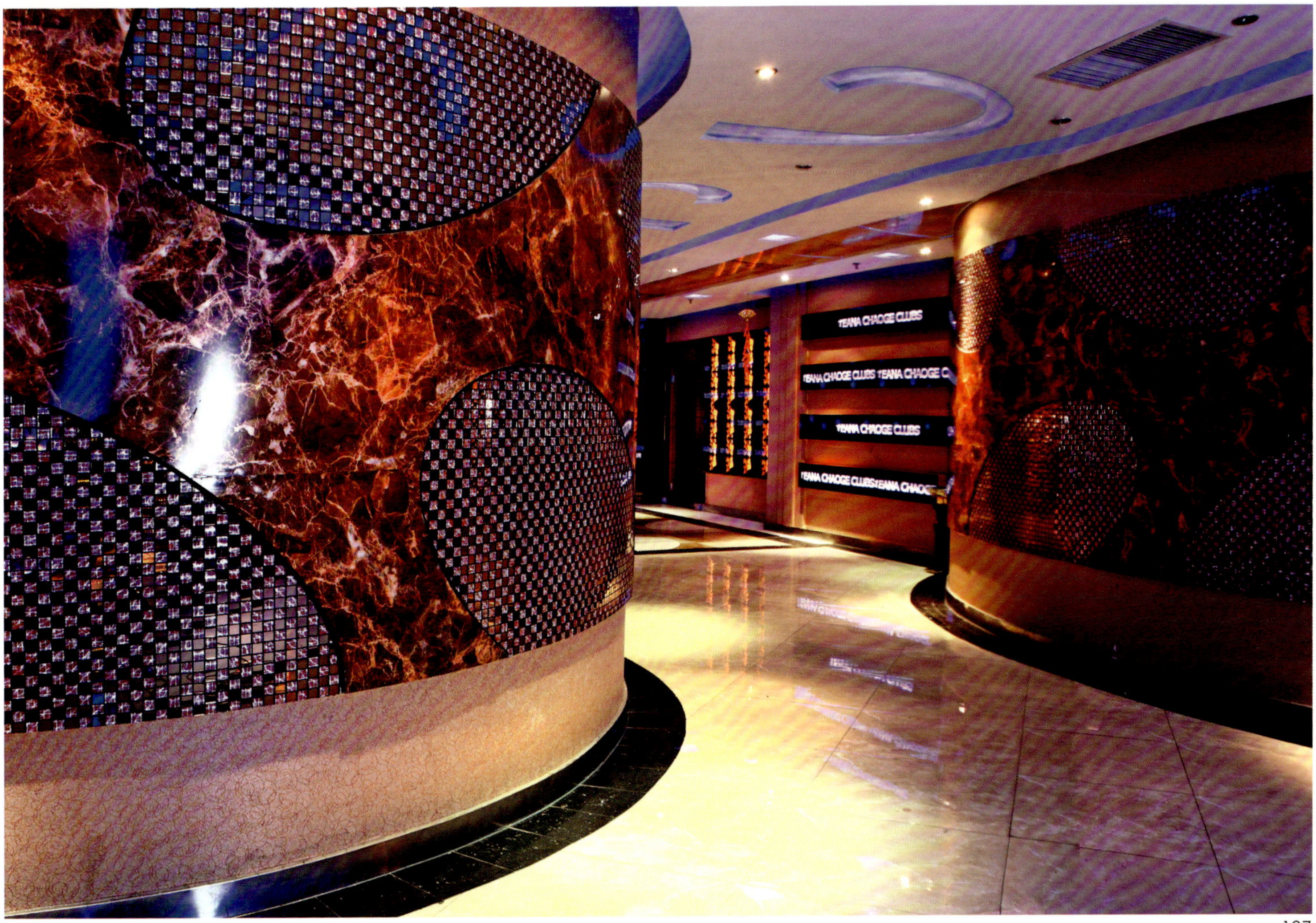
TEANA CHAOGE CLUBS
TEANA CHAOGE CLUBS

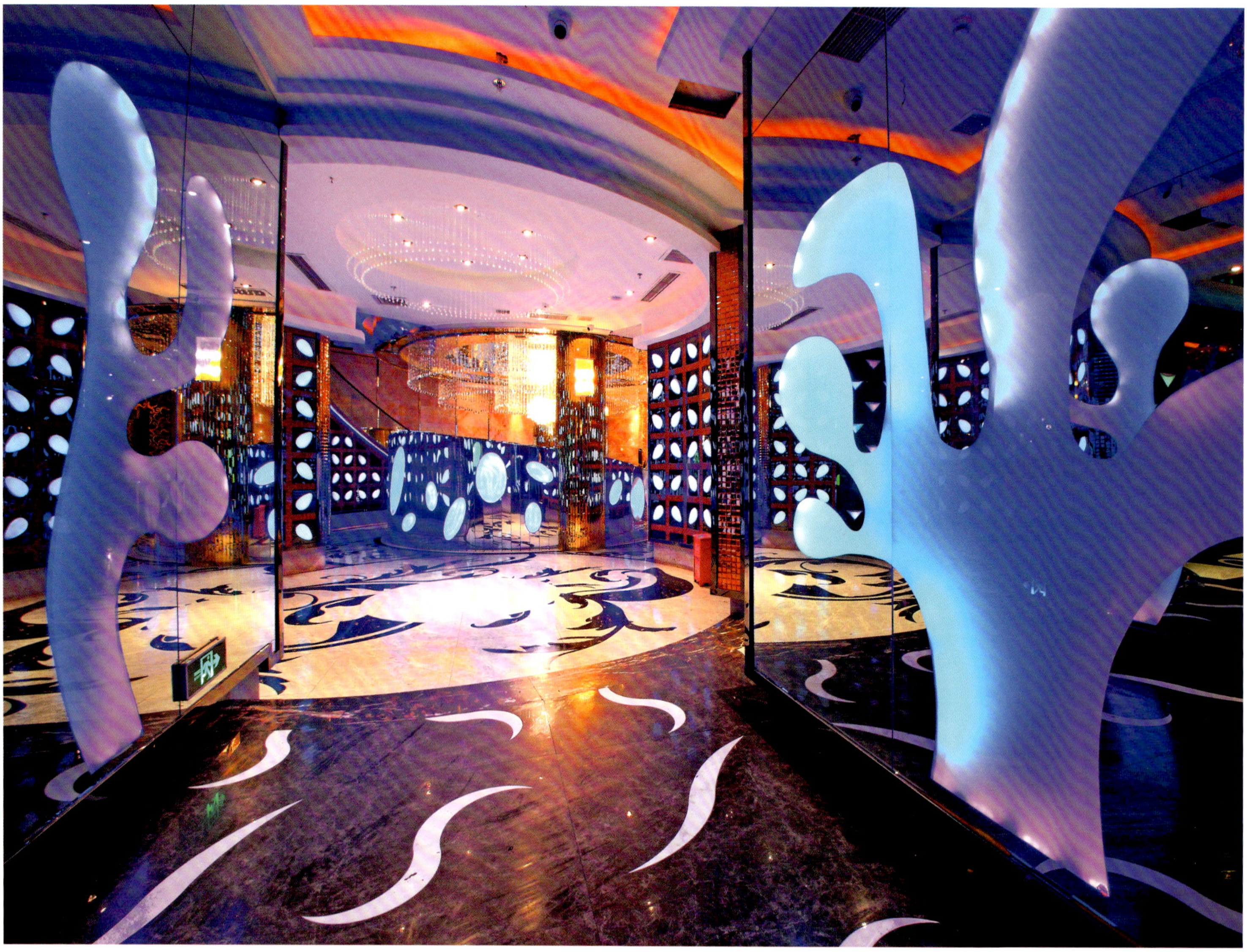

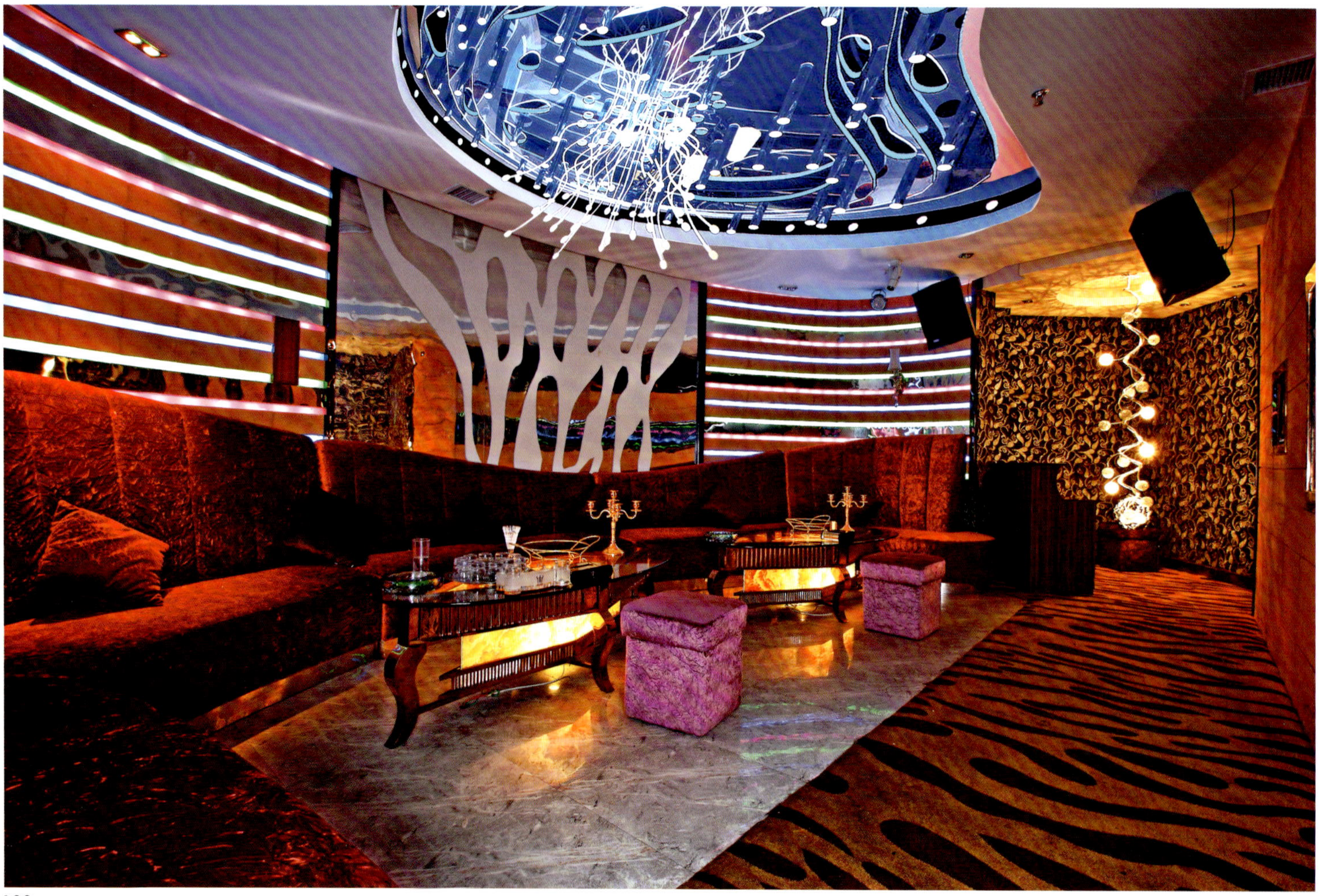

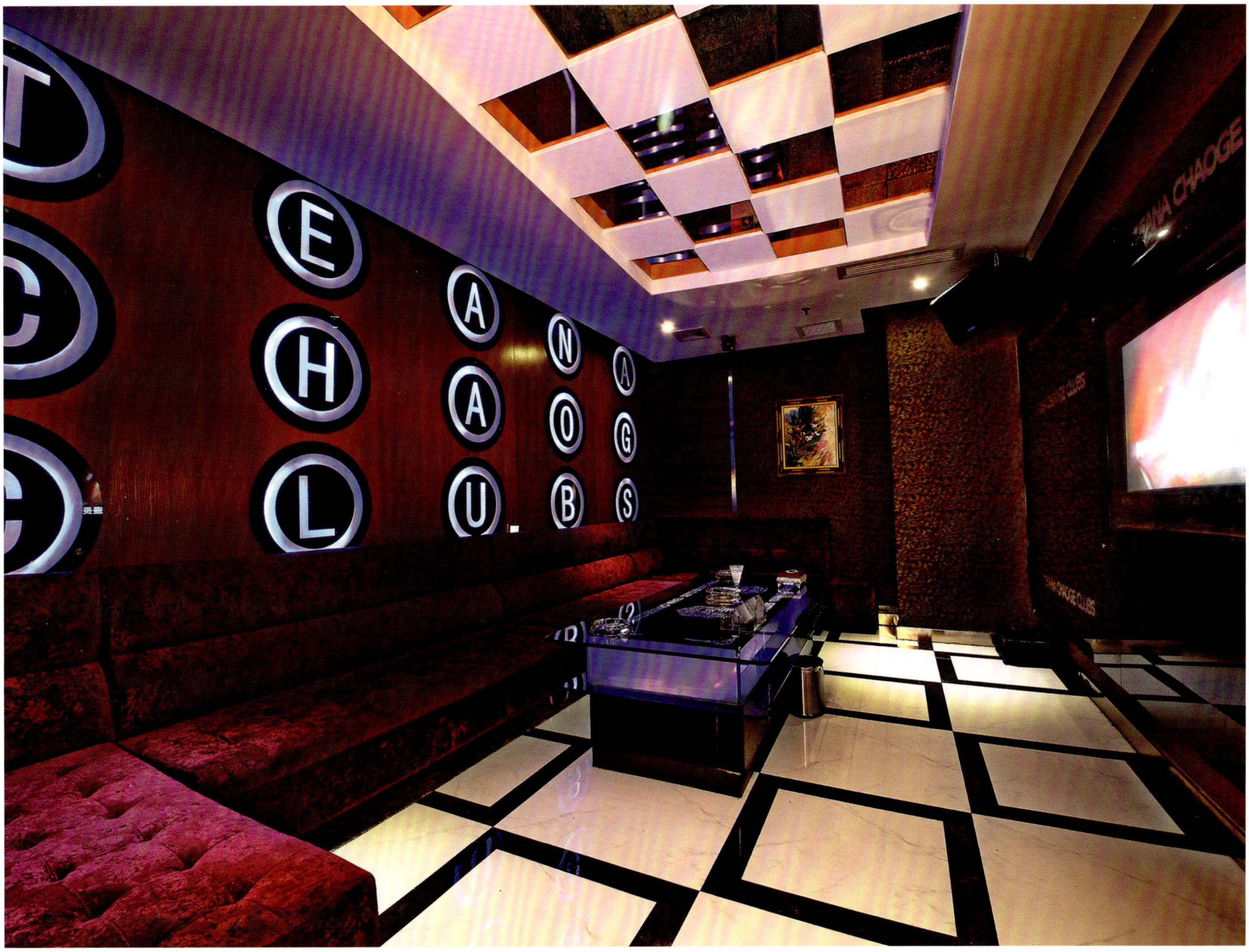

　　这是一个 KTV 项目，是面向时尚前卫一族的娱乐场所。因此，本会所的空间定义是璀璨热闹。作为夜场的设计，灯光的效果至关重要，并要在设计上出彩，要紧紧抓住消费者的心理，打造出自己的特色。本案设计师利用各种砖结合马赛克，再加上灯光的协调和衬托，凸显出空间的气势和夜场绚烂之下的艺术气息，这样的设计，才能让人印象深刻，从而吸引更多的人前来消费。

　　首先是接待大堂的设计，花瓣状的天花设计呼应着地面瓷砖的拼花造型，加之四周半围合的弧形墙设计，更加烘托出天地一体的气势，让空间充满整体感，而且各种砖和灯光的运用也做了很好的对比搭配和组合，包厢和走道也采用不同的处理方式来设计，让前来消费的顾客感受到不一样的气场和魅力。

地点	面积	设计师	设计公司	主要材料
/ 景德镇	/ 502m²	/ 邹巍、高波	/ 景德镇市东航室内装饰设计有限公司	/ 马赛克、玻璃、镜子、装饰板、光纤灯等

2012 Music Club

—— 2012 音乐会所

Nightclub is a different space for people to relax. Therefore, as a themed space of life in nightclub, it provides a unique experience, calmness and desire, extrusion and release, culture of weightlessness and hint of light…

The space planning meets the needs of overall function and operation, and on a neat and careful basic arrangement, the designers have ideas at hands, endowing the space with a luxurious and modern appearance of diversified entertainment space. The dazzling LED, psychedelic lights and charming shapes are staggered together and shine with each other, fascinating. In the design of shapes, the designers pay attention to the focusing on the contrast of decorative elements, while in lighting, they emphasize on the changes of colors. The whole atmosphere is full of modern and luxurious feeling, attracting customers to visit and consume, reaching the commercial effect of decoration and meeting the psychological needs of customers.

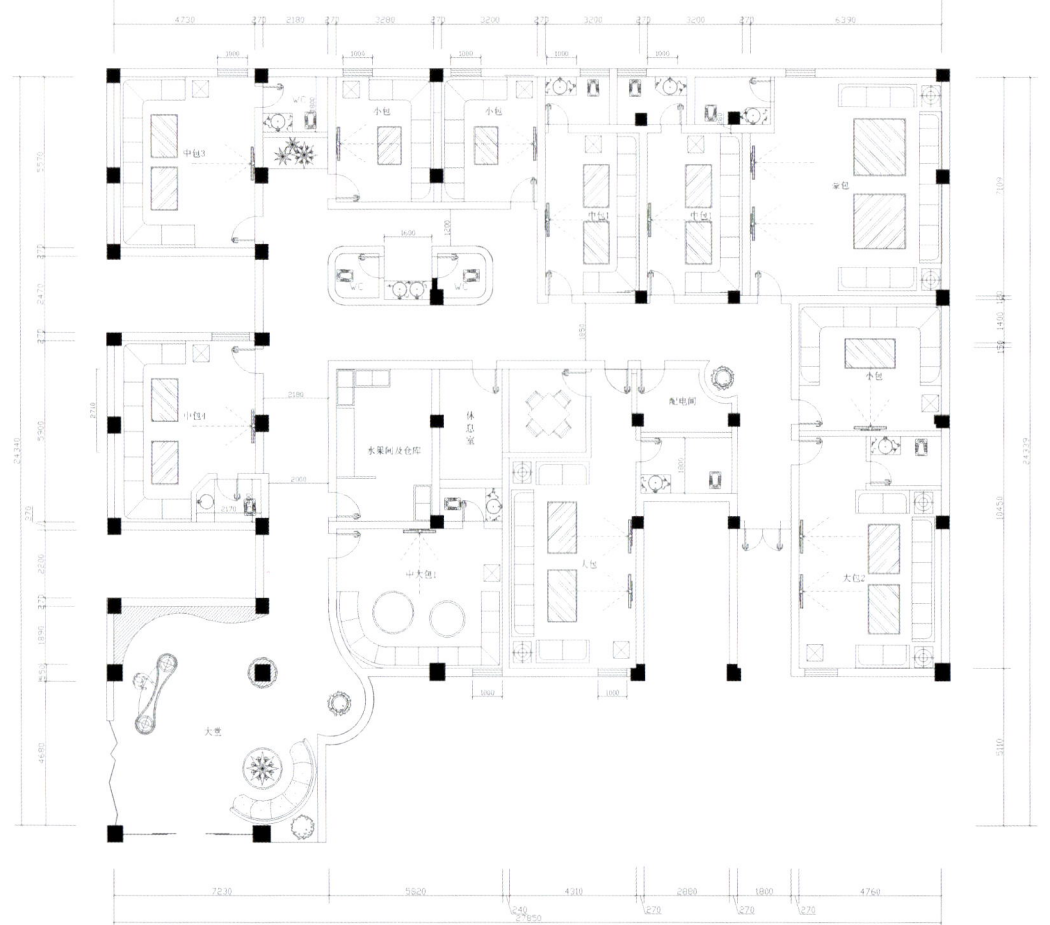

夜场是人们寻找常态之外的另一个放松个体的空间。因此，作为夜店生活的主题空间，就该给人别具风格的体验：平静和欲望，挤压和释放，失重的文化，光线的暗示……

空间的规划在满足整体功能与运营需求的情况下，在工整缜密的基础安排之上，设计师信手挥洒，赋予空间一个奢华、现代的多元化娱乐空间的外在。炫目的LED、迷幻的灯光、迷人的造型交错放光，令人神迷。在造型的处理上，设计师注意强调装饰元素的对比，而在灯光的处理上却强调色彩的变化。整体的氛围充满了现代奢华的感觉，吸引顾客驻足消费，既达到了装潢的商业效果，也满足了顾客的心理需求。

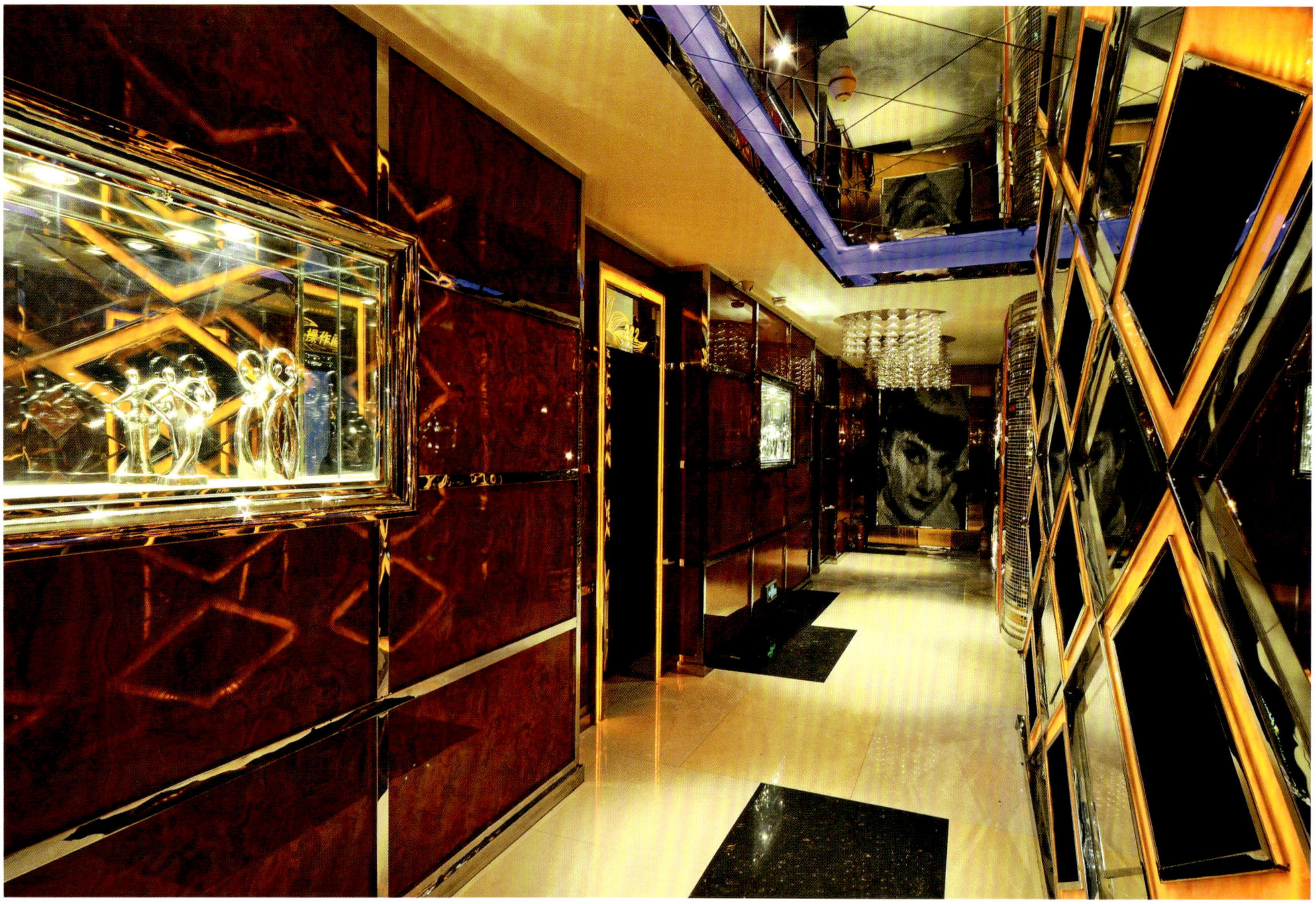

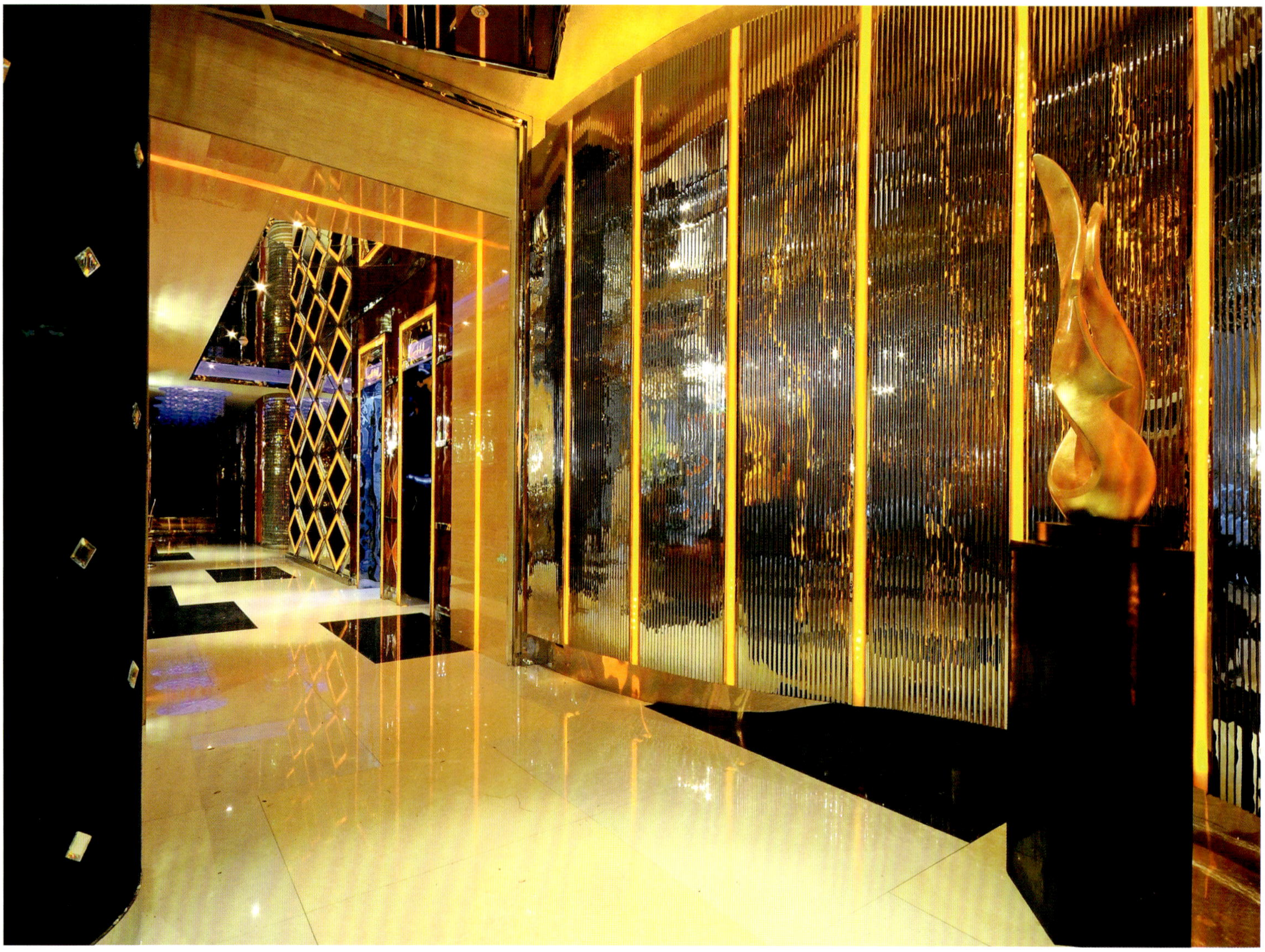

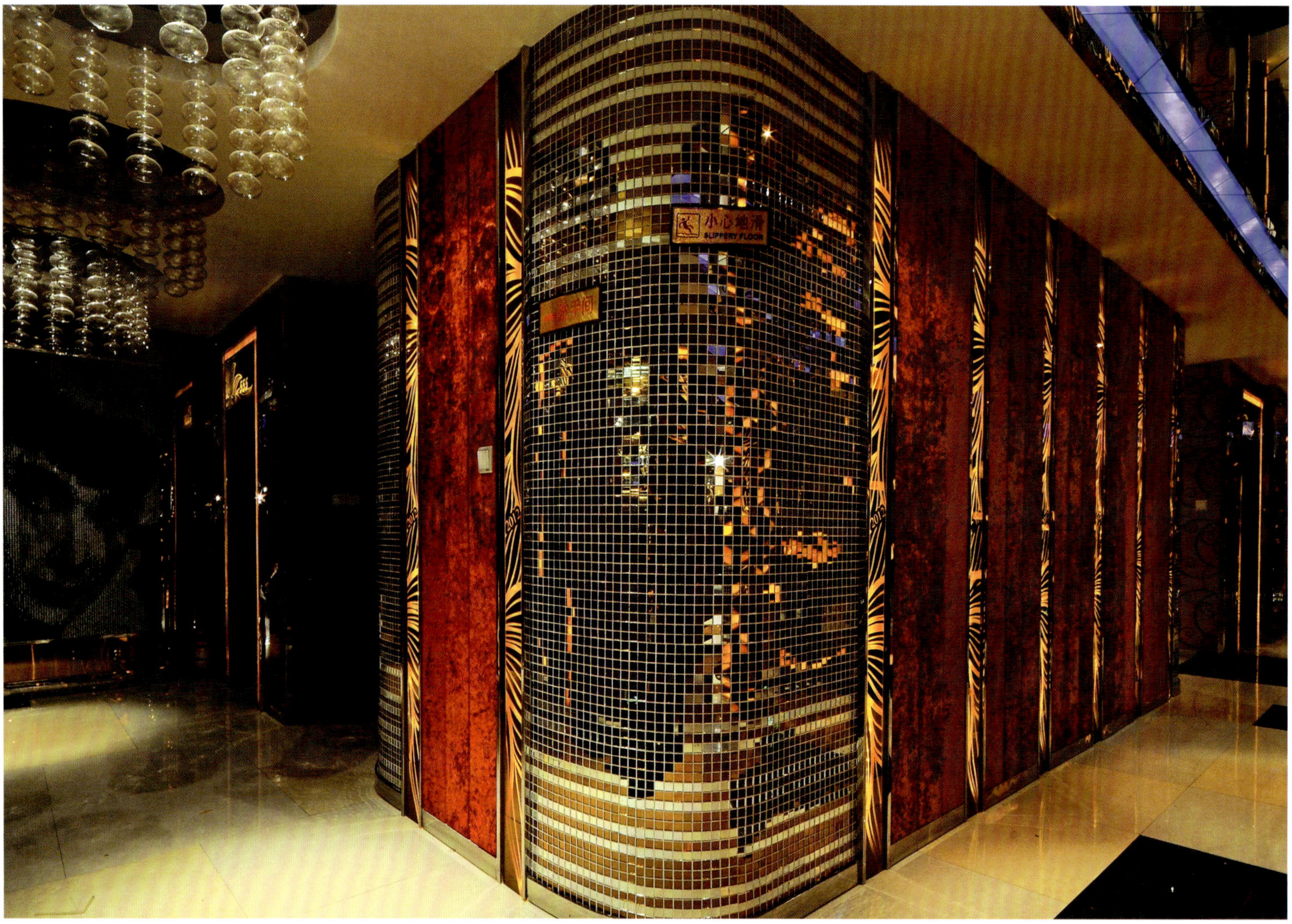
小心地滑
SLIPPERY FLOOR

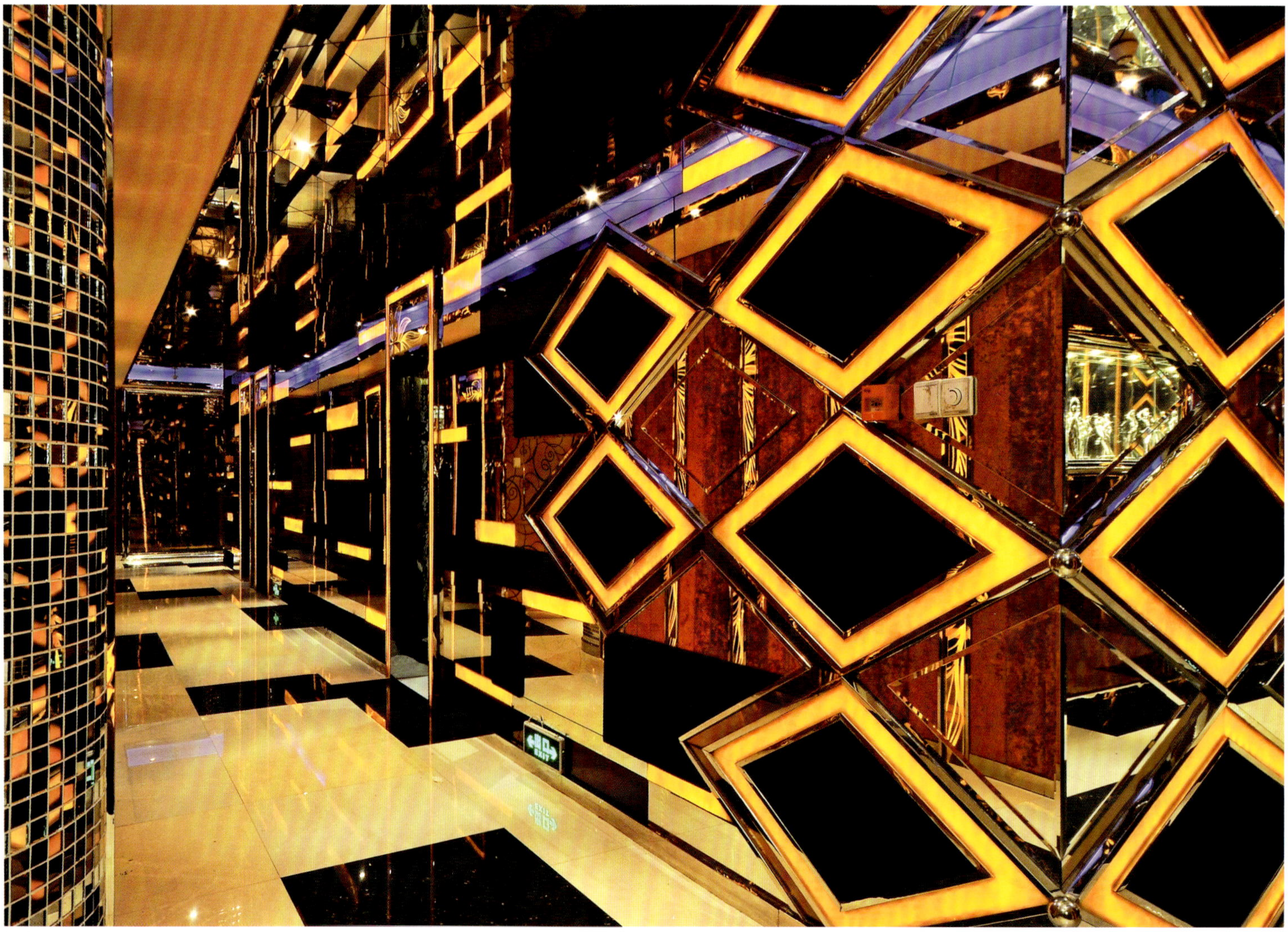

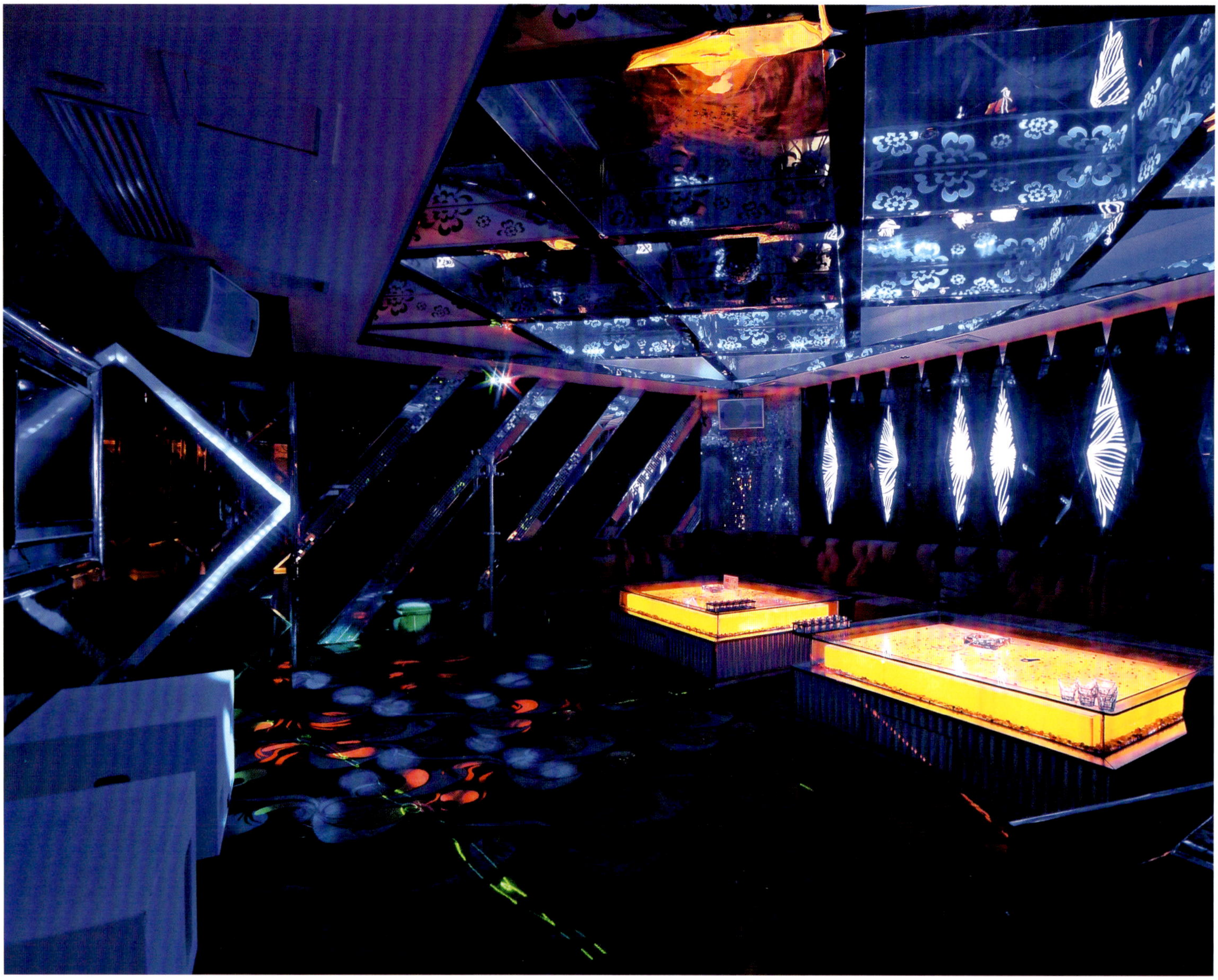

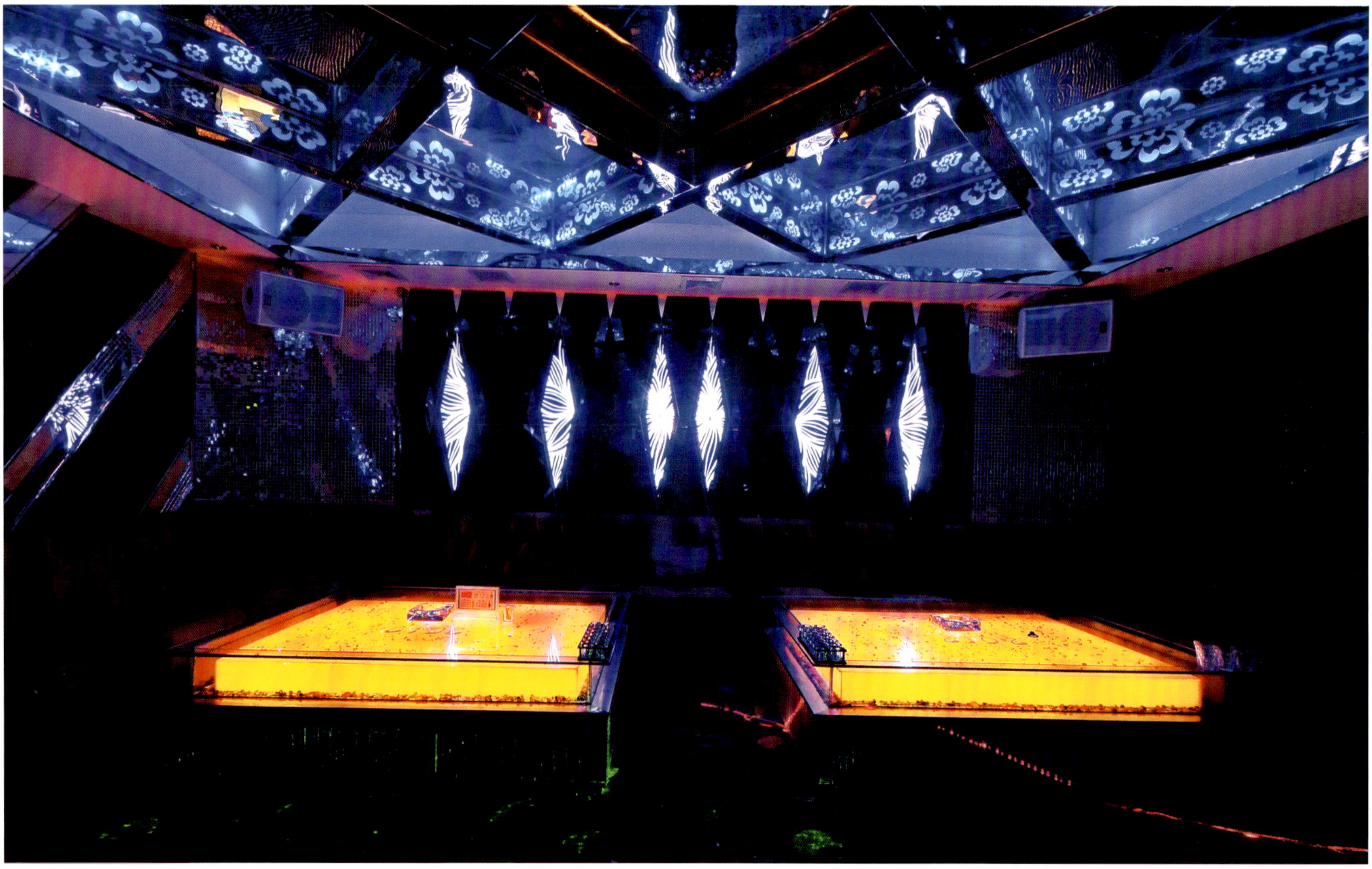

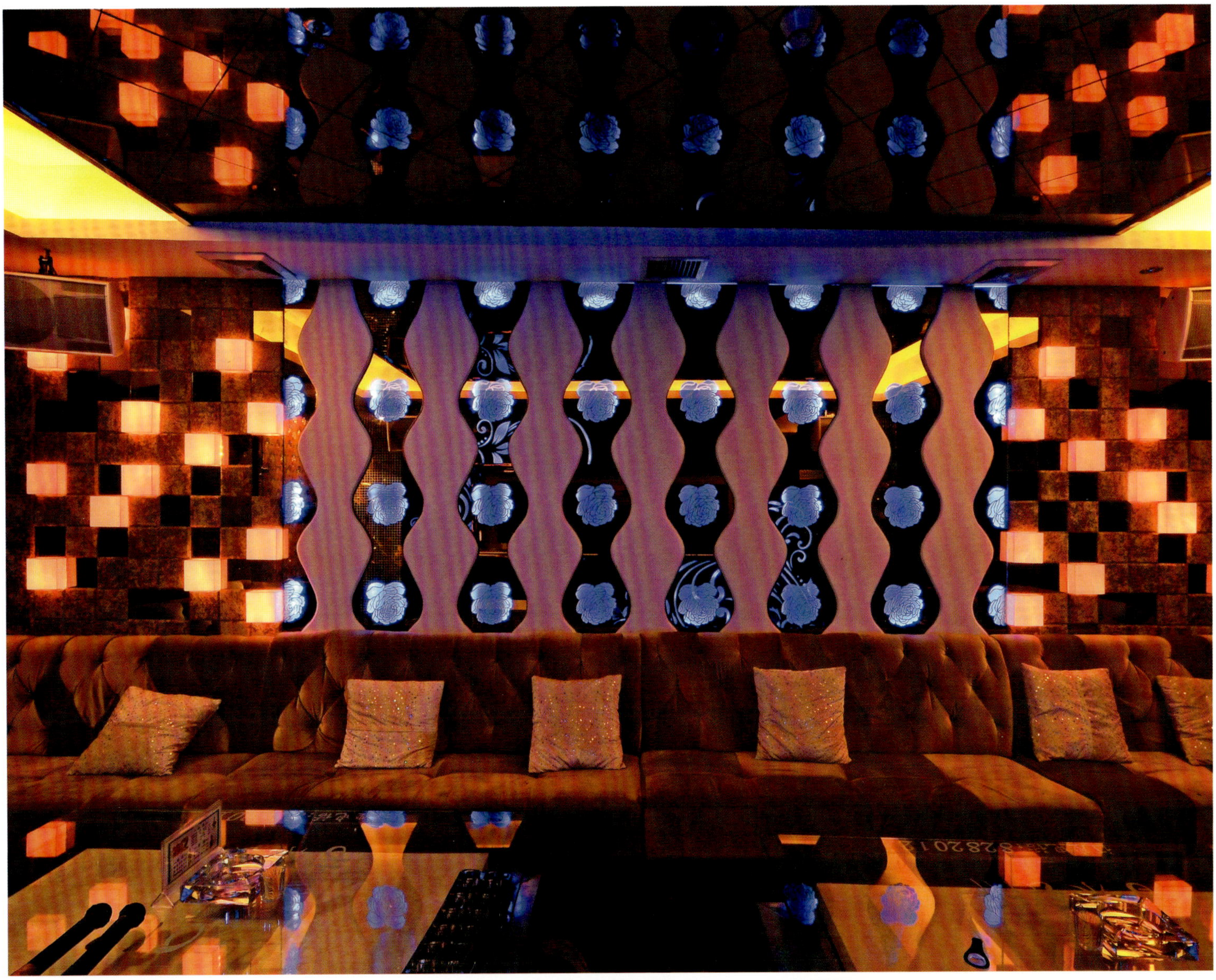

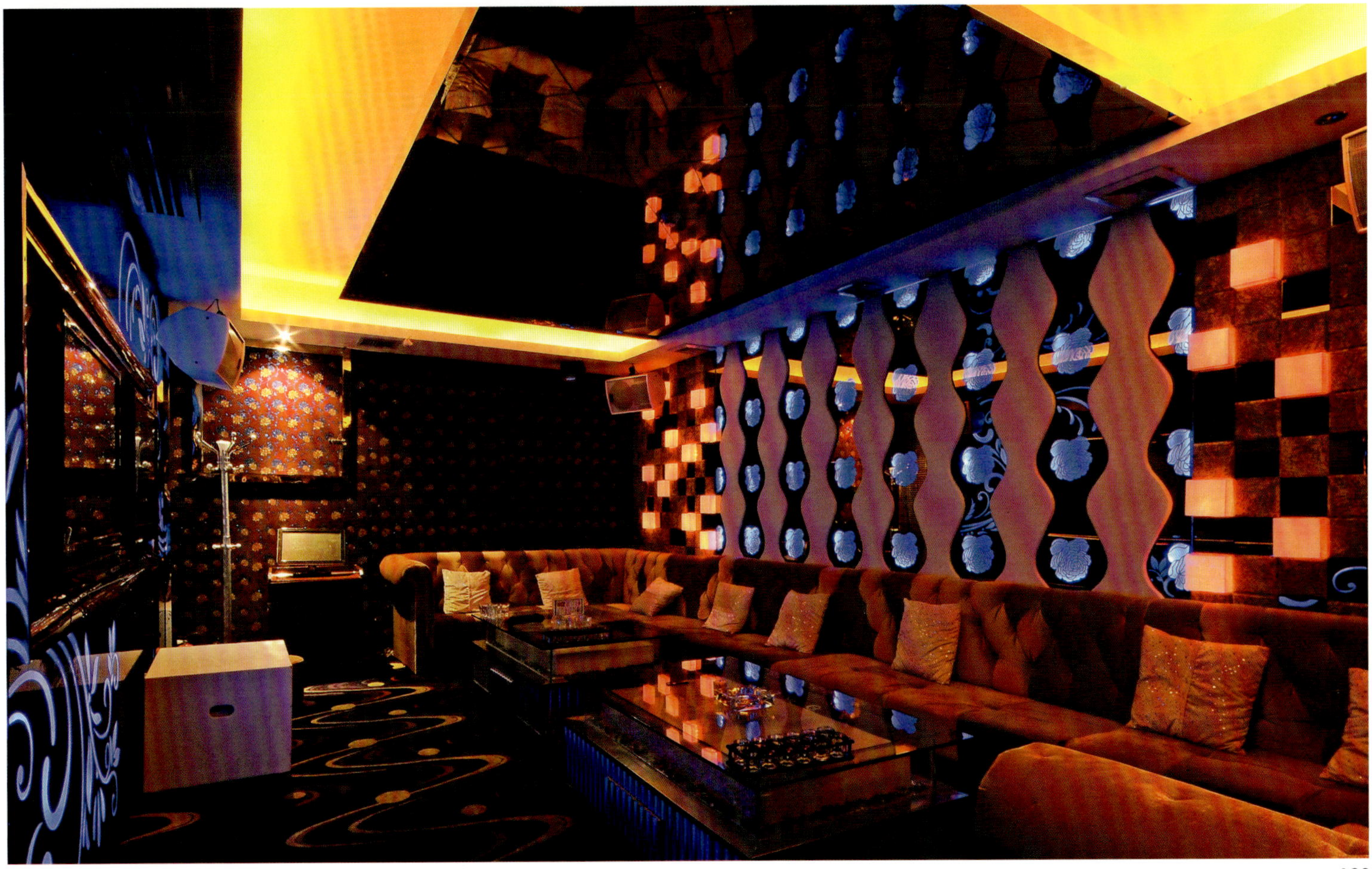

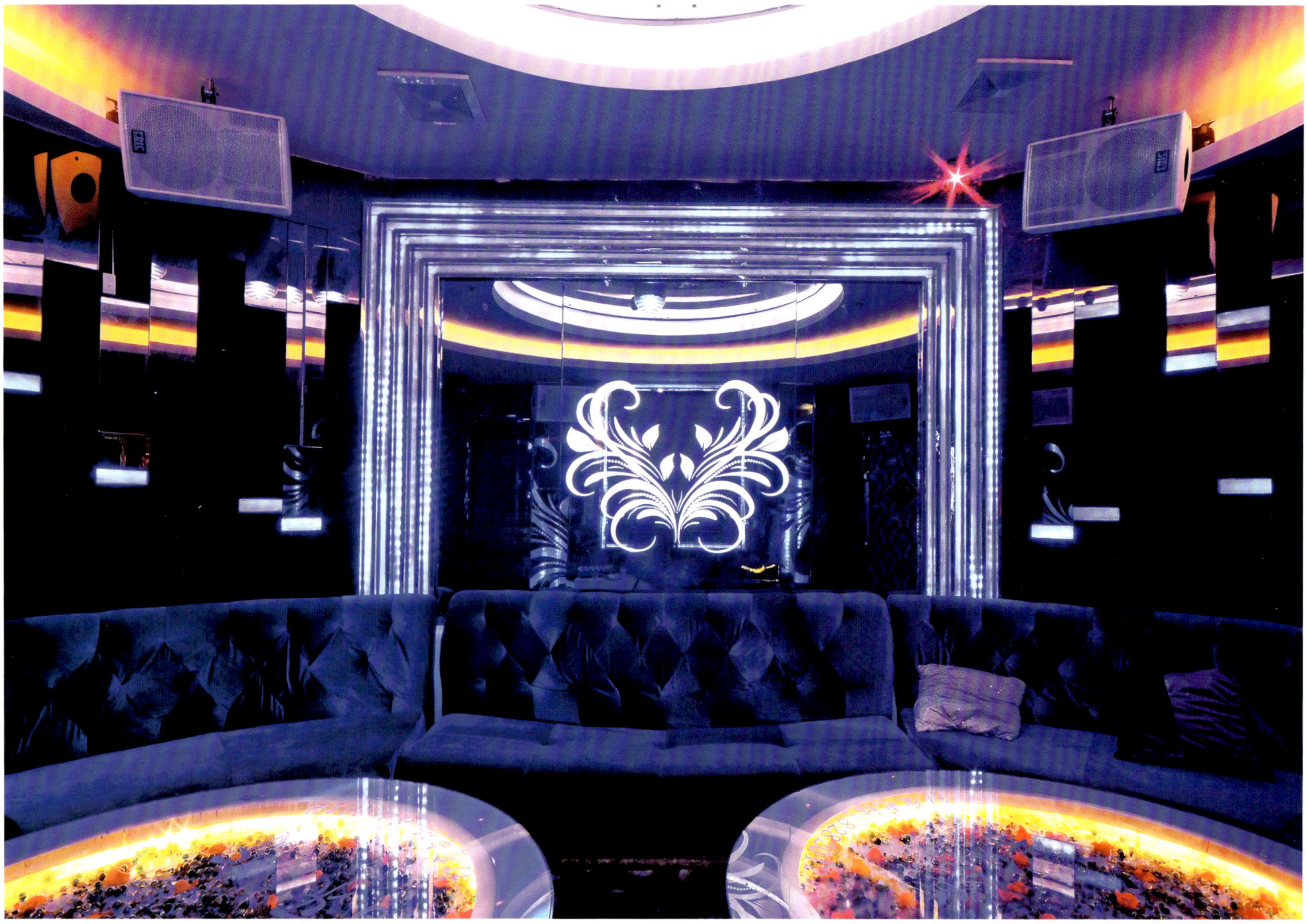

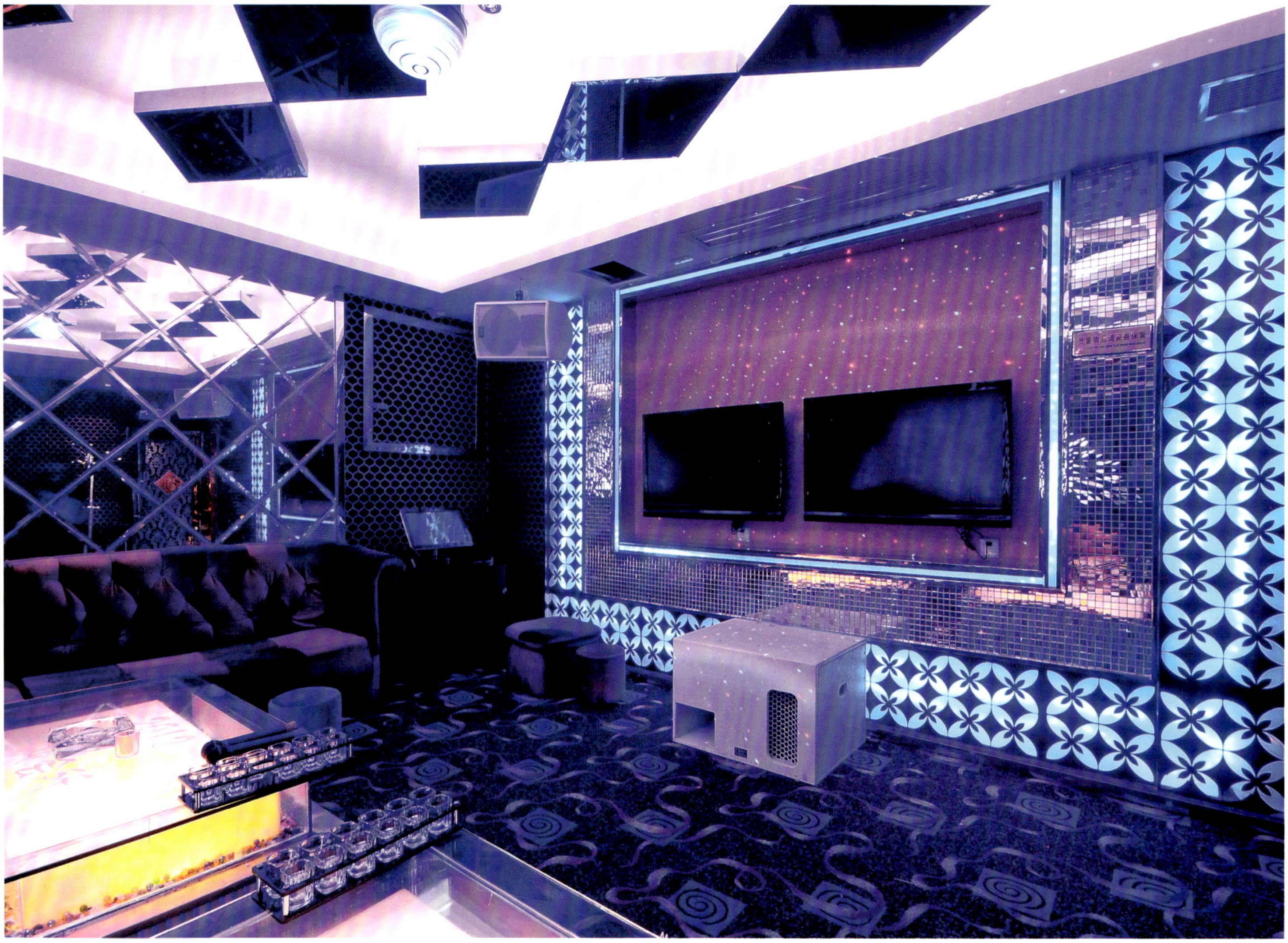

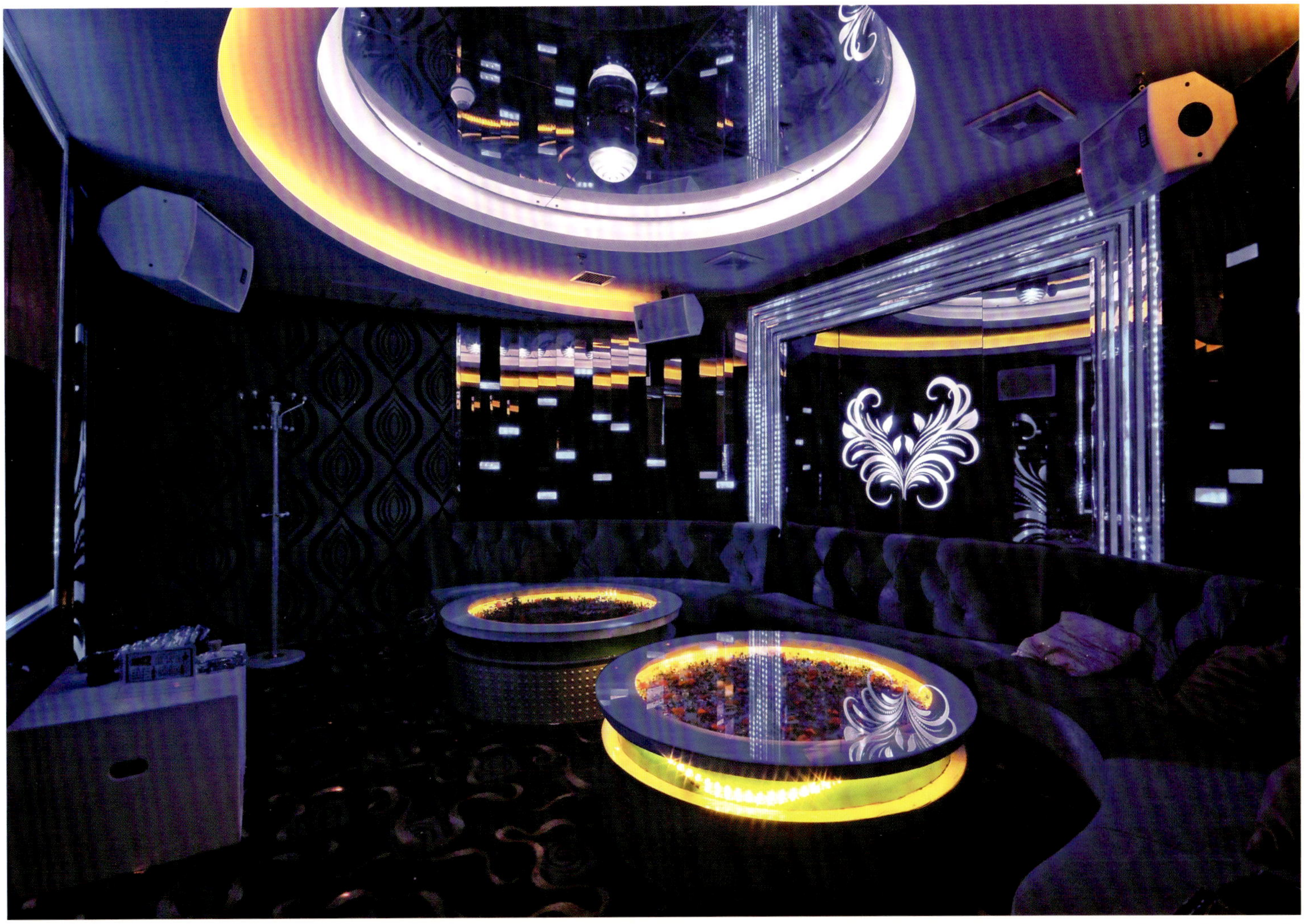

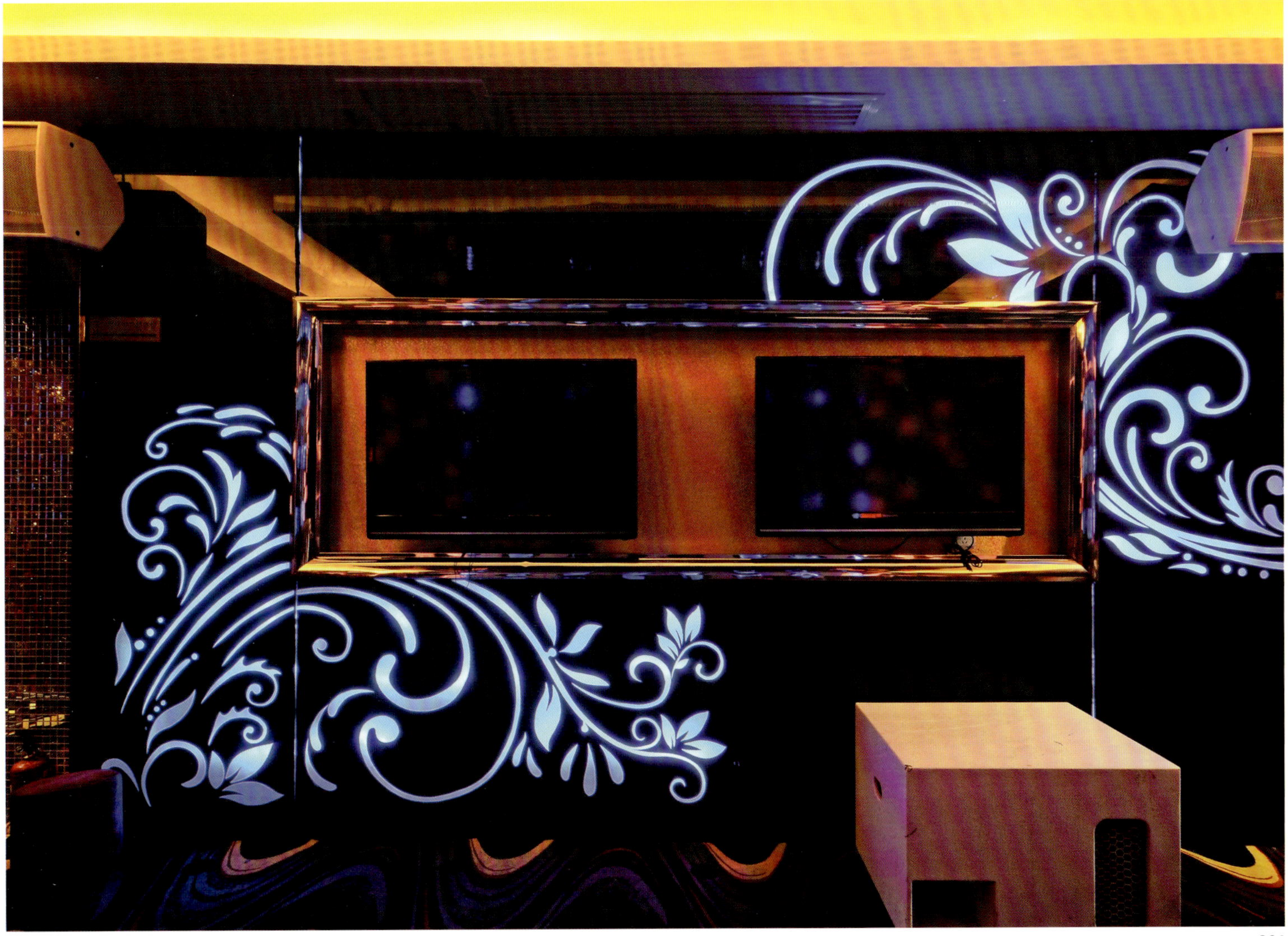

地 点	面 积	设计师	设计公司	主要材料
/ 贵州	/ 2700m²	/ 杨彦	/ 深圳品彦室内设计有限公司	/ 进口羊绒地毯、卡布奇诺砖、石材、玻璃马赛克、抛光砖、烤漆玻璃、烤漆板、亚克力透光灯片、皮革、马赛克、墙纸、布艺软包

Teana Chaoge Club

___ 天籁·朝歌

The booming development of entertainment industry has intensified the competition in this area, and nowadays the consumers no longer indulge in the luxurious and flashy style in the past and begin to pursue the connotation of sound and space. Therefore, in the design, the designer introduces comparatively pure group of colors, coupled with refreshing space layout, to match the simple and fashionable design with a low-key and luxurious atmosphere, creating a high-quality nightclub.

In the design of the appearance and the entrance of the building, the brilliant lighting model attracts the attention of passers-by, and widening the entrance with the extension created by the divergent effect of light, fascinating. The inside public areas and walkways emphasize the echoes of ceiling modeling and the floor to highlight the momentum of the space, and to create the overall sense. The ceiling is created with lighting effects, coupled with modern materials and high-tech equipments, and the ground is designed with floor tiles and carpet to create a gorgeous temperament. The private rooms are designed with comfortable sofas coupled with pure lights, expressing a gorgeous feeling in the simplicity.

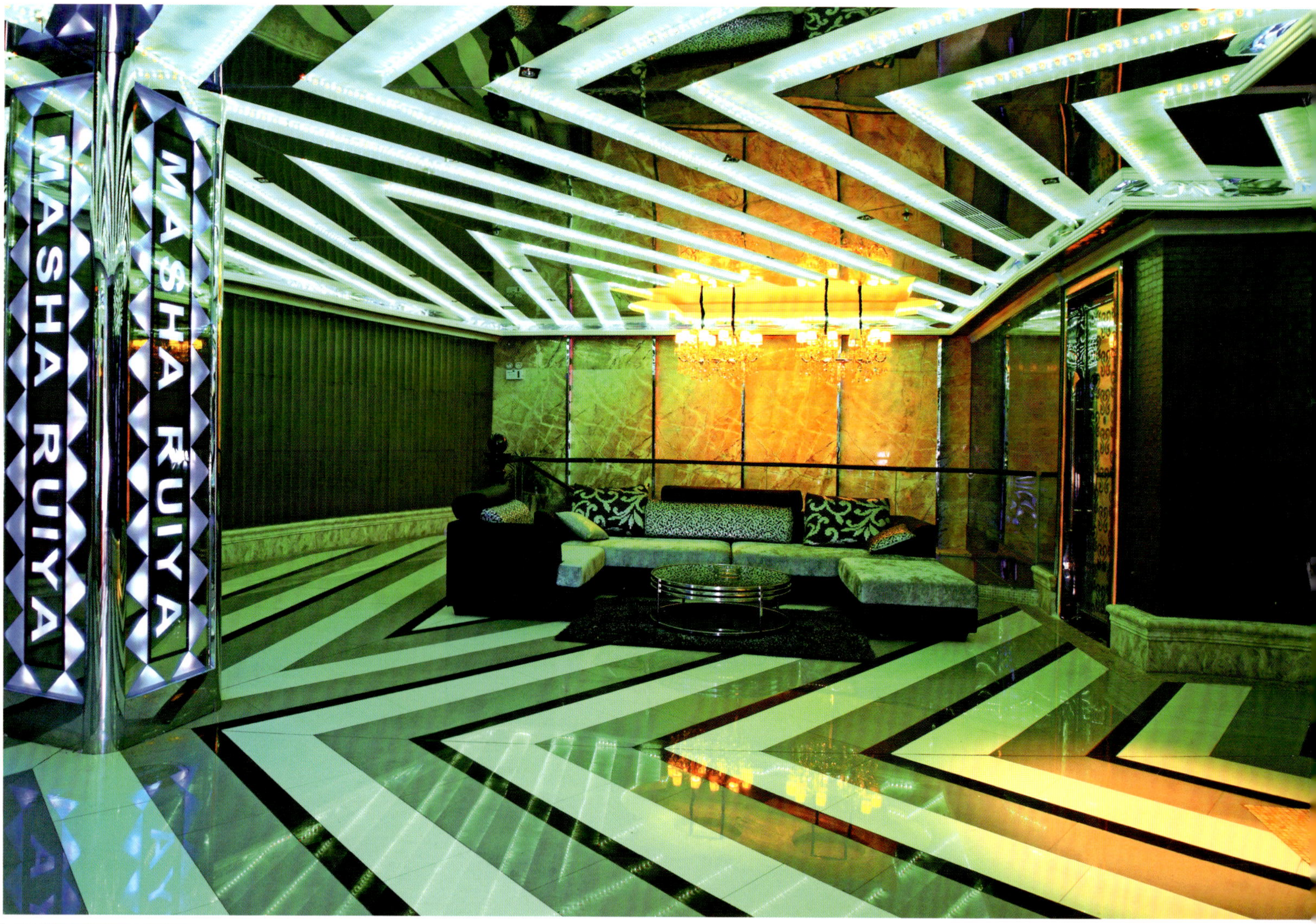
MASHA RUIYA
MASHA RUIYA

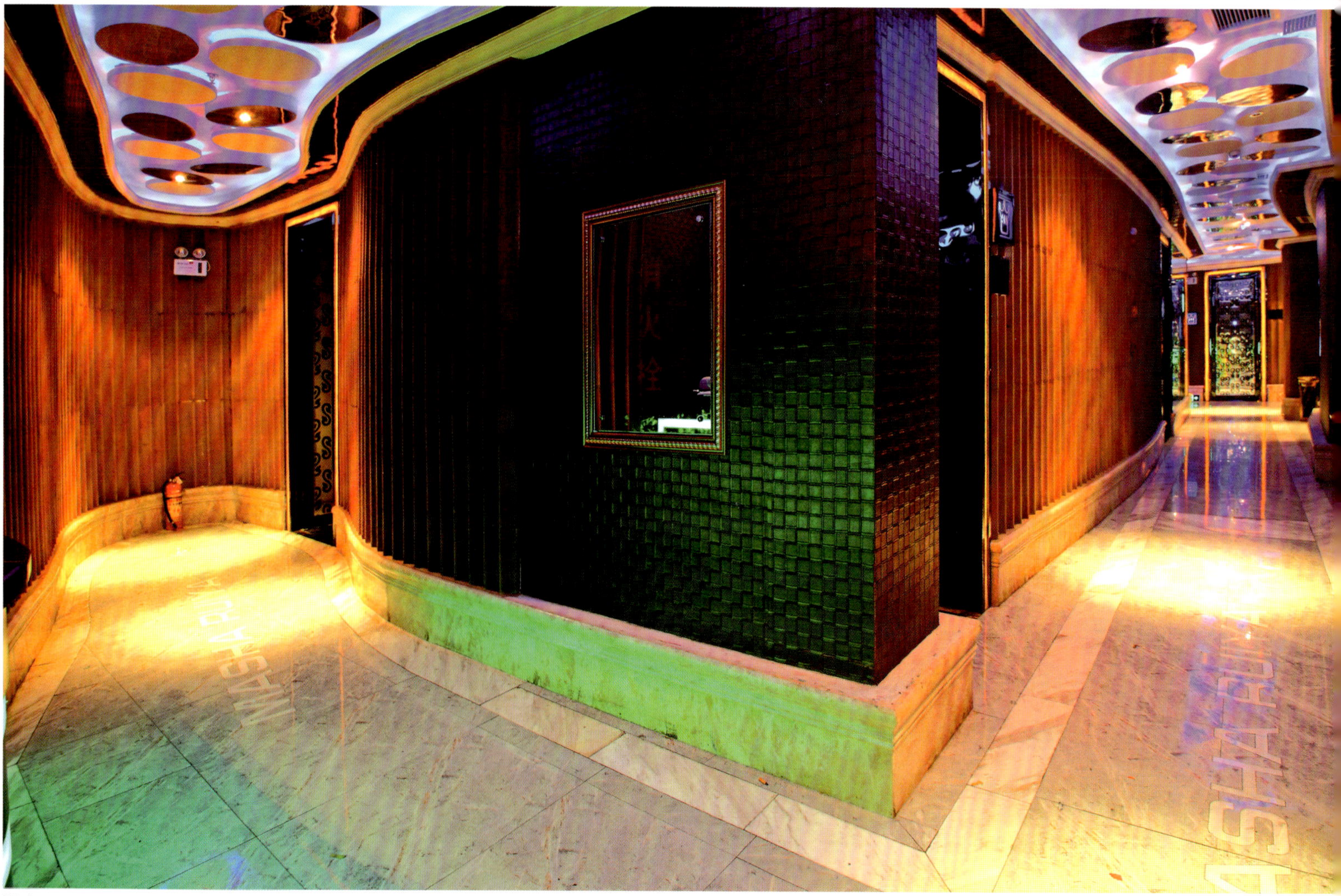

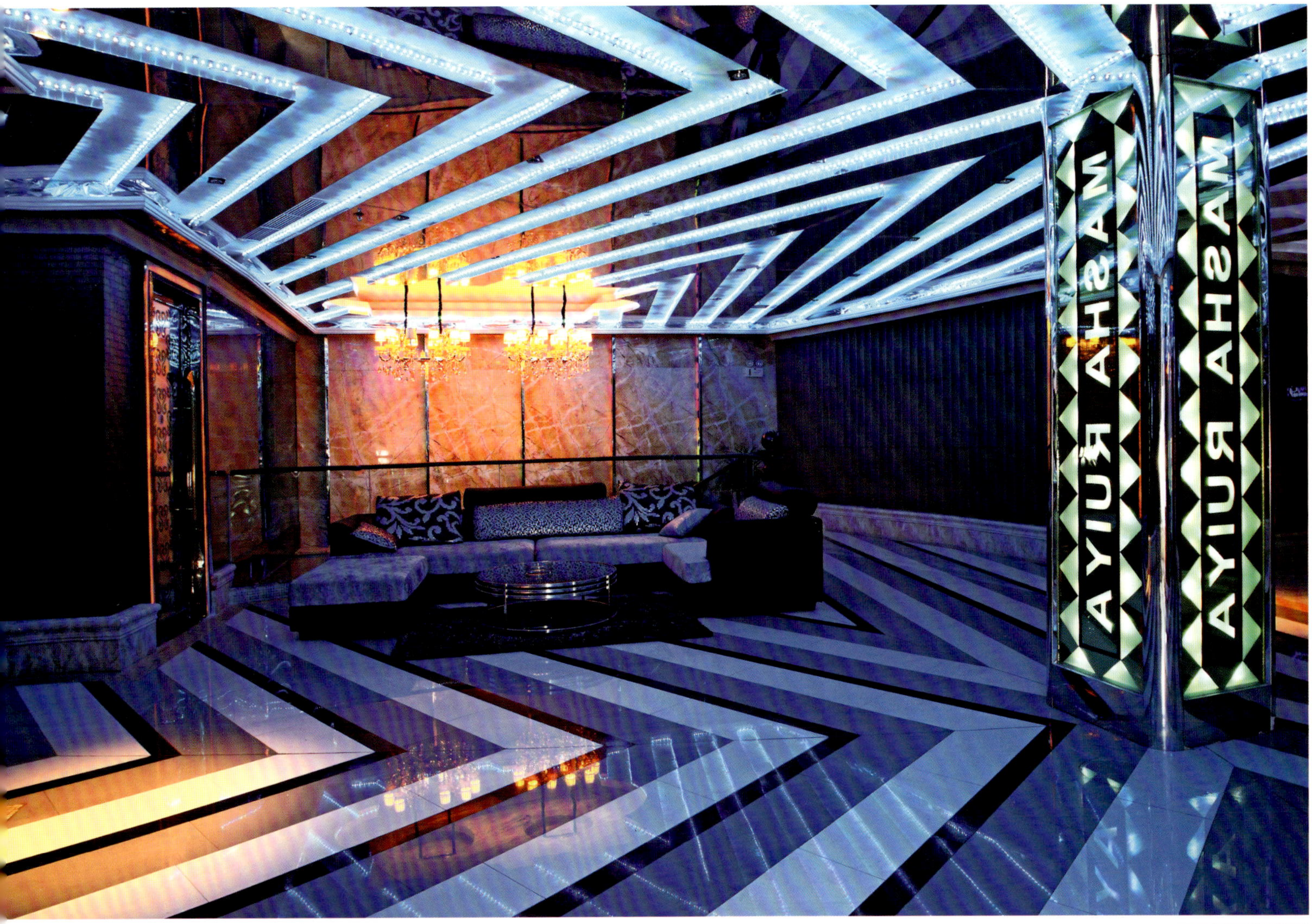

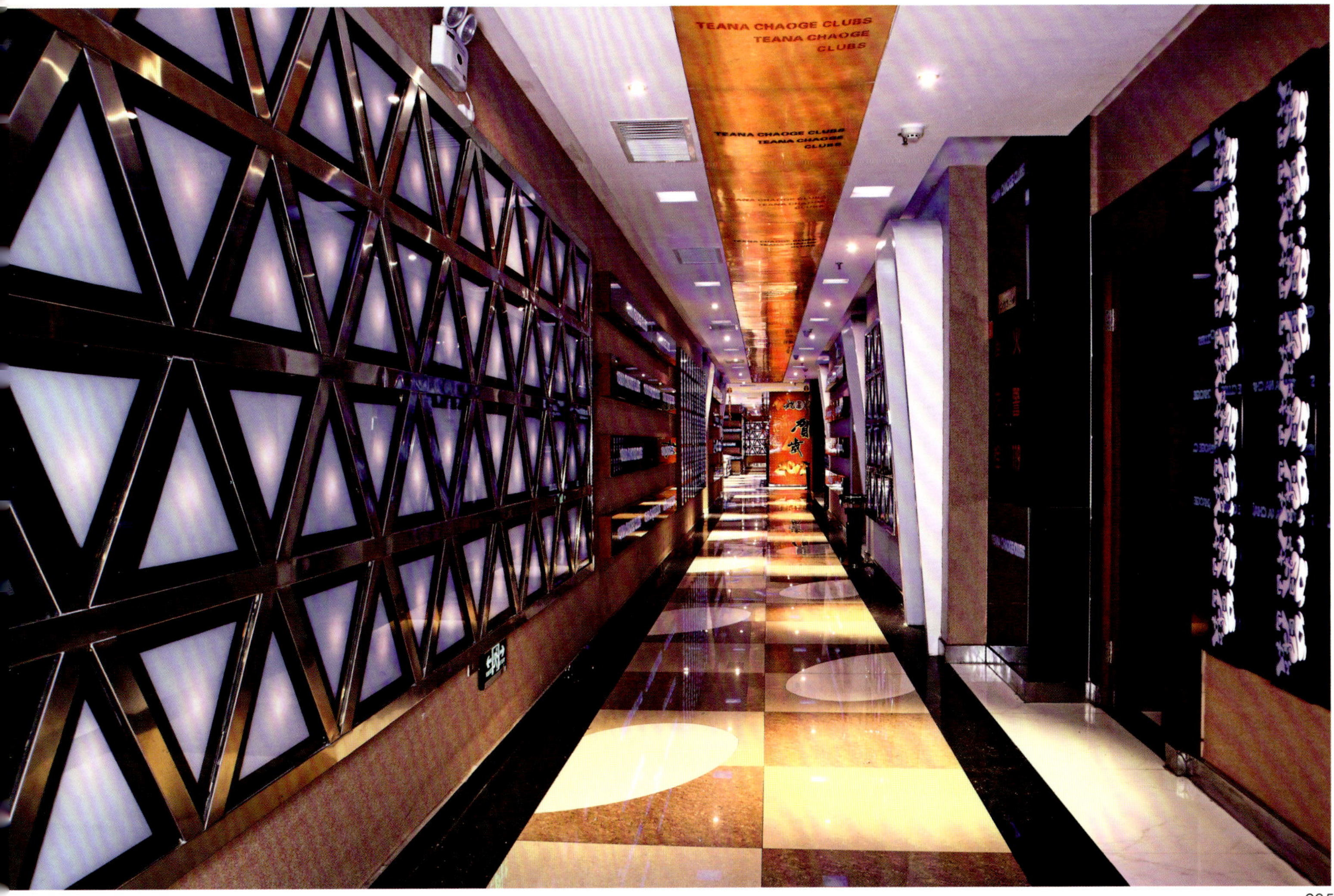

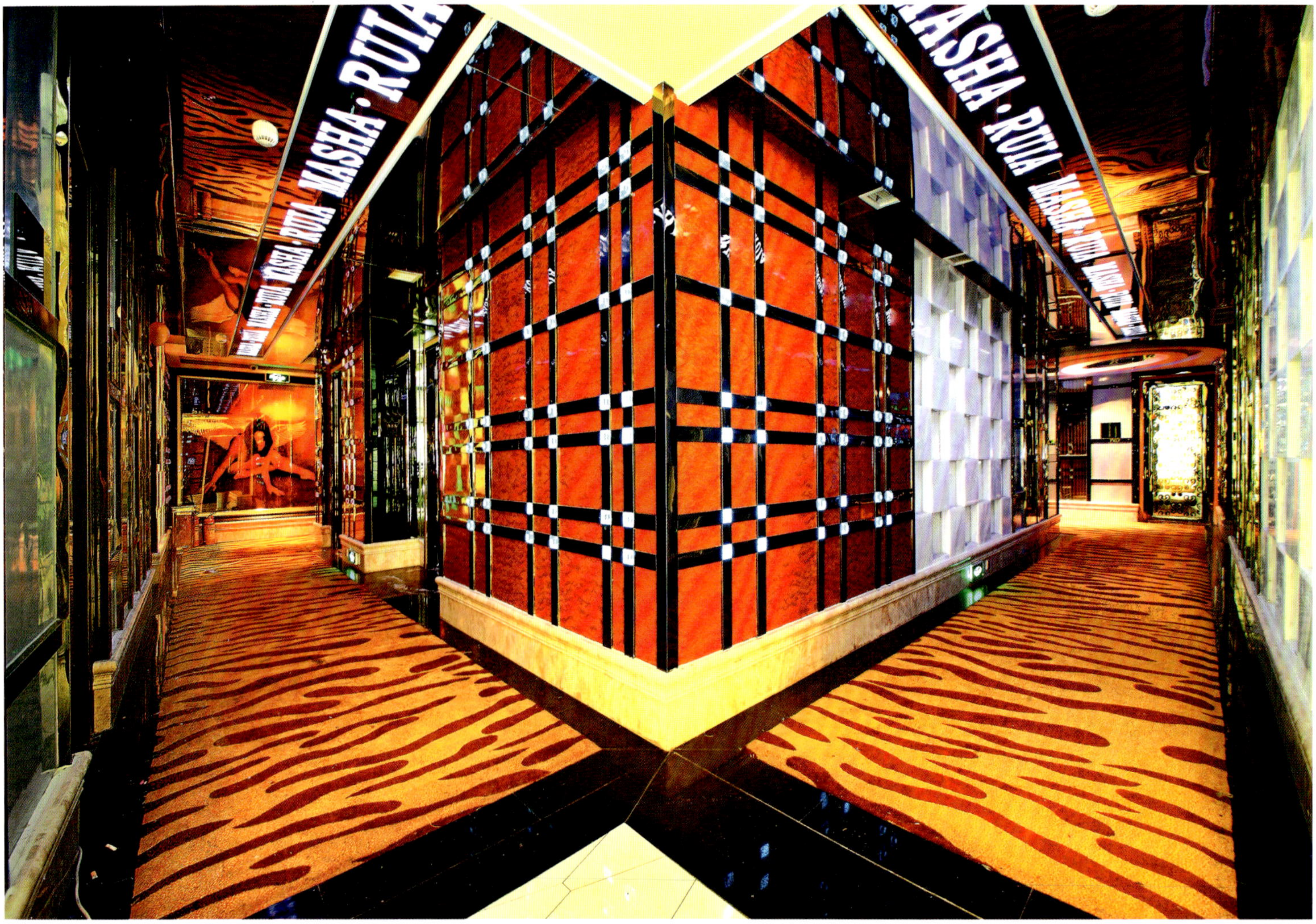

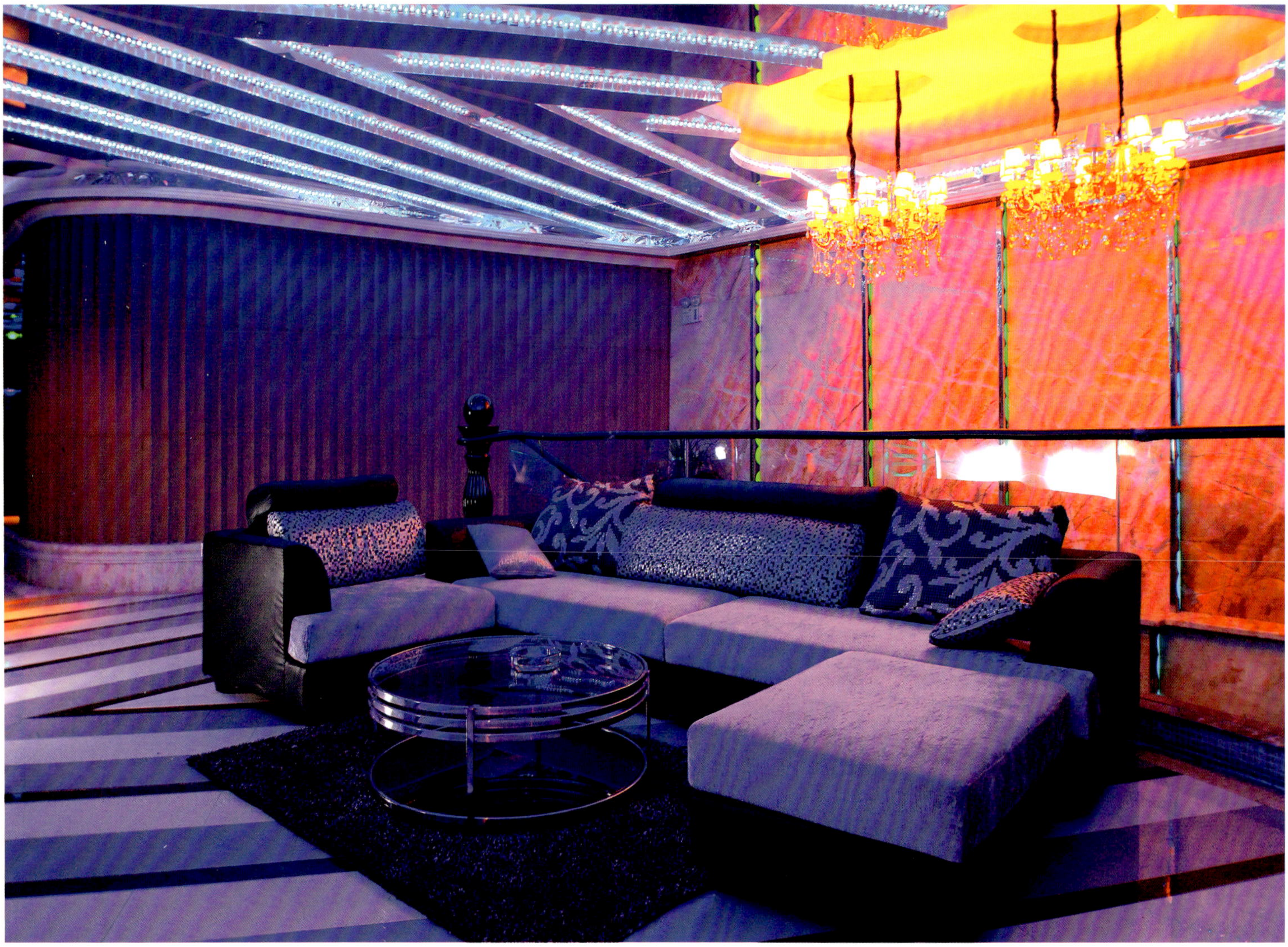

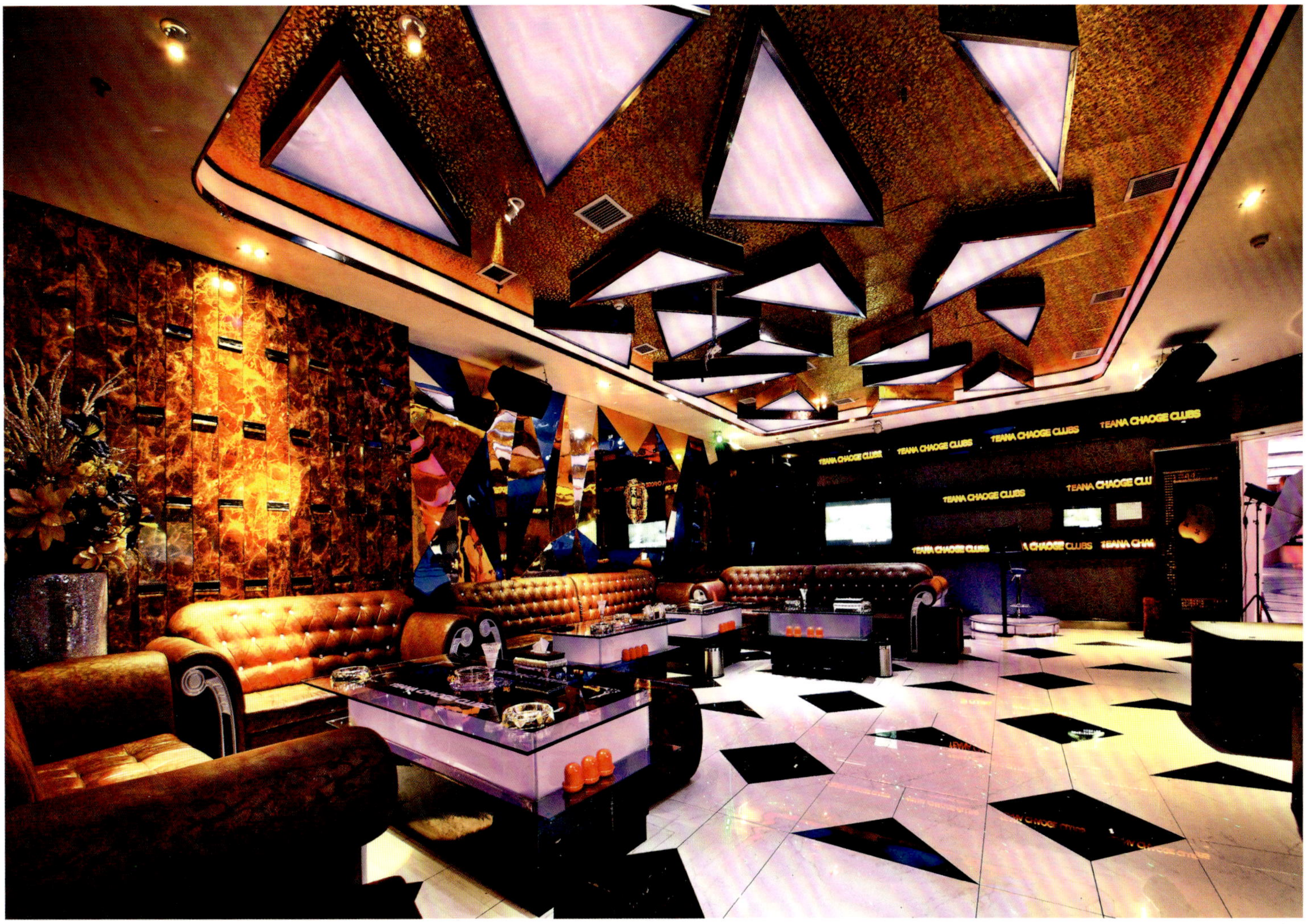
TEANA CHAOGE CLUBS

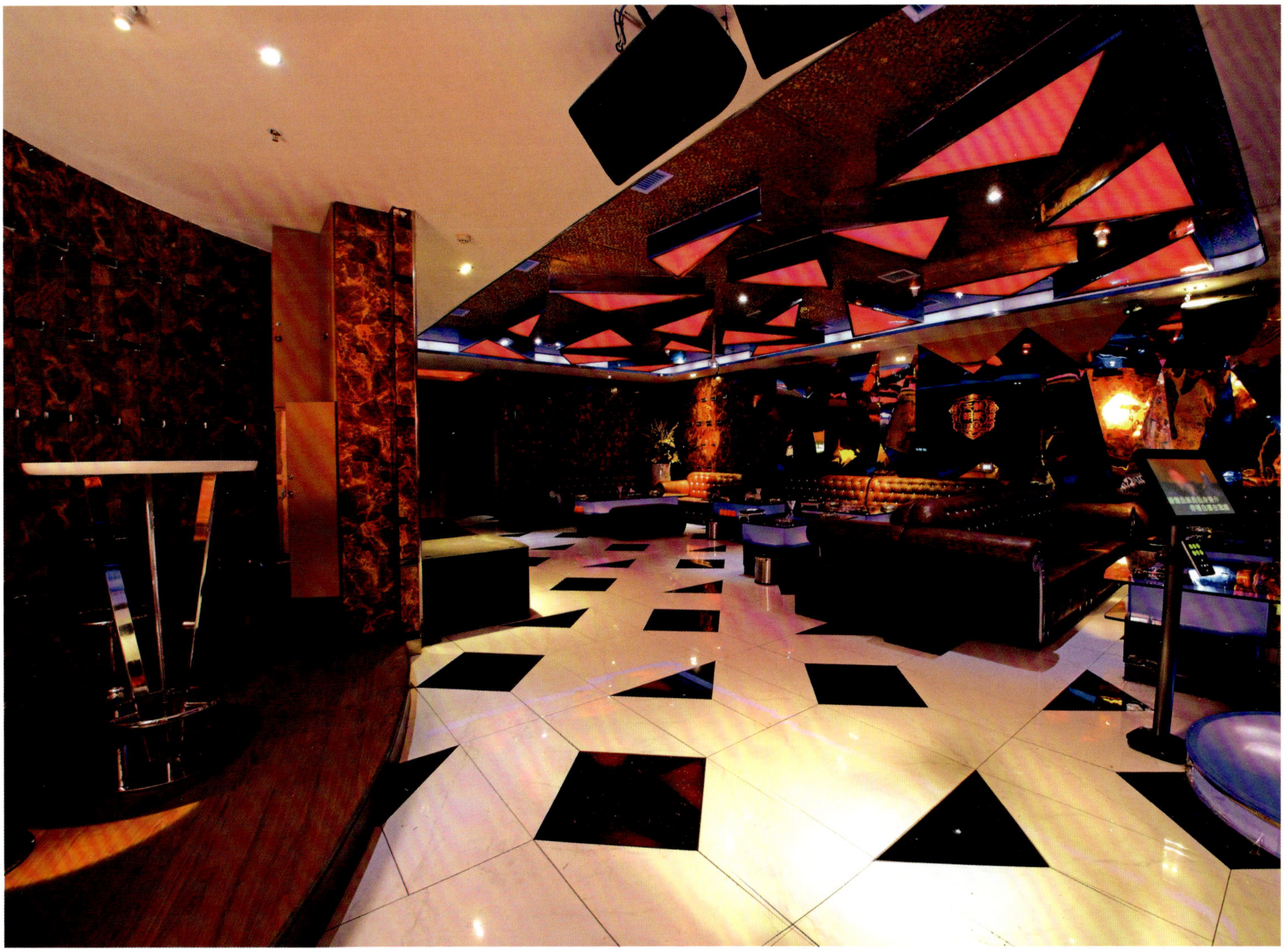

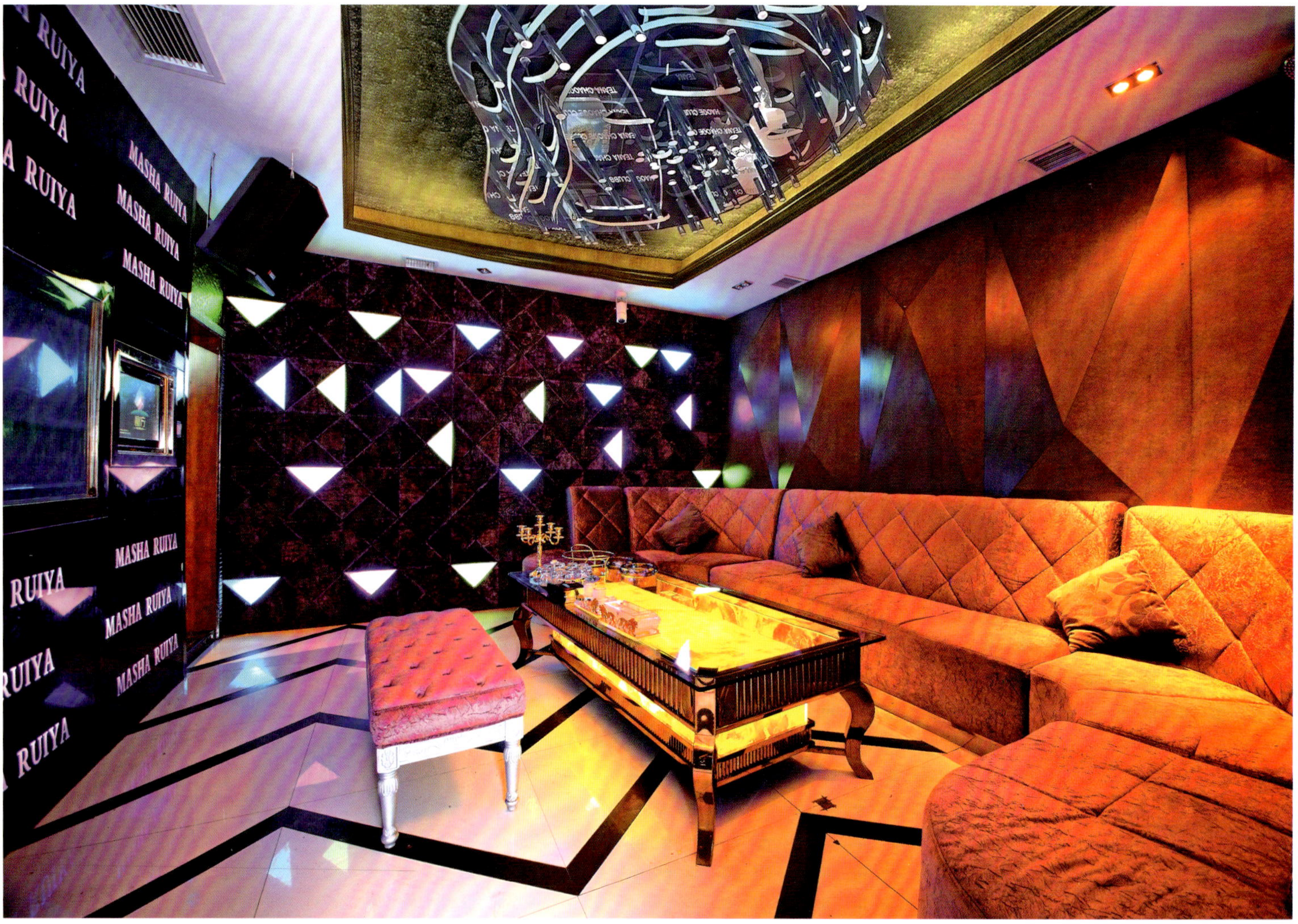
RUIYA
MASHA RUIYA

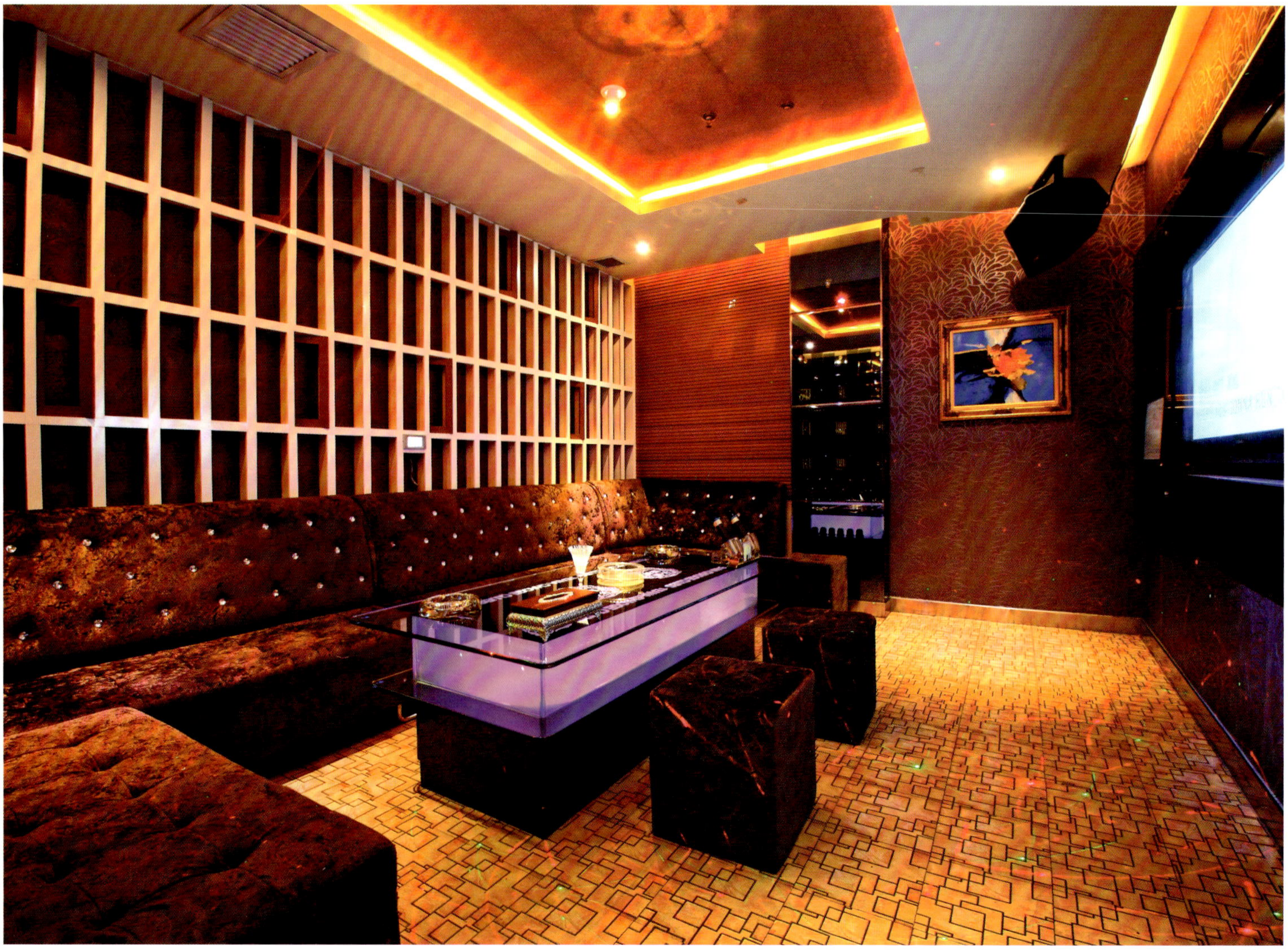

　　娱乐行业的蓬勃发展加剧了行业的竞争，加上如今的消费者不再沉迷于过去奢侈浮华的风格，开始追求音效和空间内涵，因此，在本案的设计中，设计师以比较纯粹的色彩搭配，加上清爽的空间布置，让简约时尚的设计配合低调奢华的氛围，打造出一个高品质的夜场空间。

　　在建筑外观和入口的设计上，设计师以绚丽的灯光造型吸引来往的路人，借着灯光的发散效果营造出来的延伸性将空间入口拉大，引人入胜。里面公共区域和走道则强调天花造型与地板的呼应来强调空间的气势，并创造整体感。天花板以灯光效果配合现代材料及高科技器材来打造，地面则以地面砖搭配地毯来营造华丽气质。包厢内的设计则以舒适的沙发搭配纯粹的灯光来设计，在简约中书写华丽感受。

地点	面积	设计师	参与设计	设计公司	主要材料
/ 深圳	/ 3800m²	/ 吕军	/ 陈伟贤、杨凯、魏熙政、戴永明、朱福	/ 深圳市吕氏国际室内建筑师事务所	/ 水晶、石材、马赛克、墙纸、玻璃、金属

Shenzhen Kailong International Hotel Nightclub and Club

—— 深圳凯龙国际酒店夜总会及会所

The theme of the design is "to sing and dance to extol the good times". In order to fit the public theme, the design team combines luxurious concepts of China and other countries over the world, to show the momentum and tension of the nightclub. For the special nature of nightclub, the creation of atmosphere and the interior design are equally important, and creating a luxurious space with extremely visual effect meets the market business needs of the owner and satisfies the psychological needs of customers to pursue stimulation and novelty.

The nightclubs of a similar style can be found everywhere, giving us a vulgar sense at first sight, so the designers pay attention to the introduction of colors and lighting to highlight the sense of times and art. These are better shown in materials, details, technology and intelligent aspect, "copes with shifting events by sticking to a fundamental principle," allowing us to pursue an amazing experience from the subtle changes in the popular design.

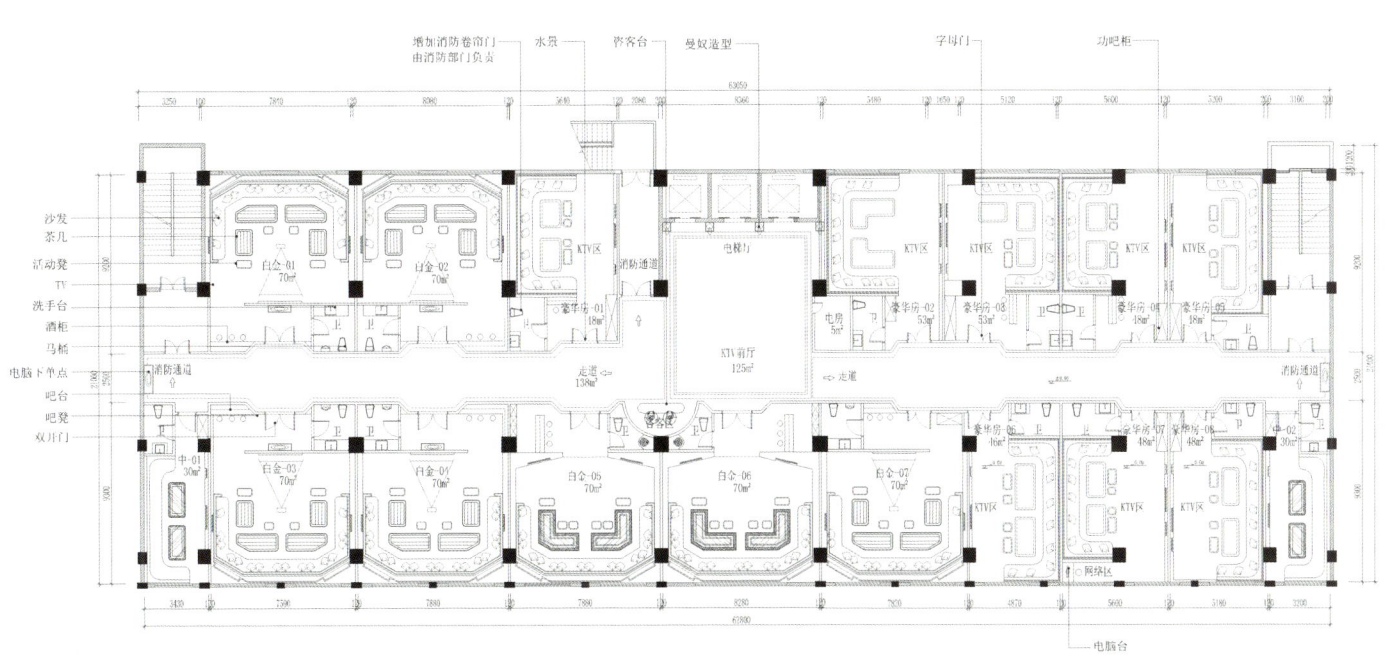

本设计的主题是"歌舞升平"。为了贴合这一张扬的主题，设计团队综合了中国及世界上其他国家的奢华理念，以此来体现夜总会的气势和张力。因为夜店性质的特殊性，氛围的营造和室内的设计同样重要，打造一个极具视觉效果的奢华空间，既能满足业主的市场商业需求，又能满足顾客寻求刺激和新奇的心理需求。

类似格调的夜总会比比皆是，而且乍然一看会给人庸俗的感觉，因此，设计师更注重色彩、灯光的运用，以突出时代感和艺术感。这些，在材料、细节、科技、智能化方面更能体现出来，"以不变应万变"，让人在通俗的设计里面追求微妙的变化带来的惊喜体验。

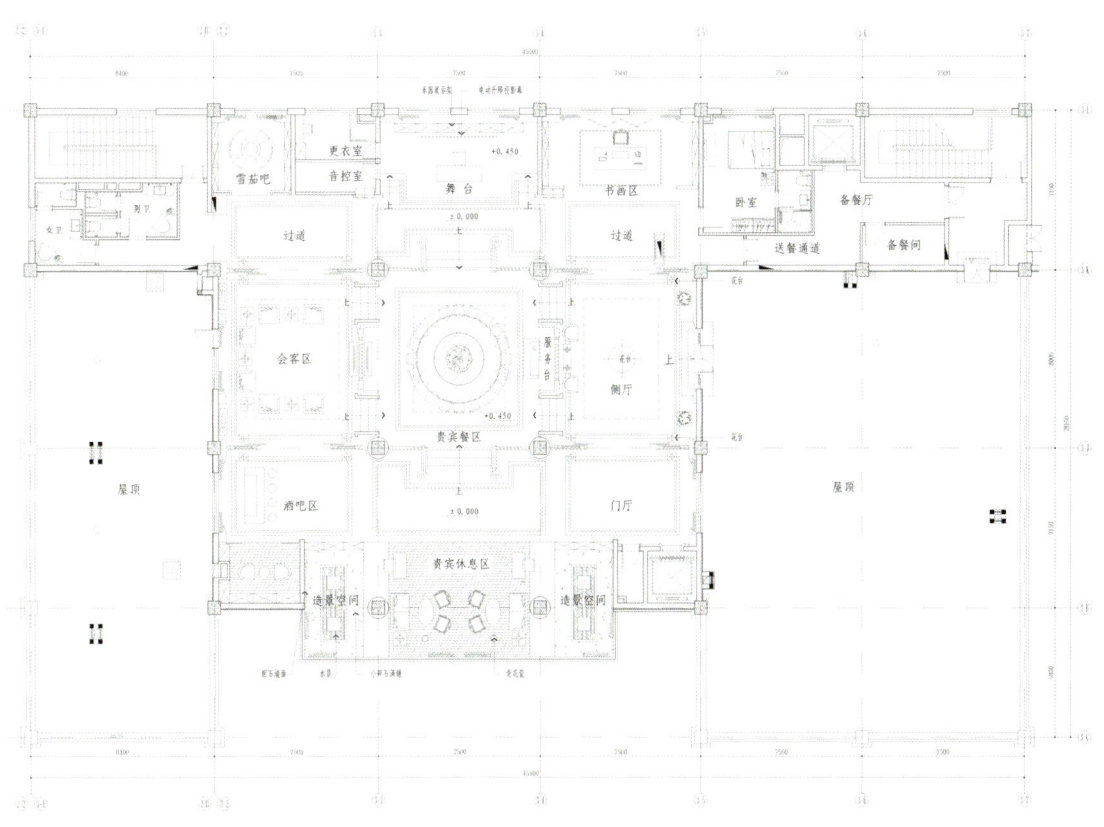

地点	面积	设计师	设计公司	主要材料
/ 成都	/ 800m²	/ 孙亮	/ 成都市优佰文化传播有限公司	/ 石材、马赛克、灰镜、钢琴烤漆、黑镜钢

Sun City KTV

___阳光都市 KTV

The project is located in the CBD of Chengdu, surrounded by office buildings, hotels and shopping malls, with a strong business atmosphere, and it is an intensive area with high-end businessmen. It is a high-end entertainment place designed for these businessmen.

The designer deliberately enlarges the public area, and some areas at corner are changed as landscape or bathroom, so the whole space is natural and generous, improving the utilization rate of the whole space. The space is decorated without too complicated or trivial shapes, but with simple and bright surfaces and lines to feature an overall feeling in favor of hotel's design techniques, showing the levels through the match of materials and furnishings. The use of materials is more clined to reflect the texture of material itself, creating a unique business atmosphere through the match of hard and soft, smooth and rough surfaces, and colors. In addition, the way of combining various light sources is used to reflect the level and contrast of light source, and a controllable switch is introduced to increase the variability of light source to meet the lighting needs in different situations.

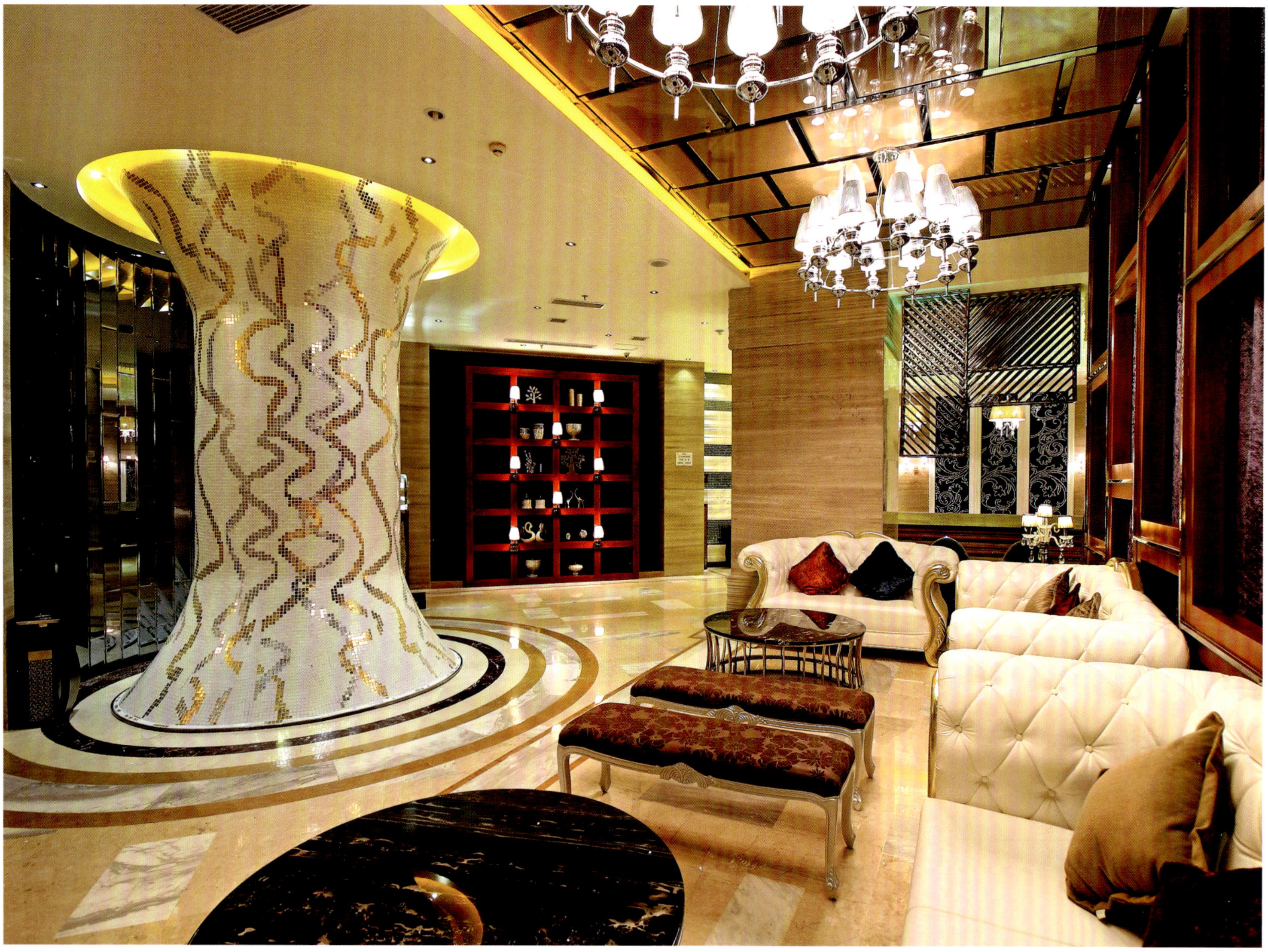

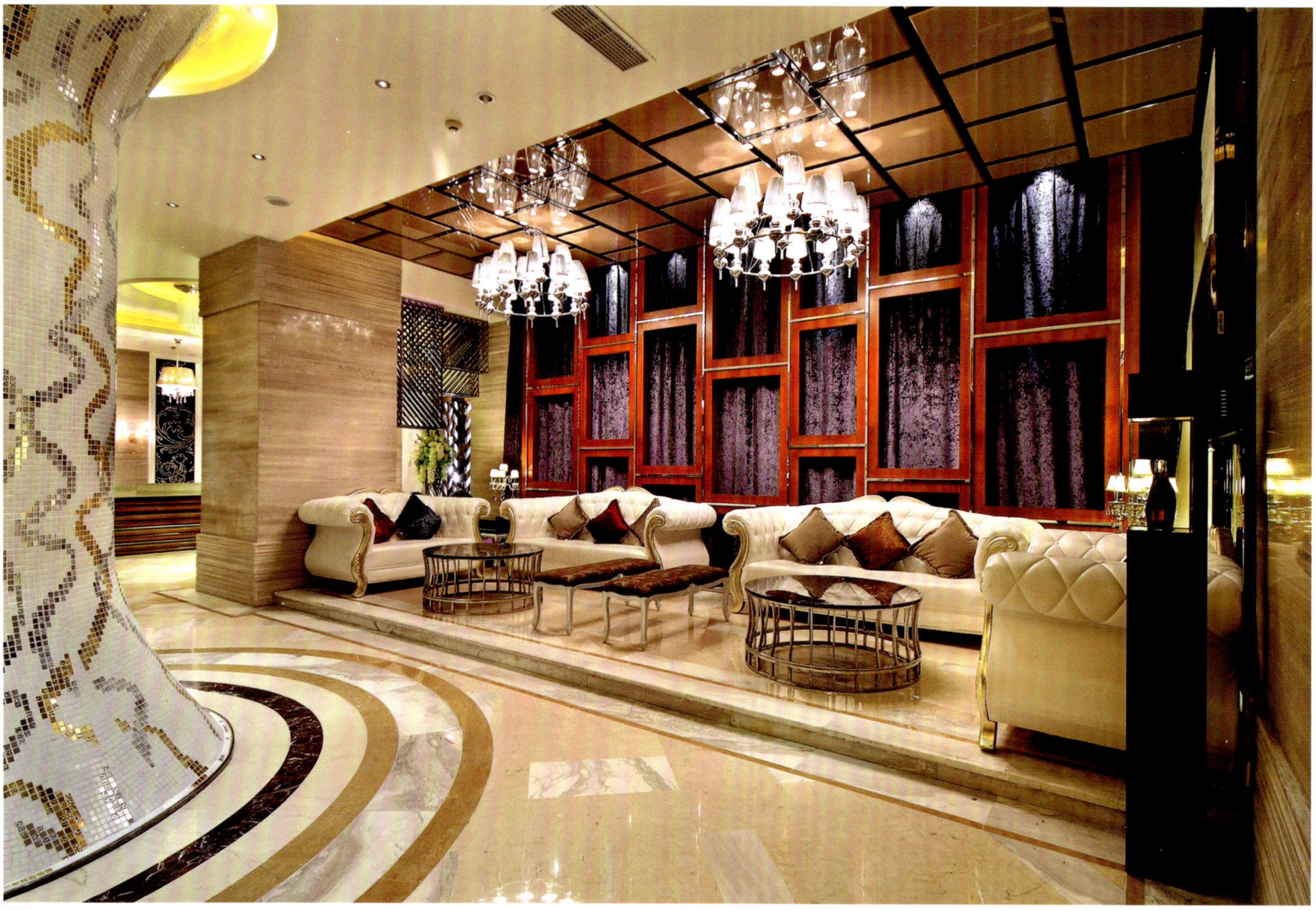

　　本案位于成都市中心CBD区域，周边写字楼、酒店、商场高楼林立，商业氛围浓厚，是高端商务人士聚集的地区。本项目便是为这些商务人士度身打造的高端娱乐场所。

　　设计师将公共区域刻意做大，把一些边角区域利用起来做景观或卫生间，使得整个空间自然、大气，又提高了整个空间的使用率。空间装饰没有过于繁杂、琐碎的造型，以简洁明快的面和线来处理，整体感觉更偏重于酒店的设计手法，通过材料和陈设的搭配来体现层次。材料的运用则更偏重于体现材料本身的质感，通过软硬搭配、光面毛面搭配和色彩搭配来营造独特的商务氛围。此外，以多种光源搭配的方式来体现光源的层次和对比，同时采用可控开关来增加光源的可变性，以满足不同情况下的灯光需求。

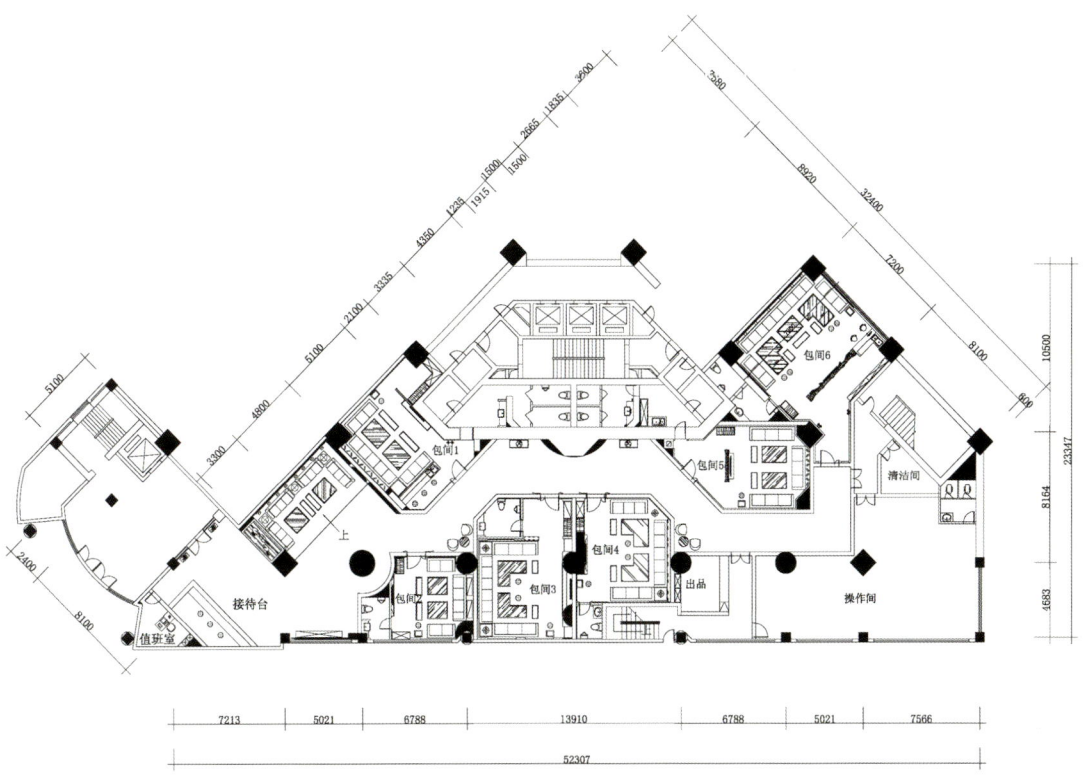

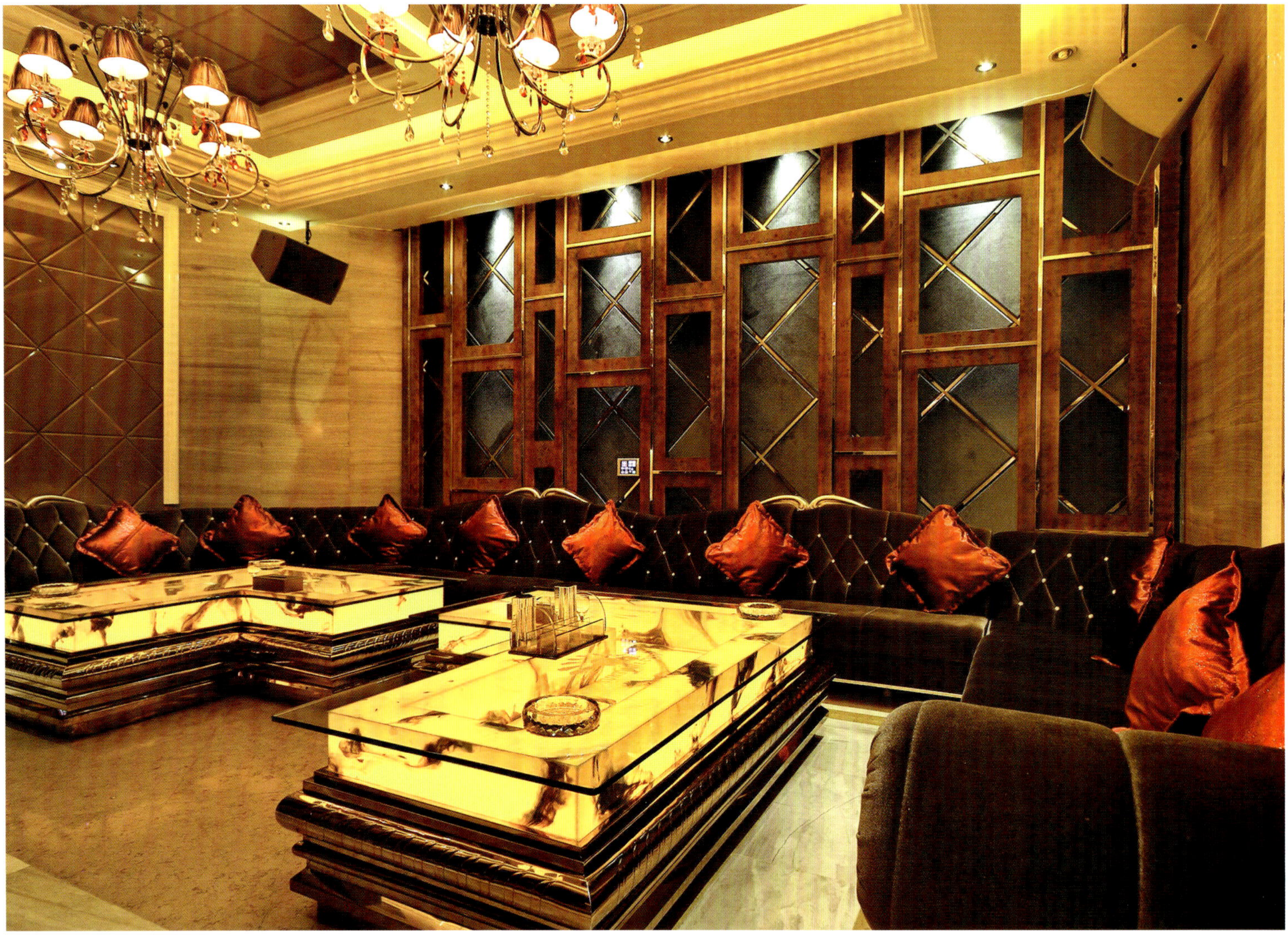

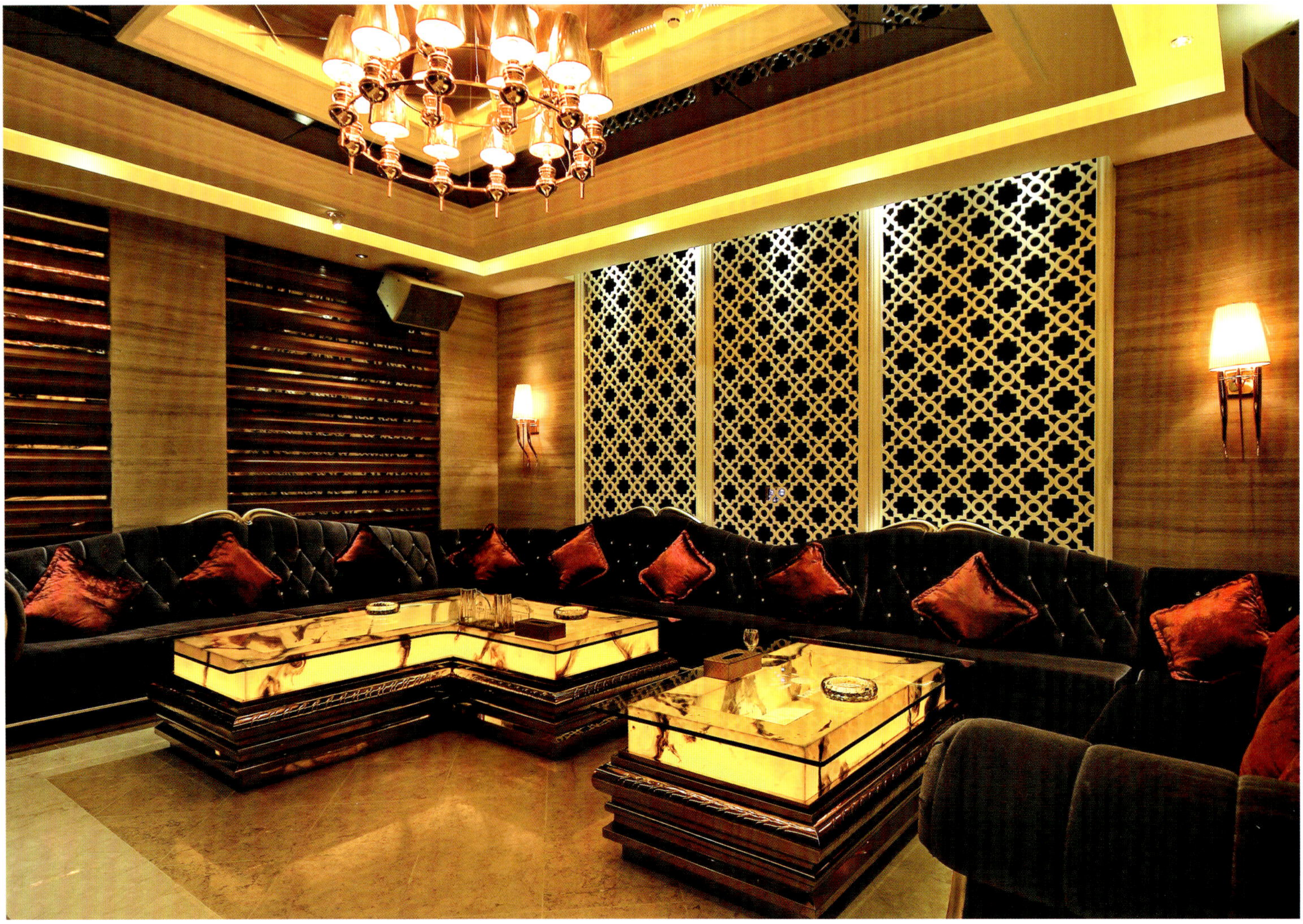

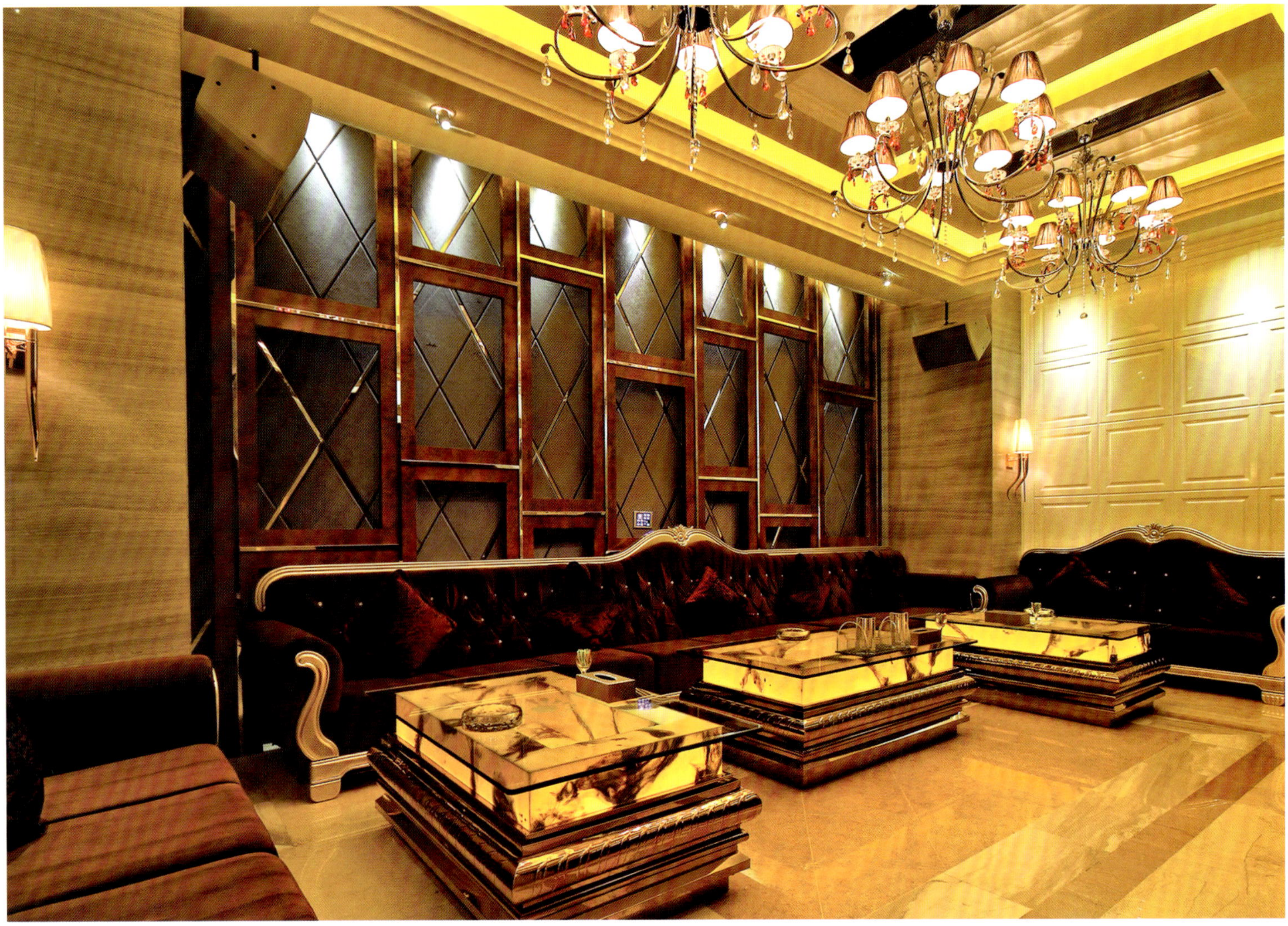

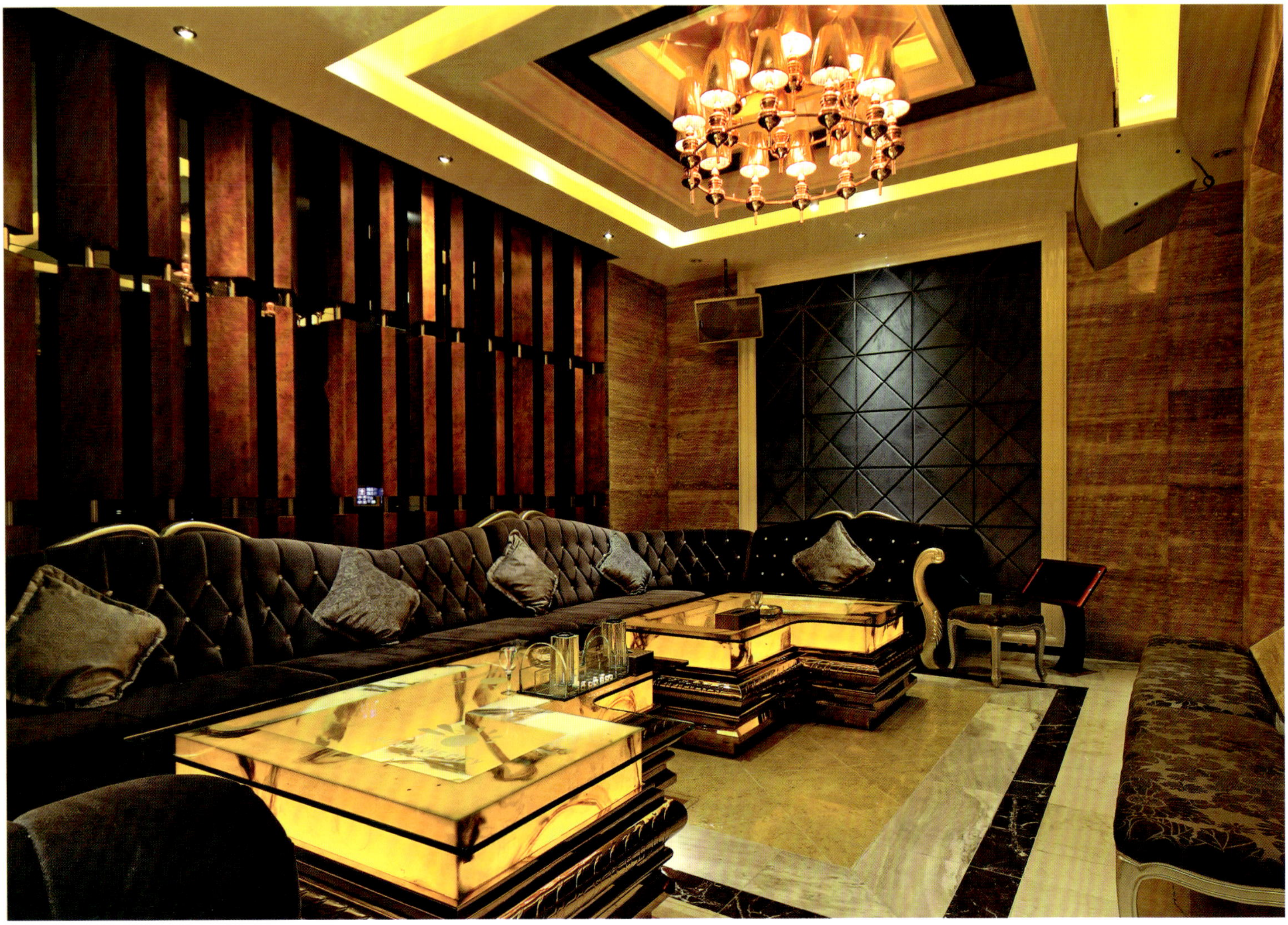

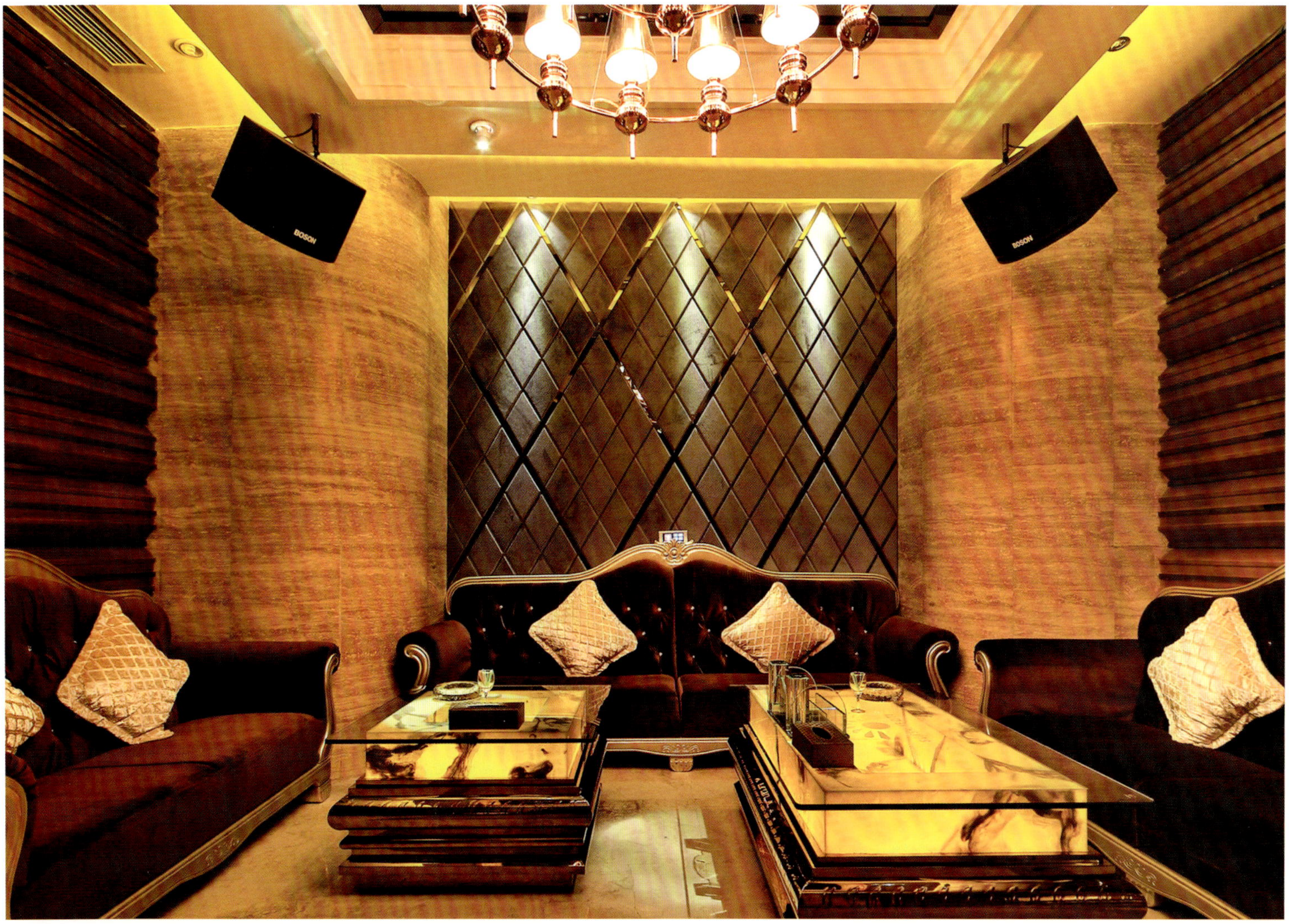
BOSON
BOSON

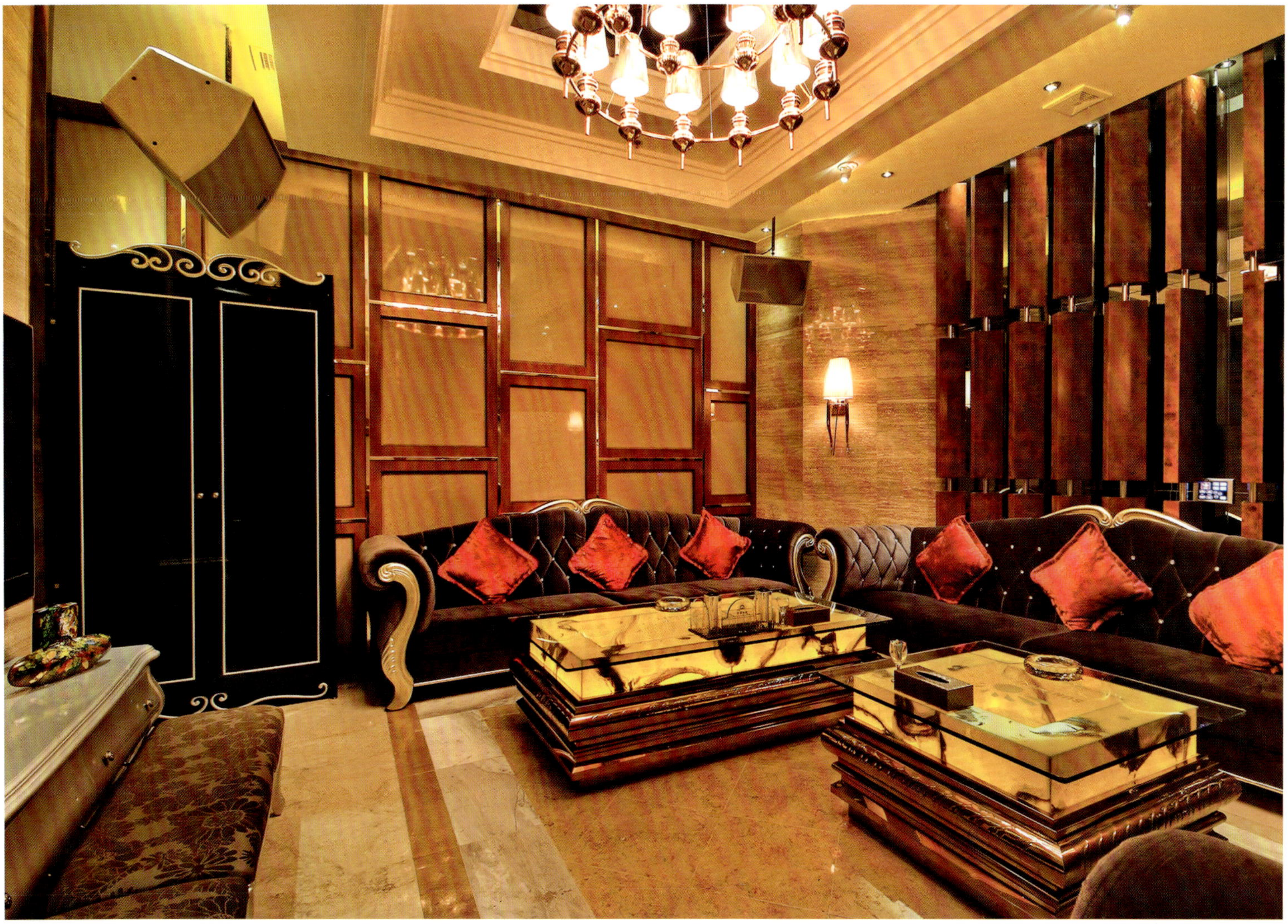

地点	面积	设计师	设计公司	主要材料
/ 深圳	/ 1500m²	/ 吴恙	/ EMC（国际）室内设计公司	/ 石材、玻璃、亚克力、皮革等

Shuidu Entertainment Club

___ 水都娱乐会所

It is a typical representative of mansion culture. The designer tries to make use of the flow and interspersion of elegant champagne gold and rich and deep coffee old gold to express a chocolate-like silky feeling and space weight, to take use of exaggerating European symbols to show a magnificent and solemn atmosphere, and to introduce the contrast of classical and modern to express the depth of history and the texture of time. As a "noisy" place, nightclub reveals the taste of red wine and cigar, like the rock of passion bursting out from the elegant appearance of vinyl records, in order to interpret the changes of ideology and enhance the taste and style to a higher level.

A high end of good taste is a real high end, and in the mutual influence of light, color and structure, the brilliant night life begins here.

　　本案是公馆文化的典型代表。设计师尝试用淡雅的香槟金色和浓郁深沉的咖啡古金色的流动与穿插，表现出巧克力般的丝滑感受和空间的重量，用夸张的欧式符号表现出华丽与庄重，用古典和现代的对比表达出历史的深度和时间的质感，让夜场这种"喧嚣"的场所流淌出红酒和雪茄的味道，就好像激情的摇滚从黑胶唱片的高雅外表里面迸发出来，以此来诠释意识形态的改变，将品位和格调上升到另一个层面。

　　有品位的高档才是真正的高档。在灯光、色彩、结构的相互影响下，绚丽的夜场生活就此拉开序幕。

本书参编人员

周锋、卢霭潮、刘宝达、欧阳亮、周强、陈哲、周美龄、雷小兰、
胡青、吴俊、方丽、段君龙、周晓琪、庄丽娟、周琴、赵丹、赵标、
闫兴宝、徐剑、王琪、黄芸、孙峰、黄宗坤、王雪松、贾春萍、
李红靖、黄静、黄康裕、杜小慧、吴俭英

参与本书翻译的人员

范连颖、王丽红